PROCEEDINGS
of the
STEKLOV INSTITUTE
of
MATHEMATICS

edited by

I. G. Petrovskiĭ and S. M. Nikol'skiĭ

number 95(1968)

ASYMPTOTIC METHODS AND STOCHASTIC MODELS IN PROBLEMS OF WAVE PROPAGATION

edited by

G. I. Petrašen′ and K. P. Latyšev

AMERICAN MATHEMATICAL SOCIETY

Providence, Rhode Island

1971

Академия Наук
Союза Советских Социалистических Республик

ТРУДЫ
ордена Ленина
МАТЕМАТИЧЕСКОГО ИНСТИТУТА
имени В. А. СТЕКЛОВА
XCV

Ответственный редактор академик И. Г. ПЕТРОВСКИЙ
Заместитель ответственного редактора профессор С. М. НИКОЛЬСКИЙ

АСИМПТОТИЧЕСКИЕ МЕТОДЫ И СТОХАСТИЧЕСКИЕ МОДЕЛИ В ЗАДАЧАХ РАСПРОСТРАНЕНИЯ ВОЛН

Сборник работ
под редакцией
Г. И. ПЕТРАШЕНЯ и К. П. ЛАТЫШЕВА

Издательство „Наука“
Ленинградское отделение
Ленинград 1968

Translated from the Russian by T. Garrett

International Standard Book Number 0-8218-1895-3

Printed in the United States of America

PROCEEDINGS OF THE STEKLOV INSTITUTE OF MATHEMATICS IN THE ACADEMY OF SCIENCES OF THE USSR

(Труды математического института им. В. А. Стеклова, т. XCV, 1968)

TABLE OF CONTENTS

INTRODUCTION

These proceedings contain new results on wave propagation which have been obtained in recent years at the Leningrad Branch of the Steklov Institute of Mathematics of the Academy of Sciences of the USSR. In subject matter the book divides into two parts.

The first part contains three papers on stochastic approaches to problems of the propagation of seismic waves. They are based on the construction of probabilistic models of seismic media in the form of a stochastic process or controlling random mechanism and the application of these models to typical problems of seismic survey. In respect of generality and the order of realization there appear to be no analogous papers in the literature.

The most important problem in seismic survey is the interpretation of fields recorded on the earth's surface consisting of "useful waves", bearing information on the structural properties of the medium in question, and intensive "wave-noise". The separation of the sum field into useful waves and wave-noise is always rather conditional and is determined by the specific practical problem confronting the seismic surveyor. In the search for geological structures waves excited on extended interfaces between media, and therefore giving sufficiently extended cophasal axes on seismograms, are considered useful. Waves whose intensity level can appreciably exceed that of useful waves belong to the field of wave-noise.

Identifying useful waves on a background of noise is essentially a statistical problem. It is, moreover, a fairly difficult problem whose solution depends first of all on the authenticity of the statistical parameters describing the properties of the noise field.

The closely related papers of L. A. Zolotuhina and L. A. Zolotuhina, K. P. Latyšev and V. N. Čugueva are devoted to a theoretical study of the properties of fields of wave-noise and a discussion of the problem of identifying a useful wave in such fields. A medium is considered which contains one extended (regular) plane interface and randomly scattered reflecting surfaces (of various lengths) distributed according to a known and sufficiently general statistical law. It is these surfaces which give rise to a field of wave-noise.

In her paper L. A. Zolotuhina discusses a stochastic model of the medium, in which the medium is regarded as the realization of a random process. She also considers probabilistic characteristics (mathematical expectation, correlation coefficients and the multivariate characteristic function of the process) of once-reflected sum waves arriving at points of the diurnal surface. Her generalization of Robbins' Theorem on the sum of a random number of random terms is of mathematical interest in its own right as a limit theorem of probability theory.

The paper by L. A. Zolotuhina, K. P. Latyšev and V. N. Čugueva investigates once and twice-reflected sum waves propagated in the given medium when the sum wave is in the form of a stochastic process with a highly complex structure. The ultimate aim of their work is to determine the conditions of recording a wave which arises on the regular interface of the medium and belongs to this complex sum wave field. For the probabilistic processes considered the authors prove limit theorems in the case of multivariate distributions enabling necessary practical results to be obtained. It is shown that under certain conditions the probabilistic process tends to normality, but in other cases it is essentially non-Gaussian. The limit laws are, however, always non-Markov and nonstationary. The relation between the problems considered by the authors and queueing problems, and the suggested generalization of problems of this type, are also of interest.

B. Ja. Gel'činskiĭ's paper, which also belongs to the probabilistic cycle, is devoted to a new formulation of problems on wave propagation in media with effective random parameters (called $\mathfrak{L}$-media). On the basis of an analysis of experimental material on the time and space distributions of fields of propagation velocities of seismic and hydroacoustic waves, and also a deep analysis of situations encountered in the geophysical interpretation of observed wave fields, the author suggests that seismic and hydroacoustic media should be considered as media with random parameters. To describe the processes of wave propagation in these media in the case of arbitrarily assigned sources the author suggests a constructive algorithm of the calculation of fields of individual waves of "arbitrary" character. Local (for example, radial) approaches to describing wave propagation in determinate media and also the use of Itô's diffusion stochastic equation are taken as the basis of this algorithm. Despite the fact that the article is principally concerned with the formulation of problems there is reason to believe that further development of the author's ideas will significantly help to resolve difficulties both in the problem of wave propagation in real seismic and hydroacoustic media and in the problem of interpreting wave fields.

The remaining papers are concerned with problems of the propagation and diffraction of waves in determinate media.

The papers by V. F. Lazutkin and I. A. Molotkov are devoted to the high-frequency asymptotic behavior of solutions of the Helmholtz equation in the case of plane regions bounded by smooth contours. The authors use the boundary layer method in whose development they themselves have been actively concerned. Lazutkin investigates the interior eigenfunction and eigenvalue problem and finds the asymptotic behavior of a subsequence of eigenfunctions whose values are concentrated in the neighborhood of a diameter of the region. Molotkov consider the problem of finding an asymptotic expansion of Green's function in the exterior problem for the Helmholtz operator when the boundary conditions are of the third kind with a variable coefficient.

L. A. Molotkov and N. V. Cepelev investigate in their papers nonstationary wave fields in two special problems. Molotkov considers the case of an elastic half-space excited by an axially symmetric source moving over the boundary surface; the aim of his paper is to describe qualitatively and quantitatively the principal parts of the wave field in the region of "comparatively high frequencies". His results are also of practical application. Cepelev studies certain questions of the propagation of acoustic waves in two half-spaces separated by an inhomogeneous transition layer. Of particular interest are his investigations of a head wave arising on the boundary of discontinuity of the wave propagation velocity gradient and of interference wave processes.

In the papers of P. V. Krauklis and of G. I. Petrašen′ and È. F. Hinen methods of investigating interference wave fields are used to enable quantitative results to be obtained only in regions of "sufficiently low frequencies". Krauklis investigates the problem in which a head wave arises in a thin high-speed elastic layer in rigid contact with two elastic half-spaces and suggests a simple method of calculating fields of screened head waves. Petrašen′ and Hinen derive and prove equations of the engineering theory of vibrations of a plate based on a study of the solutions of three-dimensional problems for an imperfectly elastic unbounded plane-parallel layer. Much attention is given to establishing conditions for the applicability of the results and discussing the problem of the exactness of describing vibratory processes in a plate within the limits of the engineering theory of vibrations.

We believe that this collection of papers will attract a wide circle of readers both among specialists interested in problems of wave propagation and among mathematicians concerned with the practical application of probability theory and mathematical statistics.

G. I. Petrašen′
K. P. Latyšev

Trudy Mat. Inst. Steklov.
95 (1968)

Proc. Steklov Inst. Math.
95 (1968)

THE FORMULATION OF WAVE PROPAGATION PROBLEMS IN MEDIA WITH EFFECTIVE RANDOM PARAMETERS

UDC 534.21; 550.834; 550.341

B. Ja. GEL'ČINSKIĬ

Abstract. In this paper we suggest a new approach to the study of wave propagation processes in solid and liquid media based on a stochastic model of a real medium called an $\mathfrak{L}$-model. The wave propagation problem formulated for this model arises from one of the ways of formalizing experimental data on the propagation of seismic and hydroacoustic waves in real media.

The proposed formalism for calculating the wave field u is based on the following assumptions. The field at the point M excited by a source at the point M_0 is defined in accordance with the law of variation of effective velocity v_e which is a functional of the form

$$v_e(M_0, M) = \phi(\mathrm{r}, \mathrm{H}(M_0, M)),$$

where the function of structure $\phi(\mathrm{r}, \mathrm{H})$ is assumed given and the vector H takes a value from the characteristic space $H^{(m)}$. The values of the functional $\mathrm{H}(M_0, M)$ are determined by trajectories joining M_0 and M in $R^{(3)}$ and $H^{(m)}$. These trajectories are found as a result of considering jointly the radial equation and a stochastic equation. The field u at M with random fixed values of H is calculated in accordance with local formulas of the corresponding determinatc problem.

The determination of the statistical characteristics of the field reduces to a consideration of a boundary value problem for the Einstein-Planck-Kolmogorov equation.

Bibliography 10 items. 1 figure.

In this paper we suggest a new approach to the study of wave propagation processes in solid and liquid media based on a stochastic model of a real medium called an $\mathfrak{L}$-model. The wave propagation problem formulated for this model arises from one of the ways of formalizing experimental data on the propagation of seismic and hydroacoustic waves in real media.

1. In seismic survey and hydroacoustics a large amount of experimental material has been accumulated on the space and time distributions of velocity fields of wave propagation (for oceans and seas).

In seismic survey the most detailed information on the magnitude of the velocity $v_p(\mathrm{r})$ of the longitudinal wave is obtained from borehole measurements by ultrasonic logging [1].[1] For each investigated area these data are in the form of a set of curves $v_p^k(z) = v_p(z, x_k, y_k)$ $(k = 1, 2, \cdots, n)$, where z is the depth; x_k, y_k are the coordinates of the kth hole, and n is the number of holes in the region.

A typical curve $v_p(z)$ is shown in the diagram. It has the form of an oscillating curve with a considerable number of phase-extrema. Velocity differentials (the ratio of velocity magnitudes at adjacent extrema) attain values of 1.5–2. Phase periods are not constant. Cases are very often met when up to ten phases are packed into a length λ of the seismic wave. A simultaneous study of the whole set of curves (even in the most straightforward regions formed of sedimentary rocks with a gentle, almost horizontal dip) shows that the characteristics of only a relatively small number of phases remain almost invariant from hole to hole. These phases correspond to the so-called reference points—extended key horizons or benches. Moreover, only averaged parameters vary smoothly from hole to hole: these are the so-called mean and sometimes the mean-interval velocities.

The parameters of the overwhelming majority of phases—their location, magnitudes, periods and number—vary rapidly from hole to hole. These phases are generally assumed to correspond to little-aged layers or intercalations.[2] Of all the various properties of these beds the most important for our purposes are the following.

In the direction of $\operatorname{grad} v_p(\mathrm{r})$ the velocity changes rapidly in magnitude along the length of the wave. In the medium, along with the micro-inhomogeneity, a medium-scale bedding exists whose characteristic dimensions in the direction of $\operatorname{grad} v_p(\mathrm{r})$ are of order λ or one tenth of λ. The isosurfaces of $v_p(\mathrm{r})$ are not "confined along the strike" (in the direction perpendicular to $\operatorname{grad} v_p(\mathrm{r})$) and have a highly complex configuration. As a rule we encounter many layers whose dimensions along the strike are of order one tenth of the distance from the source to the observation point, so that over this distance an order of ten such layers are packed successively.

In hydroacoustics a considerable amount of material has also been obtained on the sound velocity field $c(\mathrm{r})$ taken at various times in the same water areas. A statistical study of the fields $c(\mathrm{r})$ is, however, complicated by the fact that in the ocean there is time as well as space variation. It is very difficult to take

---

1) In individual cases ultrasonic logging also enables the velocity $v_s(z)$ of the transverse wave to be determined.

2) A similar pattern holds for other forms of logging data (electrical, nuclear, cavitygrams, and so on).

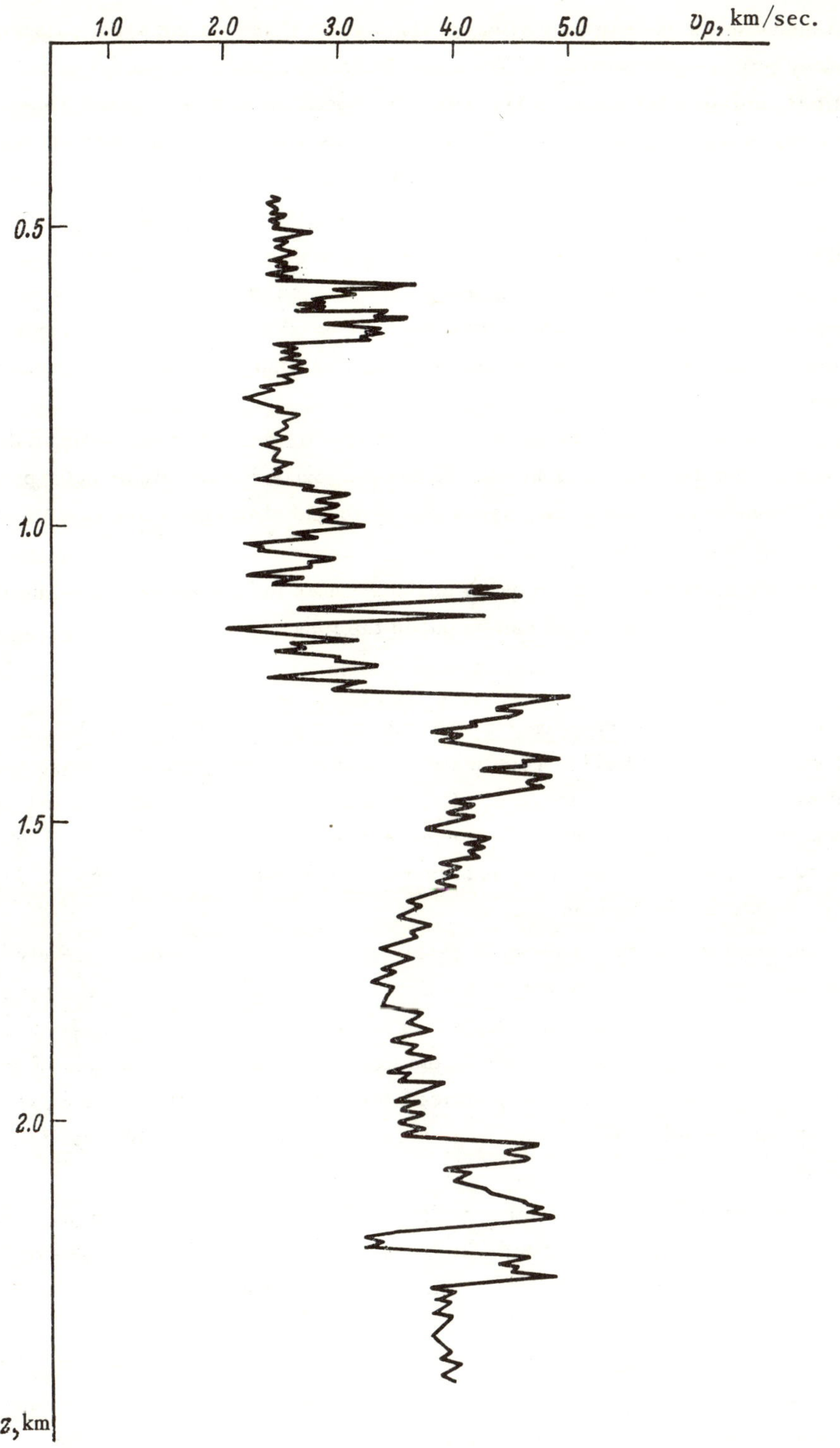
1.0
2.0
3.0
4.0
5.0
v_p, km/sec.
0.5
1.0
1.5
2.0
z, km

continuous measurements of vertical temperature profiles (of the speed of sound) at many points simultaneously. Nevertheless, on the basis of theoretical considerations and the evaluation of a number of nonsynchronous measurements, a whole hierarchy of inhomogeneities (globules) of various sizes, including medium-scale inhomogeneities, is assumed to exist in the ocean. At the same time, it is observed that there is a regular variation of the velocity of sound with depth and in some cases horizontally also.

Thus on land and in the ocean a regular variation of the velocity field of wave propagation complicated by the presence of inhomogeneities of various orders is observed. The necessity to describe this kind of complex field naturally arises.

For brevity we shall speak of one velocity $c(\mathbf{r})$ only, without distinguishing the cases of solid and liquid media. We assume that $\mathbf{r}$ is the radius-vector of the point in space (x, y, z) in the case of the earth and of the point (x, y, z, t) in the case of the ocean.

For the most general description of the velocity field it is usually assumed that the refractive index $n(\mathbf{r})$ can be put in the form

$$n(\mathbf{r}) = f(\mathbf{r}) + \pi(\mathbf{r}), \tag{1}$$

where $f(\mathbf{r})$ is a determinate function defining the regular part of the field $n(\mathbf{r})$ and $\pi(\mathbf{r})$ is a random field of coordinates and time. Doubts frequently arise concerning this additive decomposition of the velocity field into systematic and random parts mainly on account of the presence of medium-scale beddings and inhomogeneities in the medium. It is usually not clear to which of the terms in (1) the medium-scale inclusions relate.

But even if this question can be resolved in some way or other it is very difficult in practice to carry out such a decomposition with respect to the set of values $c(\mathbf{r}_k)$ measured at a bounded set of points $\mathbf{r}_k (k = 1, 2, \cdots, n)$. We do not know, for example, of a single case in the literature when the parameters of the functions $f(\mathbf{r})$ and $\pi(\mathbf{r})$ could in practice be determined for earth or ocean, given assumptions regarding the properties of these functions sufficiently close to reality.

2. Following the above brief description of the most important (for our purposes) properties of velocity fields, it is natural to say how the processes of wave propagation in media characterized by such fields as $c(\mathbf{r})$ are studied theoretically.

So far there has been no sucess in considering even approximately the problem of wave propagation in such media. We therefore have to resort to far-reaching simplifications. It is usual to consider either determinate problems for various sufficiently complex functions $c(r)$ or wave propagation problems in a medium with random inhomogeneities, assuming that certain characteristics of the random component of the velocity field (or wave field) are in some sense small in comparison with the corresponding characteristics of the determinate part.

In the first case the wave field is, as a rule, studied by means of approximate local methods [2,3] since it has not yet been found possible to construct or investigate an exact solution of the problem for sufficiently complex functions $c(r)$. These methods are widely known and we do not dwell on them here. We merely note that mass calculations in practice take account only of the first dominant term of local decompositions. This is explained by the fact that, even in the case of horizontally bedded heterogeneous media consisting of a relatively small number of layers, calculations of the subsequent terms in radial series are necessarily very laborious. Attempts to define the field more exactly in terms of a superposition of normal (surface) waves has also in a number of cases been ineffective, since this requires the summation of a large number of oscillating terms. From the practical point of view, the advisability of calculating subsequent approximations (in very precise calculations) is dubious because the parameters of the real medium are approximated very roughly.

Incidentally we mention here the method of synthetic seismograms developed for seismic survey in the past decade [1]. In this method the real medium is represented by a horizontally bedded system consisting of layers of equal time. With this approximation we obtain formulas for a field caused by a normal incident wave which are very convenient for computer calculation. Calculations of wave fields for systems consisting of hundreds and thousands of layers can be carried out relatively simply by this means. Programs have already been compiled and calculations performed for these systems. In these models up to about ten layers are stacked on the length of a wave. Apparently the authors of the method suppose that, with this approximation of the real medium, a more correct account is taken of the medium-scale bedding existing in it.

In our opinion the use of synthetic seismograms to calculate medium scale inhomogeneities is no better than other methods, and at the same time the approach is not very constructive from a practical point of view. This view stems from the fact that the dynamic system is described by hundreds of parameters which are chosen with respect to the logging data from one of the bore holes. This choice is basically random in character because the medium-scale inhomogeneities are not confined along the strike. The calculated synthetic seismogram is therefore the realization of a random process. However, the determination of the stable characteristics of a seismogram (for example, the mean value requires a sorting out of the variants which in the case of hundreds of parameters is unrealizable.

In considering wave propagation processes in media with random inhomogeneities simplifications of another type are made [4-8]. First, it is assumed that $n(\mathbf{r})$ can be put in the form

$$n(\mathbf{r}) = n_0(\mathbf{r})[1 + \mu(\mathbf{r})], \quad (2)$$

where $n_0(\mathbf{r})$ is a determinate (mean) component and $\mu(\mathbf{r})$ is a random field for which

$$\overline{\mu(\mathbf{r})} = 0, \quad \overline{(\mu(\mathbf{r}))^2} \ll 1. \quad (3)$$

Secondly, $n_0(\mathbf{r})$ is assumed to be a very simple function; in the overwhelming majority of papers it is assumed that

$$n_0(\mathbf{r}) = \text{const}.$$

Certain papers only, such as [7], [8], include a study of the cases

$$n_0(\mathbf{r}) = n_0(z).$$

Thirdly, problems are considered concerning the incidence of a plane or spherical wave on the boundary of a stochastic medium. As far as we are aware, no consideration has been given to the problem of a point source in a heterogeneous medium for which the mean value n_0 depends on the coordinates (even if only on z).

Solutions of problems are constructed either by using perturbation theory for small-scale inhomogeneities or by Rytov's method (the method of smooth perturbations) in the case of large scale inhomogeneities. For perturbation theory to be applicable it is necessary that the stray field should be small in comparison with the incident. In the second case it is apparently necessary that at least the fluctuations of the amplitude should be small.

It is therefore clear that in these problems for stochastic media no account is taken of the presence in many real media of medium-scale inhomogeneities, large fluctuations of $c(\mathbf{r})$ $(n(\mathbf{r}))$ and the complex determinate dependence on the coordinates.

From the descriptions of experimental data given below it will be seen that in many media large fluctuations in amplitude are observed. The methods we have just discussed do not "work" in such conditions.

3. We now describe experimental data obtained in the study of the propagation of nonstationary seismic and acoustic waves and the results of comparing these data with calculations.

Processes of wave propagation are studied experimentally by means of multichannel (multipoint) logging. On the basis of the interpretation of a set of multichannel readings alone sufficiently reliable information on the fundamental properties of the wave field and their connections with the structure of a region can be obtained. Experience in interpreting such data enables the following conclusions to be made.

In the case of "high" frequencies the waves are not propagated to the "necessary" distances and we have a confused (noninterpretable) wave field. For given distances it is nearly always possible to choose an interval of "medium" frequencies corresponding to an interpretable wave picture which is not so complex. The interpretation of the wave field is a nonformalized (partly intuitive) process.

As a result of the interpretation the basic part of the recorded field is decomposed into fields of separate waves. To each wave there corresponds a moving surface Σ along which the field varies smoothly. These waves are usually identified as body, head (lateral) and surface (normal) waves. If these fields are considered qualitatively the impression is created that in the process of propagating medium-frequency waves it is as if the parameters of the medium are averaged and an effective wave field is formed whose principal part can be sufficiently well described by local (and, in particular, radial) formulas.

For a quantitative comparison of observational and calculated data the experimental material is decomposed into homogeneous sets by taking account of all geological (oceanological) and geophysical (hydroacoustic) information relating to the region and a review of the wave fields obtained. As a result, the whole region is split up into separate homogeneous blocks. Attempts are made to choose these blocks so that each of them corresponds to one law of formation or a single structure.

An important step is the so-called parametrization of the model, by which is meant the choice of the form of the function $\phi(\mathrm{r}; H_1, H_2, \cdots, H_m)$, where $H_1, H_2, \cdots, H_m$ are parameters, approximating the field $c(\mathrm{r})$ in each block (structure). In choosing the form of $\phi(\mathrm{r}; H_1, H_2, \cdots, H_m)$ functions are sought which accurately reflect the basic regularities of $c(\mathrm{r})$ in the block in question and which correspond as far as possible to simple calculation formulas for the wave field.

Thus, for example, in using reflected waves to carry out seismic surveys in regions composed of sedimentary rocks the area in question can very often be split approximately into sectors with horizontal bedding. In each sector a piecewise linear function $\phi(z\ H_1, H_2, \cdots, H_m)$ is chosen so that each jump in the velocity corresponds either to a key reflecting horizon or to a change in the sedimentation regime, and the magnitudes of the velocity gradients characterize the specific geological periods of these regimes. Thus the form of the function $\phi(z; H_1, H_2, \cdots, H_m)$ and the choice of parameters $H_i (i = 1, 2, \cdots, m)$ of strata depths, velocity jumps and gradients reflect the basic regularities in the variation of the field $c(\mathrm{r})$. Such a parametrization of the model takes no account of medium-scale bedding because the number m of parameters H_i used to describe the model is tens or hundreds of times smaller than the number of parameters needed for a detailed description of the ultrasonic logging data (or the number of layers in synthetic seismogram models). The chosen function $\phi(z; H_1, H_2, \cdots, H_m)$ therefore describes the averaged effective model of the medium.

After determining the form of $\phi(\mathrm{r}; H_1, H_2, \cdots, H_m)$ the values of the components $H_i (i = 1, 2, \cdots, m)$ of the vector H are chosen with the aim that they should correspond to averaged values of local logging data (bathythermographs) or to individual essential characteristics of datum marks (key horizons). In a successful parametrization of a model the values of H can usually be chosen so that the results of the calculations of wave fields using local formulas correspond qualitatively for the chosen effective model to the set of experimental data for the block in question. A quantitative comparison of calculated and experimental data will, however, reveal substantial divergences which cannot be explained by additive noise.

In order to obtain agreement between calculation and experiment it is necessary to take different effective models to describe waves of different types. Even for a single wave the effective model is not the same for different frequencies. Nevertheless, even if these requirements are met, the calculated and experimental fields diverge appreciably, the deviations containing both a systematic and a random component. This is evidently explained by the fact that in averaging medium-scale inhomogeneities it is impossible to obtain a single effective model for different waves and frequencies and in the case of a single wave when the source and detector are arbitrarily located.

In interpreting the divergences of calculated and experimental values it can be assumed that the effective parameters change in the propagation process. In order to give a simple explanation of this assumption we describe the wave propagation process in an effective model by means of the following scheme with finite difference steps.

Let the form of the function $\phi(\mathrm{r}; H)$ be fixed and the vector $\mathrm{H}(H_1, H_2, \cdots, H_m)$ assume a value from the m-space $H^{(m)}$. Then to each effective model in $H^{(m)}$ there corresponds a point which we call the H-point.

Suppose that a source is located at the point $M_0(\mathrm{r}_0)$ and that the initial value H_0 of H is assigned on the basis of a priori information. It is natural to assume that in a neighborhood of M_0 a wave is propagated as if it were in the same effective model with parameter H_0. In a small neighborhood of the source we may on this hypothesis associate with the wave the propagation time $\tau(M_0, M, H_0)$, the direction of the ray with unit vector $n(M_0, M, H_0)$ and the wave field $u(M_0, M, H_0)$, which are defined with respect to the corresponding formulas for a determinate medium characterized by the velocity $\phi(r; H_0)$. After passing along a small segment of the path

$$\Delta s_1 = \Delta s_1 n(M_0, M_0, H) \tag{4}$$

the wave reaches the point $M_1(r_1)$, where

$$r_1 = r_0 + \Delta s_1, \tag{5}$$

in which under the action of a heterogeneous medium the effective vector changes by a jump of magnitude ΔH_1, so that to the points M_0 and M_1 there corresponds an effective medium with parameter

$$H_1 = H_0 + \Delta H_1.$$

The wave passes along a path from M_1 to a neighboring point M_2 with radius vector

$$r_2 = r_1 + \Delta s_2 \equiv r_0 + \sum_{k=1}^{2} \Delta s_k,$$

where $\Delta s_2 = \Delta s n(M_0, M_1, H_1)$ and $n(M_0, M_1, H_1)$ is the unit tangent vector to the ray at M_1 corresponding to a nonrandom medium with vector H_1.

The wave field $u(M_0, M, H_1)$ at M_2 is calculated by formulas corresponding to a determinate distribution of the magnitude of the velocity $\phi(r; H_1)$. The vector H receives an increment ΔH_2 at M_2 and the wave moves in an effective medium with vector H_2 given by

$$H_2 = H_0 + \sum_{k=1}^{2} \Delta H_k.$$

Thus the effective parameters defining the propagation process change by jumps.

The model of the medium in which the propagation process takes place in accordance with this scheme will be called an $\mathfrak{L}$-model (or $\mathfrak{L}$-medium).

To proceed from this intuitive picture of the propagation process in an $\mathfrak{L}$-medium to a quantitative description, it is first necessary to define the law of variation of H. Various models of effective media will be obtained depending on the assumptions about this law.

Models with a nonrandom law of variation of H will correspond to a determinate $\mathfrak{L}$-medium. We note incidentally that in one form or another the processing of kinematic data in seismic surveys uses individual representations corresponding to determinate $\mathfrak{L}$-model. In stochastic models, however, it is natural to assume, at least at first, that the law of variation of H is Markov. Media in which the effective vector H is a multivariate Markov procees will be called Markov $\mathfrak{L}$-media.

If, however, we consider the analogy between the walk of the H-point in the space $H^{(m)}$ and Brownian motion, a more particular assumption regarding the law of variation of the effective parameters can be made by taking H to be an m-dimensional random diffusion process. The corresponding $\mathfrak{L}$-model is called a diffusion model. Only models of this kind are considered in this paper; thus for brevity we shall omit the word "diffusion" from now on.

4. We now formalize the presentation and attempt to do this at a physical level of rigor. We limit ourselves to considering body and lateral (head) waves only. This restriction is obviously of no significance.

The suggested procedure for calculating wave fields is the same for both solid and liquid media, so we shall describe a single formalism without distinguishing expressions corresponding to seismic and hydroacoustic waves.

The field $u(\mathrm{r}_0, \mathrm{r}, \omega)$ at the point $M(\mathrm{r})$ excited by a stationary source at the point $M_0(\mathrm{r}_0)$ is defined with the help of the following formalism.

I. It is assumed that the basic part of the field in the region in question is a superposition

$$u(\mathbf{r}_0, \mathbf{r}, \omega) = \sum_i u_i(\mathbf{r}_0, \mathbf{r}, \omega) \tag{6}$$

of effective fields $u_i(\mathrm{r}_0, \mathrm{r}, \omega)$ of separate waves. The summation in (6) is over all waves arriving at M from the source in accordance with the distribution of values of effective speed defined in II. As already mentioned, cases are considered in which a finite number of body or head waves arrive at M.

II. The wave field $u_i(\mathrm{r}_0, \mathrm{r}, \omega)$ is defined at an arbitrary point $M(\mathrm{r})$ in accordance with the law of variation of effective velocity $v_{\mathrm{e}}^{(i)}$ which is a functional of the form

$$v_{\mathrm{e}}^{(i)}(M_0, M, \omega) = \phi(\mathrm{r}, \mathrm{H}^{(i)}(M_0, M, \omega)), \tag{7}$$

where the function $\phi(\mathrm{r}, \mathrm{H}^{(i)})$ determining the structure of the medium is assumed given and $\mathrm{H}^{(i)}$ is the vector with components $H_1^{(i)}, H_2^{(i)}, \cdots, H_m^{(i)}$.

We call $\phi(\mathrm{r}, \mathrm{H}, \omega)\,[\phi(\mathrm{r}, \mathrm{H}(M_0, M, \omega)]$ the function (functional) of structure, H the vector of the characteristic parameters (or simply the characteristic vector), and $H^{(m)}$ the phase space (or the space of characteristic parameters).

For example, in the case of an n-layered axially symmetric system with a piecewise linear law of variation of velocity the function of structure has the form

$$\varphi(\mathbf{r}, \mathbf{H}) = \varphi(z, \mathbf{H}) = \sum_{k=1}^{n} H_{1+3k}[1 + H_{2+3k}(z - H_{3k-3})] \times [\varepsilon(z - H_{3k-3}) - \varepsilon(z - H_{3k})], \tag{8}$$

where $\epsilon(z)$ Is Heaviside's function;

$$H_{2+3k} = \frac{1}{v_k}\frac{\partial v}{\partial z}, \tag{9}$$

$H_{1+3k} = v_k$ is the velocity of wave propagation on the upper boundary of the kth layer; and $H_{3k-3} = z_k$ is the depth of the upper boundary of the kth layer. The number of components of the characteristic vector H for this n-layered system is $m = 3n$. If, however, on certain boundaries the velocity and its gradient are continuous the dimension of the phase space $H^{(m)}$ is correspondingly reduced.

III. The values of the functional $\mathrm{H}(M_0, M, \omega)$ are determined by trajectories joining the points M_0 and M in $H^{(m)}$ and $R^{(3)}$. These trajectories are found by considering the equation

$$ds^{(i)}(M) = \mathbf{n}^{(i)}(s, \mathbf{r}, \mathbf{H}^{(i)}(s))\, ds^{(i)}(M) \tag{10}$$

jointly with the stochastic Langevin-Itô equation

$$d\mathbf{H}^{(i)}(M, \omega) = \mathbf{a}(s^{(i)}, \mathbf{r}, \mathbf{H}^{(i)}(s), \omega)\, ds^{(i)}(M) + \sum_{k=1}^{m} \mathbf{b}_k(s^{(i)}, \mathbf{r}, \mathbf{H}^{(i)}(s), \omega)\, dw_k(M), \tag{11}$$

where s is the arc length M_0M; ω is the frequency;

$$\mathbf{n}^{(i)} = \frac{\operatorname{grad} \tau^{(i)}}{|\operatorname{grad} \tau^{(i)}|}, \tag{12}$$

and $\tau^{(i)} = \tau^{(i)}(M_0, M, \mathrm{H}^{(i)})$ is a solution of the equation

$$(\operatorname{grad} \tau^{(i)})^2 = \frac{1}{[\varphi(\mathbf{r}, \mathbf{H}^{(i)})]^2} \tag{13}$$

for the ith wave propagated in a determinate medium for a fixed value of the vector $\mathrm{H}^{(i)}$. Equation (11) defines an m-dimensional diffusion process H (see [9], Russian p. 486). In (11) a and b_k ($k = 1, 2, \cdots, m$) are given vector functions from $H^{(m)}$ and $R^{(3)}$, and $w_k(M)$ ($k = 1, 2, \cdots, m$) are independent one-dimensional processes of a Brownian motion. As we know (see [9], Russian p. 521), the vector a defines the "drift" of the parameter which takes place for Brownian motion in the phase space and the vectors b_k characterize the diffusion properties of the multivariate process $\mathrm{H}^{(i)}$.

The solutions of the system (10)–(11) are stochastic integrals and can be calculated using the procedure described below with initial conditions

$$\tau^{(i)}(M_0)=0,\ \frac{\partial\tau^{(i)}(M_0)}{\partial x}=\alpha_1,$$

$$\frac{\partial\tau^{(i)}(M_0)}{\partial y}=\alpha_2,\quad \mathbf{H}(M_0,M_0)=\mathbf{H}_0. \tag{14}$$

First of all, a step

$$\Delta\mathbf{s}_1^{(i)}=\mathbf{n}(\alpha_1,\alpha_2,\mathbf{H}_0)\Delta s_1^{(i)}(M_0) \tag{15}$$

is made in the direction (α_1, α_2) along a ray from the point M_0 to the point $M_1(\mathbf{r}_1)$, where

$$\mathbf{r}_1=\mathbf{r}_0+\Delta\mathbf{s}_1^{(i)}. \tag{16}$$

At $M_1(\mathbf{r}_1)$ the vector $\mathbf{H}^{(i)}$ receives an increment $\Delta\mathbf{H}^{(i)}(M_1)$ determined from (11). The direction of $\Delta\mathbf{s}_2^{(i)}$ at M_1 is determined by (12) and (13) for a fixed value of the vector $\mathbf{H}_1^{(i)}$ given by

$$\mathbf{H}_1^{(i)}(M_1)=\mathbf{H}_0^{(i)}+\Delta\mathbf{H}^{(i)}(M_1). \tag{17}$$

The wave then arrives at the point $M_2(\mathbf{r}_2)$, where

$$\mathbf{r}_2=\mathbf{r}_0+\sum_{l=1}^{2}\Delta\mathbf{s}_l^{(i)} \tag{18}$$

and so on. Finally, after N steps the ray reaches the point $M(\mathbf{r})$, where

$$\mathbf{r}=\mathbf{r}_0+\sum_{l=1}^{N}\Delta\mathbf{s}_l^{(i)}. \tag{19}$$

The vector $\mathbf{H}^{(i)}$ at this point is given by

$$\mathbf{H}^{(i)}(M)=\mathbf{H}_0^{(i)}+\sum_{l=1}^{N}\Delta\mathbf{H}^{(i)}(M_l). \tag{20}$$

If we now pass to the limit as $N\to\infty$ or $\Delta s\to 0$, where

$$\Delta s=\max_l \Delta s^{(i)}(M_l), \tag{21}$$

the value of $\mathbf{H}(M)$ can be found from the formula

$$\begin{aligned}\mathbf{H}^{(i)}(M)-\mathbf{H}_0^{(i)}&=P-\lim\sum_{l=1}^{N}\Delta\mathbf{H}^{(i)}(M_l)\\ &=P-\lim\sum_{l=1}^{N}\sum_{k=1}^{m}\mathbf{b}_k(s_{l-1}^{(i)},\mathbf{r}_{l-1},\mathbf{H}^{(i)}(s_{l-1}^{(i)}),\omega)[w_k(M_l)-w_k(M_{l-1})]\\ &\quad+\int_{M_0}^{M_0}\mathbf{a}(s^{(i)},\mathbf{r},\mathbf{H}^{(i)}(s),\omega)\,ds(M).\end{aligned} \tag{22}$$

Remark. The procedure suggested here of considering jointly the trajectories in the phase space and in $R^{(3)}(x, y, z)$ differs from Itô's classical procedure in that, in particular, the vectors a and $\mathbf{b}_k (k = 1, 2, \cdots, m)$ depend on the radius vector r. A rigorous mathematical justification of the procedure would therefore require further investigation. We shall not concern ourselves with this here in view of our stated intention to formalize the procedure at a physical level of rigor.

IV. In accordance with the scheme described, diffusion of the vector H occurs during the propagation of waves. If no further conditions are introduced, the values of H as s increases are "scattered" over the whole phase space.[1] This more or less corresponds to a uniform distribution of states of the $\mathfrak{L}$-model over the whole space $H^{(m)}$, which is equivalent to absorption of the wave. For certain poorly ordered media such a description of the wave propagation process is obviously satisfactory. But, as already mentioned, in many real media there are regularities bringing order into the structure. In order that our formalism should reflect the existence of these regularities we need to consider (10) and (11) together with additional conditions.

In a number of cases we can introduce a determinate force $\mathbf{F}(s, \mathbf{H}, \widetilde{\mathbf{H}}(s), \omega)$ obstructing the scatter of the H-points over the phase space. On the right side of (11) we must then add an impulse

$$\pi = \int \mathbf{F}(s, \mathbf{H}(s) \widetilde{\mathbf{H}}(\mathbf{r}), \omega)\, ds. \tag{23}$$

It may of course be assumed from the start that the drift vector $\mathbf{a}(s, \mathbf{r}, \mathbf{H}, \omega)$ contains a term corresponding to (23). But singling out a separate term in (11) corresponding to the force F is physically more intuitive and in practice more convenient.

In other situations it may be more appropriate to consider a boundary value problem with conditions of the "reflecting or absorbing walls" type. It is known [9,10] that in such cases it is convenient to pass to the Einstein-Fokker-Planck-Kolmogorov equation

$$\frac{\partial p(s, \mathbf{r}, \mathbf{H}, \omega)}{\partial s} = -\sum_{j=1}^{m} \frac{\partial}{\partial H_j}(a_j(s, \mathbf{r}, \mathbf{H}, \omega)\, p(s, \mathbf{r}, \mathbf{H}, \omega))$$
$$+\frac{1}{2} \sum_{j,\, l=1}^{m} \frac{\partial^2}{\partial H_j \partial H_l}(b_{jl}(s, \mathbf{r}, \mathbf{H}, \omega)\, p(s, \mathbf{r}, \mathbf{H}, \omega)) \tag{24}$$

for the density $p(s, \mathbf{r}, \mathbf{H}, \omega)$ of the direction of H.

1) From now on we omit the index i.

In (24) $a_j (j = 1, 2, \cdots, m)$ is the jth component of the drift vector $\mathbf{a}$ and the coefficient $b_{jl} (j, l = 1, 2, \cdots, m)$ is an element of the matrix $\{b_{jl}\}$ of local dispersions related to the components of the vectors $\mathbf{b}_k (k = 1,2, \cdots, m)$ by

$$b_{jl} = \sum_{k=1}^{m} b_k^j b_k^l \quad (j, l = 1, 2, \ldots, m). \tag{25}$$

We need to add to (24) the initial condition

$$p(s, \mathbf{r}, \mathbf{H}(s)\,\omega)\,|_{s=0} = p_0(\mathbf{H}_0, \omega) \tag{26}$$

and the boundary condition

$$p(s, \mathbf{r}, \mathbf{H}(s), \omega) = 0 \quad \text{on} \quad S \tag{27}$$

or

$$\frac{\partial}{\partial n} p(s, \mathbf{r}, \mathbf{H}(s), \omega) = 0 \quad \text{on} \quad S, \tag{27'}$$

where S is a surface in $H^{(m)}$ given by

$$f[\mathbf{H}(s)] = 0. \tag{28}$$

V. After constructing the trajectories in $H^{(m)}$ and $R^{(3)}$ and determining the magnitudes of $\mathbf{H}(M)$ along them, it is natural to proceed to the calculation of the wave field $u(\mathbf{r}_0, \mathbf{r}, \mathbf{H}(M), \omega)$.

We assume that at an arbitrary point $M_k(\mathbf{r}_k)$ with a fixed value of the vector $\mathbf{H}(M_k) = \mathbf{h}_k$ the field u is determined by

$$u(\mathbf{r}_0, \mathbf{r}_k, \mathbf{H}(M_k), \omega) = \Phi(s, \mathbf{r}_k, \mathbf{h}_k, \omega), \tag{29}$$

where $\Phi(s, \mathbf{r}_k, \mathbf{h}_k, \omega)$ is a local solution of the problem corresponding to the wave in question for a determinate medium characterized by the velocity

$$c(\mathbf{r}, \omega) = \varphi(\mathbf{r}, \mathbf{h}_k, \omega). \tag{30}$$

The expression for the function $\Phi(s, \mathbf{r}_k, \mathbf{h}_k, \omega)$ is determined by a local method [2,3] whose choice depends on the distribution of the velocity $c(\mathbf{r}, \omega)$, the location of the source and the point M_k, and the type of wave being studied.

The stochastic characteristics of the wave field u at an arbitrary point $M_k(\mathbf{r}_k)$ can be found by two different methods.

In the first method the calculations are as follows. We first construct a solution of the boundary value problem (24), (26), (27) for the density $p(s, \mathbf{r}_k, \mathbf{H}(s), \omega)$ of the probability distribution of the characteristic vector. We then use known formulas to calculate the moments of the function $\Phi(\mathbf{r}_0, \mathbf{r}_k, \mathbf{H}(s), \omega)$. For example, its expectation is found from the expression

$$M\Phi(s, \mathbf{r}_k, \mathbf{H}(s), \omega) = \int_{(V)} \Phi(s, \mathbf{r}_k, \mathbf{H}(s), \omega)\, p(s, \mathbf{r}_k, \mathbf{H}(s), \omega)\, d\mathbf{H}, \tag{31}$$

where V is the domain in $H^{(m)}$ bounded by the surface S (28). And, finally, we average with respect to s (or $\mathbf{r}_0$).

Calculations in the second method are based on Itô's formula for the differentiation of a complex function (see [9], Russian p. 501). If we limit consideration to those media for which $\Phi(s, \mathbf{r}_k, \mathbf{h}, \omega)$ (nonrandom) has the necessary regularity properties (or on the surfaces of discontinuity of its derivatives with respect to s, $\mathbf{h}$ satisfies conjunction conditions), then the process $\eta(M) = \Phi(s, \mathbf{r}, \mathbf{H}(s), \omega)$ will also be determined by the Langevin-Itô equation (see [9], Russian p. 504)

$$d\eta(M) = \left[\frac{\partial\Phi(s, \mathbf{r}, \mathbf{H}(s), \omega)}{\partial s} + \sum_{l=1}^{m} \frac{\partial}{\partial H_l}(\Phi(s, \mathbf{r}, \mathbf{H}(s), \omega)\, a_l(s, \mathbf{r}, \mathbf{H}(s), \omega))\right.$$

$$+ \frac{1}{2} \sum_{i, j, l}^{m} \frac{\partial^2}{\partial H_i \partial H_j}[\Phi(s, \mathbf{r}, \mathbf{H}(s), \omega)\, b_l^{(i)}(s, \mathbf{r}, \mathbf{H}(s), \omega)]\, ds(M)$$

$$+ \sum_{j, l=1}^{m} \frac{\partial}{\partial H_j}[\Phi(s, \mathbf{r}, \mathbf{H}(s), \omega)\, b_l^{(j)}(s, \mathbf{r}, \mathbf{H}(s), \omega)]\, dw_l(M), \tag{32}$$

where a_l and $b_l^{(j)}$ $(j, l = 1, 2, \cdots, m)$ are components of the drift and diffusion vectors in (11).

The wave field is now studied on the basis of equation (32) or of the corresponding equation for the probability distribution analogous to (24). But here it should be noted that finding stochastic integrals for (32) or the probability distribution is a very difficult problem, since the coefficients of the equation contain derivatives of $\Phi(s, \mathbf{r}, \mathbf{H}(s), \omega)$.

The most promising method of solving problems for complex $\mathfrak{L}$-media will obviously be the modelling of equation (32) or (11) on a computer. As we know, Itô's equation can be modelled relatively simply on digital machines.

5. Therefore in describing an $\mathfrak{L}$-medium the following characteristics of the model must be given: the form of the function of structure (7), of the vectors of drift $\mathbf{a}(s, \mathbf{r}, \mathbf{H}, \omega)$ and diffusion $\mathbf{b}_k(s, \mathbf{r}, \mathbf{H}, \omega)$ $(k = 1, 2, \cdots, m)$, and any additional condition from IV. The position of the source and the initial value $\mathbf{H}_0$ of the characteristic vector, or the initial probability distribution $p_0(H_0)$, must also be known.

A full investigation of the wave propagation problem includes the calculation of stochastic integrals of equations (11) and (32) with initial data (14) and additional conditions from IV, or finding a solution of the boundary value problem for the diffusion equation (24), and subsequently determining with their help the statistical characteristics of the wave field (29).

Such an investigation of the wave propagation process can be very difficult, but in the solution of practical problems it is possible to restrict oneself to determining only the first two moments of the wave and parametric fields, which, as a rule, are calculated more simply.

In conclusion, let us formulate the basic aspects of this paper and make a number of remarks.

The paper contains a number of conclusions obtained as a result of generalizing experimental data on wave propagation in real media and relating to: the decomposition of the field of vibrations into fields of separate waves each of which can be described by local formulas; the possibility of introducing into this description an averaged model characterized by the function of structure; and the variation of effective parameters of the model in the propagation process.

On the basis of the formalization of these experimental aspects a stochastic model of the medium is suggested for which the problem can be formulated. This formulation does not follow uniquely and strictly logically from the given experimental data. Within this framework other variants of the $\mathfrak{L}$-model can be proposed.

Furthermore, we regard the formalism described in this paper for a selected $\mathfrak{L}$-model as an initial rough scheme for calculating the wave field. In future work on its development and on the formulation of its aspects on a strict mathematical level individual details will be refined or changed. Obtaining an $\mathfrak{L}$-model as a result of averaging the solution of the problem in a classical formulation for a medium with a sufficiently complex structure would be of very great importance in proving the suggested procedure.

We hope also to evaluate the reliability of the suggested model by comparing the results of calculations for an $\mathfrak{L}$-medium with experimental data.

The author is grateful to G. I. Petrašen' and M. I. Freĭdlin for helpful discussions on the paper.

BIBLIOGRAPHY

[1] *Problems of seismic survey*, Gostoptehizdat, Moscow, 1962. (Russian)

[2] A. S. Alekseev, V. M. Babič and B. Ja. Gel'činskiĭ, *The ray method of calculating the intensity of wave fronts*, Problems of the Dynamic Theory of the Propagation of Seismic Waves, vol. 5, Izdat. Leningrad. Univ., Leningrad, 1961, pp. 3–25. (Russian)

[3] G. D. Maljužinec and L. A. Vaĭnšteĭn, *Transverse diffusion during diffraction on an impedance cylinder of large radius.* I, II, Radiotehn. i Èlektron. 6 (1961), 1247–1258, 1489–1495 = Radio Engrg. and Electronics 6 (1961), 1106–1117, 1324–1330.

[4] L. A. Černov, *Wave propagation in a medium with random inhomogeneities*, Izdat. Akad. Nauk SSSR, Moscow, 1958. (Russian)

[5] V. I. Tatarskiĭ, *Theory of fluctuation phenomena in the propagation of waves in a turbulent atmosphere*, Akad. Nauk SSSR Inst. Fiz. Atmosfery, Izdat. Akad. Nauk SSSR, Moscow, 1959. (Russian) MR 23 #B1572.

[6] S. M. Rytov, *Introduction to statistical radiophysics*, "Nauka", Moscow, 1966. (Russian)

[7] N. G. Denisov, *On fluctuations of amplitude and phase of a wave passing across a layer with random inhomogeneities*, Izv. Vysš. Učebn. Zaved. Radiotehnika 2 (1959), no. 2, 316. (Russian)

[8] F. G. Bass and S. I. Hankina, *On the theory of electromagnetic wave propagation in a heterogeneous medium with fluctuations of the dielectric constant*, Izv. Vysš. Učebn. Zaved. Radiotehnika 3 (1960), no. 2, 216. (Russian)

[9] I. I. Gihman and A. V. Skorohod, *Introduction to the theory of random processes*, "Nauka", Moscow, 1965; English transl., Saunders, Philadephia, Pa. 1969. MR 33 #6689.

[10] E. B. Dynkin, *Markov processes*, Fizmatgiz, Moscow, 1963; English transl., Vols. I, II, Die Grundlehren der math. Wissenschaften, Bände 121, 122, Academic Press, New York; Springer-Verlag, Berlin, 1965. MR 33 #1886; 1887.

Trudy Mat. Inst. Steklov.
95 (1968)

Proc. Steklov Inst. Math.
95 (1968)

PROBABILISTIC CHARACTERISTICS AND ASYMPTOTIC DISTRIBUTIONS OF ONCE-REFLECTED SUM WAVES IN A STOCHASTIC MODEL OF A STRATIFIED MEDIUM APPLIED TO SEISMIC PROSPECTING

UDC 519.217

L. A. ZOLOTUHINA

Abstract. A stochastic model of a plane-parallel medium is considered in which reflected waves are propagated. The basic probabilistic characteristics of a once-reflected sum wave (mathematical expectation, correlation coefficients, multivariate characteristic function) are calculated. The multivariate asymptotic distribution of the process investigated is obtained by generalizing Robbins' theorem on the sum of a random number of random functions.

Bibliography 8 items.

The problem of interpreting seismic data is usually solved by deterministic methods. This approach, however, ignores the effect of noise. This effect can be taken into account if the problem is solved by using methods from the theory of probability and mathematical statistics. A basic difficulty in the statistical approach lies in the correct description of noise, especially if the noise is related to waves from noninterpretable objects. In this case the description of noise must be based on ergodic considerations and the distribution of noise depends essentially on the model of its origin. The necessity of a probabilistic approach to problems of seismic survey is considered in more detail in [1], where several stochastic models are suggested. In this paper, on the basis of a stochastic model of the medium, we calculate the basic probabilistic characteristics of a once-reflected sum wave arising in the medium and realized by a random process. These characteristics are the mathematical expectation, correlation coefficients and the multivariate characteristic function. The multivariate asymptotic distribution of the process is also found and is obtained by generalizing Robbins' theorem on the sum of a random number of random variables. Two illustrative examples are discussed.

Since the fundamental contribution to noise is provided by waves from noninterpretable objects, it is important to investigate the behavior of waves reflected from various inhomogeneities, the boundaries of strata situated in the mass of the earth's crust. We shall consider the following probabilistic model of a stratified medium. Boundaries of inhomogeneities (strata) in the form of straight line segments are randomly situated in a vertical section of the earth which is a half-plane bounded by the diurnal surface. In what follows we shall simply call them boundaries. All the boundaries are parallel to each other and to the diurnal surface. Each boundary is characterized by its length l, depth z and coordinate of origin ξ. We assume that the boundary origins ξ form a random flow; that is, if we denote by $S(x)$ the number of boundary origins in the interval $(0, x]$ then $S(x)$ is a random process which takes integral values; the function $S(x)$ does not decrease as x increases.[1] We assume that the boundary lengths l are independent, identically distributed random variables with distributions functions $F(x) = P(l > x)$, $-\int_0^\infty x dF(x) = s < \infty$, and not dependent on $S(x)$. The depths z are also independent, identically distributed random variables with distribution function $G(z)$.

Analogous situations arise in queueing theory if the flow of origins is considered as a flow of calls, the boundary lengths as conversation lengths and the depths as serv . The flow of calls is arbitrary, the distribution of conversation lengths is arbitrary and the server is randomly chosen with an arbitrary distribution function $G(z)$ (for continuous $G(z)$ the number of servers is infinite).

One of the problems of seismic survey is the study of the medium. This cannot be carried out directly but we can, however, study the waves propagated in this medium caused by an explosion and recorded on seismic detectors. The characteristics of the medium can be assessed from the properties of the waves. Since in the proposed model a medium is the realization of a random process, the problem of studying the medium can be considered from the point of view of the probabilistic properties of waves propagated in such media. The source of the explosion and the seismic detectors are assumed to lie along the diurnal surface. We introduce the following coordinate system: the origin is at the point of the explosion and the x-axis is a straight line dividing the half-plane and perpendicular to the z-axis. Let $(2b_1, 0)$, $(2b_2, 0), \cdots, (2b_K, 0)$ be the coordinates of the detectors in this system. We consider once-reflected waves and assume that the reflections obey the rules of geometric optics. The waves reflected from the boundaries intersection the straight line $x = b$ will arrive at the detector situated at a distance $2b$ from the source. If the boundary situated at depth z intersects $x = b$, then at time t a wave reflected from this boundary will arrive at the point $2b$, the wave being of the form

---

1) In [1] $S(x)$ is taken to be a Poisson flow.

$$y(t,\ b) = Af\left(t - \frac{2}{v}\sqrt{b^2 + z^2}\right)$$

$$= \begin{cases} 0, & \text{if}\quad t - \frac{2}{v}\sqrt{b^2 + z^2} \leqslant 0, \\ Af_1\left(t - \frac{2}{v}\sqrt{b^2 + z^2}\right), & \text{if}\quad t - \frac{2}{v}\sqrt{b^2 + z^2} > 0, \end{cases} \qquad (1)$$

where the delay time $\tau = 2\sqrt{b^2 + z^2}/v$ is easily calculated from geometric considerations; f_1 is a determinate function; A is the amplitude of the reflected wave, a random variable independently and identically distributed for different boundaries with distribution function $G_A(x)$; v is the velocity of wave propagation, a constant; and z is the depth of the reflecting boundary.[1)]

If $x = b$ intersects $N(b)$ boundaries ($N(b)$ random) the sum wave $U(t,\ b)$ arriving at the point $2b$ at time t can be put in the form

$$U(t,\ b) = \sum_{i=1}^{N(b)} y_i(t,\ b),$$

where $y_i(t,\ b)$ is calculated in accordance with (1) for $z = z_i$. If $N(b) = 0$ then $U(t,\ b) = 0$.

Since the $y_i(t,\ b)$ $(i = 1, \cdots, N(b))$ are functions of independent and identically distributed random variables z_i and A_i, the random variables $y_i(t,\ b)$ for fixed t and b are also independent and identically distributed. Therefore $U(t,\ b)$ for fixed t and b is the sum of a random number of random terms which are independent and identically distributed. In calculating the probabilistic characteristics of $U(t,\ b)$ we require the probabilistic characteristics of $y(t,\ b)$ and $N(b)$:

$$a = Ey(t,\ b) = \int_{-\infty}^{\infty}\int_{0}^{H} Af\left(t - \frac{2}{v}\sqrt{b^2 + z^2}\right) dG_A(A)\, dG(z),$$

$$d^2 = Ey^2(t,\ b) = \int_{-\infty}^{\infty}\int_{0}^{H} A^2 f^2\left(t - \frac{2}{v}\sqrt{b^2 + z^2}\right) dG_A(A)\, dG(z),$$

$$c^2 = \operatorname{var} y(t,\ b) = d^2 - a^2.$$

The characteristic function of $y(t,\ b)$ is

$$\Psi(\theta) = Ee^{i\theta y(t,\ b)} = \int_{-\infty}^{\infty}\int_{0}^{H} e^{i\theta Af\left(t - \frac{2}{v}\sqrt{b^2+z^2}\right)} dG_A(A)\, dG(z).$$

---

1) Since the observations are made in a finite time interval $(0,\ T)$, waves from very deep boundaries will not reach the seismic detector and therefore the flow of boundaries can always be considered in the band $(0,\ H)$ parallel to the x-axis and $G(z)$ assumed given in the interval $(0,\ H)$, where $0 < H < \infty$.

The quantities a, d, c and $\Psi(\theta)$ depend on t and b, but for brevity these arguments are omitted.

In terms of queueing theory $N(b)$ can be interpreted as the number of conversations in progress at time b. But finding the law of distribution of the number of conversations in progress at a certain time is one of the basic problems of queueing theory. The distribution of $N(b)$ is completely determined by the flow and the service and is calculated in queueing theory under various assumptions regarding the character of the flows and service. We give some known results from this field.

Example 1. If the flow of calls is simple with parameter λ, the distribution of conversation lengths is arbitrary and the number of servers infinite, then under conditions of stationarity [2,3]

$$P(N=k)=e^{-\lambda s}\frac{(\lambda s)^k}{k\,!}, \tag{2}$$

where $s=\int_0^\infty x dF(x)<\infty$ (for a stationary flow the argument b can be omitted. This case will be the main illustrative example throughout the remainder of the paper.

It is also possible to consider the case in which the distances between calls are independent and identically distributed random variables with an arbitrary distribution law, the conversation lengths are also governed by an arbitrary distribution law, and the number of servers is infinite. For this case Takács [3] has found an integral equation which must be satisfied by the generating function for the number of conversations in progress at a fixed point of time.

We assume from now on that the distribution of $N(b)$ is known. We use the following notation, omitting the argument b:

$$\begin{gathered}\omega_k=P(N=k),\\ EN=\alpha,\\ EN^2=\beta^2,\\ \operatorname{var} N=\gamma^2=\beta^2-\alpha^2,\\ \mathcal{F}(\theta)=E\left\{e^{i\theta\frac{N-\alpha}{\gamma}}\right\}\end{gathered}$$

is the characteristic function of the normalized random variable $(N-\alpha)/\gamma$. The moments of $U(t,b)$ are then easily calculated:

$$\begin{gathered}EU(t,b)=\alpha a,\\ EU^2(t,b)=\alpha c^2+\beta^2a^2,\\ \sigma^2=\operatorname{var} U(t,b)=\alpha c^2+\gamma^2a^2.\end{gathered}$$

The characteristic function of the normalized random variable $w = (U(t, b) - EU(t, b))/\sigma$ is

$$\varphi(\theta) = Ee^{i\theta w} = e^{-\frac{i\alpha a\theta}{\sigma}} \sum_{k=0}^{\infty} \omega_k \left[\Psi\left(\frac{\theta}{\sigma}\right)\right]^k.$$

Example 1 (continuation). Since $P(N = k) = e^{-\lambda s}(\lambda s)^k/k!$, we have

$$EN = \alpha = \lambda s, \quad \operatorname{var} N = \gamma^2 = \lambda s, \quad \sigma^2 = \lambda s d^2 \tag{3}$$

and then

$$\varphi(\theta) = e^{-i\frac{\sqrt{\lambda s}\,a\theta}{d}} e^{-\lambda s\left(1-\Psi\left(\frac{\theta}{\sqrt{\lambda s}\,d}\right)\right)}.$$

We now assume that the distribution of the origins, and therefore the distributions of N, depends on the parameter λ. The distribution of $U(t, b)$ will then also depend on λ and we can seek the asymptotic distribution of $U(t, b)$ as $\lambda \to \infty$.

The following result due to Robbins [4] is known.

Theorem 1. *Let $U = \Sigma_{i=1}^{N} y_i$, where y_i are independent identically distributed random variables with characteristic function $\Psi(\theta)$, and let $Ey_i = a$, $Ey_i^2 = d^2$ and $\operatorname{var} y_i = c^2$ $(0 \leqslant c^2 < \infty)$; let N be a random variable not depending on y_i whose distribution function depends on λ, where $EN = \alpha$, $EN^2 = \beta^2$ and $\operatorname{var} N = \gamma^2$; and let $\mathfrak{F}(\theta)$ be the characteristic function for $(N - \alpha)/\gamma$ (α, β^2, $\mathfrak{F}(\theta)$ depend on λ).*

Further, let $\operatorname{var} U = \sigma^2$, and let $\phi(\theta)$ be the characteristic function of $(U - EU)/\sigma$.

Then, if as $\lambda \to \infty$

1) $\sigma^2 \to \infty$,

2) $\gamma = o(\sigma^2)$,

we have

$$\varphi(\theta) = \mathcal{F}(\delta\theta)\, e^{-\frac{\theta^2(1-\delta^2)}{2}} + o(1), \tag{4}$$

where

$$\delta = \frac{\gamma a}{\sigma} = \left(\frac{\gamma^2 a^2}{\alpha c^2 + \gamma^2 a^2}\right)^{1/2} \quad (0 \leqslant \delta \leqslant 1).$$

In the conditions of the theorem it is not assumed that $\lim_{\lambda\to\infty} \mathfrak{F}(\theta)$ exists. If, however, the random variable $(N - \alpha)/\gamma$ has a limiting distribution function $K(x)$ so that $\lim_{\lambda\to\infty} \mathfrak{F}(\theta) = \mathfrak{F}_1(\theta)$ exists and if $\nu = \lim_{\lambda\to\infty} c^2\alpha/a^2\gamma^2$ also exists, then (4) can be rewritten in the form

$$\lim_{\lambda\to\infty}\varphi(\theta)=\mathscr{F}_1\left(\frac{\theta}{(1+\nu)^{1/2}}\right)e^{-\frac{\left(\theta\left(\frac{\nu}{1+\nu}\right)^{1/2}\right)^2}{2}}, \tag{5}$$

that is, $w=(U(t,\ b)-EU(t,\ b))/\sigma$ has a limiting distribution function given by

$$H(x)=\mathscr{K}(x(1+\nu)^{1/2})*H_0\left(x\left(\frac{1+\nu}{\nu}\right)^{1/2}\right). \tag{6}$$

Here $*$ is the sign of convolution and $H_0(x)$ is the normal distribution with parameters (0, 1).

Example 1 (continuation). Since in this case N has a Poisson distribution with parameter λs, its distribution as $\lambda\longrightarrow\infty$ is asymptotically normal and conditions 1) and 2) of Theorem 1 are satisfied (see (3)). Hence, as we see from (5) or (6), the distribution of $U(t,\ b)$ is also asymptotically normal as $\lambda\longrightarrow\infty$. There also exist more precise asymptotic expansions for the distribution of $U(t,\ b)$ when N has a Poisson or normal distribution [5].

We note that for $a\neq 0$ condition 2) of Theorem 1 is always satisfied and can be omitted. If, however, $a=0$ condition 2) is in fact necessary, and in this case is equivalent to the condition $\gamma/\alpha\longrightarrow 0$ as $\lambda\longrightarrow\infty$.

The results stated apply to univariate characteristics of the process $U(t,\ b)$ for fixed t and b. Since $U(t,\ b)$ represents a random field, in order to describe it completely we need to know the multivariate distributions for arbitrary sets of the parameters t and b.

Let $U_j=U(t_j,\ b_j)$ $(j=1,\cdots,\ k)$ be the sum wave arriving at the point $2b_j$ at time t_j, where $t_j\in(0,\ T]$, and let U_j $(j=1,\cdots,\ k)$ be written in the order of increasing b_j; that is, $b_1\leq b_2\leq\cdots\leq b_k$ (the relation $b_j=b_{j+1}$ means that $U(t_j,\ b_j)$ and $U(t_{j+1},\ b_{j+1})$ are calculated for the same value of b). If there are N_j boundaries in the section $x=b_j$, then

$$U_j=\sum_{l=1}^{N_j}y_l(t_j,\ b_j)=\sum_{l=1}^{N_j}A_lf\left(t-\frac{2}{v}\sqrt{b_j^2+z_l^2}\right)\quad(j=1,\ \ldots,\ k).$$

Let us consider the vector

$$\mathrm{U}=(U_1,\ \ldots,\ U_k)=\left(\sum_{l=1}^{N_1}A_lf\left(t_1-\frac{2}{v}\sqrt{b_1^2+z_l^2}\right),\right.$$

$$\left.\sum_{l=1}^{N_2}A_lf\left(t-\frac{2}{v}\sqrt{b_2^2+z_l^2}\right),\ \ldots,\ \sum_{l=1}^{N_k}A_lf\left(t_k-\frac{2}{v}\sqrt{b_k^2+z_l^2}\right)\right).$$

We see that each component of U is the sum of a random number of independent random variables, but the components themselves (that is, the U_j) are mutually dependent, for: 1) $N_1, \cdots, N_k$ are dependent; 2) there are also dependent terms among those which form the components, since some boundaries may be observed in several sections.

We call the region between the lines $x = b_i$ and $x = b_{i+1}$ $(i = 1, \cdots, k-1)$ the ith region, the region to the left of the line $x = b_1$ the zero region and the region to the right of $x = b_k$ the kth region.

Let n_n $(i = 1, \cdots, k-1)$ be the number of boundaries which appear in the ith region (if $b_i = b_{i+1}$ the ith region equals 0 and therefore $n_i = 0$), and let μ_{ij} $(i = 0, \cdots, k-1;\ i \leq j \leq k)$ be the number of boundaries whose origins lie in the ith region and whose ends lie in the ith region (if $b_i = b_{i+1}$ then $\mu_{ij} = 0$ for $i \leq j \leq k$ and $\mu_{li} = 0$ for $l = 0, \cdots, i-1$).

It is obvious that

$$n_i = \mu_{ii} + \ldots + \mu_{ik} \quad (i = 1, \ldots, k). \tag{7}$$

Then N_i $(i = 1, \cdots, k)$ can be put in the form

$$\begin{aligned} N_1 &= \mu_{01} + \mu_{02} + \ldots + \mu_{0k}, \\ N_2 &= \mu_{02} + \ldots + \mu_{0k} + \mu_{12} + \mu_{13} + \ldots + \mu_{1k} \\ &\cdots\cdots\cdots\cdots\cdots\cdots \\ N_i &= \mu_{0i} + \ldots + \mu_{0k} + \mu_{1i} + \ldots + \mu_{1k} + \ldots + \mu_{i-1\,i} + \ldots + \mu_{i-1\,k} \\ &= \sum_{j=0}^{i-1} \sum_{l=i}^{k} \mu_{jl}, \\ &\cdots\cdots\cdots\cdots\cdots\cdots \\ N_k &= \mu_{0k} + \ldots + \mu_{k-1\,k}. \end{aligned} \tag{8}$$

Therefore U can be rewritten in the form

$$\begin{aligned} \mathrm{U} = \Bigg(&\sum_{l=1}^{k} \sum_{\nu=0}^{\mu_{01}} A_\nu^{(0l)} f\left(t_1 - \frac{2}{v}\sqrt{b_1^2 + (z_\nu^{(0l)})^2}\right), \ldots, \\ &\sum_{j=0}^{i-1} \sum_{l=i}^{k} \sum_{\nu=0}^{\mu_{jl}} A_\nu^{(jl)} f\left(t_i - \frac{2}{v}\sqrt{b_i^2 + (z_\nu^{(jl)})^2}\right), \ldots \\ &\ldots, \sum_{j=0}^{k-1} \sum_{\nu=0}^{\mu_{jk}} A_\nu^{(jk)} f\left(t_k - \frac{2}{v}\sqrt{b_k^2 + (z_\nu^{(jk)})^2}\right)\Bigg), \end{aligned}$$

where $z_i^{(jl)}$ is the depth of the ith boundary out of those μ_{jl} boundaries starting in the jth and ending in the lth region; and $A_i^{(jl)}$ is the amplitude of the wave generated by this layer. With this notation the only dependent terms are those with identical indices j, l, i. But the random variables $\mu_{01}, \mu_{02}, \cdots, \mu_{k-1\,k}$ themselves are dependent. Knowing the joint distribution of $\mu_{01}, \mu_{02}, \cdots, \mu_{k-1\,k}$ we can write down the characteristic function of U.

Example 1 (continuation). We show how the joint distribution of $\mu_{01}, \cdots, \mu_{k-1\,k}$ is obtained in the case of a simple flow of boundary origins and an arbitrary distribution of boundary lengths under stationary conditions. Since a simple flow is characterized by the absence of an aftereffect, N_1 and $n_1, \cdots, n_k$ are independent. In accordance with (7) and (8) each n_i and N_1 can be decomposed into the sum of terms of the form μ_{il} $(l = i + 1, \cdots, k)$ and by the independence of l_i the chains

$$\begin{gathered}\mu_{01}, \mu_{02}, \cdots, \mu_{0k}, \\ \mu_{12}, \cdots, \mu_{1k}, \\ \cdot\ \cdot\ \cdot\ \cdot\ \cdot \\ \mu_{k-1\,k}\end{gathered}$$

are independent. It is therefore sufficient to write the distribution of each of the chains separately.

Let us calculate the distribution of the first chain $\mu_{01}, \cdots, \mu_{0k}$. First of all, we find

$$P\left(\mu_{01}=r_{01}, \ \ldots, \ \mu_{0k-1}=r_{0k-1} \,|\, N_1=\sum_{i=1}^{k}\mu_{0i}=r\right).$$

Since the boundary lengths are independent and identically distributed, we have

$$P(\mu_{01}=r_{01}, \ \ldots, \ \mu_{0\,k-1}=r_{0\,k-1} \,|\, N_1=r)$$
$$=\frac{r\,!}{r_{01}\,!\ \ldots\ r_{0\,k-1}\,!\ (r-r_{01}-\ \ldots\ -r_{0\,k-1})\,!}\, p_1^{r_{01}}\ \ldots\ p_k^{r-r_{01}-\cdots-r_{0\,k-1}},$$
$$\sum_{i=1}^{k} p_i = 1,$$

where p_i $(i = 1, \cdots, k - 1)$ is the probability that a boundary with origin in the zero region has its end in the ith region. In the case in question this is the probability that the end of a boundary which intersects the line $x = b_1$ lies in the ith region. In terms of queueing theory p_k is the probability that a conversation we found to be in progress arbitrarily chosen moment of time will not have finished after the elapse of time $b_k - b_1$. As was shown by A. Ja Hinčin (see [2], p. 204),

$$p_k = \frac{1}{s} \int_{b_k - b_1}^{\infty} F(u)\, du = \varepsilon(b_k - b_1),$$

and p_i $(i = 1, \cdots, k-1)$, the probability that a conversation in progress at time b_1 will finish in the interval (b_i, b_{i+1}) is given by

$$p_i = \varepsilon(b_i - b_1) - \varepsilon(b_{i+1} - b_1).$$

Therefore

$$\begin{aligned} &P(\mu_{01} = r_{01}, \ldots, \mu_{0\,k-1} = r_{0\,k-1} \mid N_1 = r) \\ &= \frac{r!}{r_{01}! \ldots r_{0\,k-1}!\,(r - r_{01} - \ldots - r_{0\,k-1})!} (\varepsilon(b_k - b_1))^{r - r_{01} - \cdots - r_{0\,k-1}} \\ &\times (\varepsilon(b_{k-1} - b_1) - \varepsilon(b_k - b_1))^{r_{0\,k-1}} \ldots (1 - \varepsilon(b_2 - b_1))^{r_{01}}. \end{aligned} \tag{9}$$

Using (2) and (9) and setting $r_{0k} = r - r_{01} - \cdots - r_{0\,k-1}$, we can easily write down the distribution of the first chain in the form

$$\begin{aligned} &P(\mu_{01} = r_{01}, \ldots, \mu_{0\,k-1}, = r_{0\,k-1}, \mu_{0k} = r_{0k}) \\ &= P(\mu_{01} = r_{01}, \ldots, \mu_{0\,k-1} = r_{0\,k-1}, N_1 = r) \\ &= e^{-\lambda s} \frac{1}{r_{01}! \ldots r_{0k}!} [\lambda s \varepsilon(b_k - b_1)]^{r_{0k}} \\ &\times [\lambda s(\varepsilon(b_{k-1} - b_1) - \varepsilon(b_k - b_1))]^{r_{0\,k-1}} \ldots [\lambda s(1 - \varepsilon(b_2 - b_1))]^{r_{01}}. \end{aligned} \tag{10}$$

Let us write down the distribution of $\mu_{i\,i+1}, \cdots, \mu_{ik}$ $(i = 1, \cdots, k-1)$. Suppose that r boundaries appear in the ith region and the origins of these boundaries have coordinates $\xi_1, \cdots, \xi_r$ along the x-axis. It can be shown that the ξ_i $(i = 1, \cdots, r)$ are uniformly distributed in the interval (b_i, b_{i+1}). Then

$$\begin{aligned} &P(\mu_{i\,i+1} = r_{i\,i+1}, \ldots, \mu_{ik} = r_{ik}) \\ &= \sum_{r = r_{i\,i+1} + \cdots + r_{ik}}^{\infty} P(n_i = r) \int_{b_i}^{b_{i+1}} \cdots \int_{b_i}^{b_{i+1}} \frac{d\xi_1 \ldots d\xi_r}{(b_{i+1} - b_i)^2} \\ &\times P(\mu_{i\,i+1} = r_{i\,i+1}, \ldots, \mu_{ik} = r_{ik} \mid n_i = r, \xi_1, \ldots, \xi_r). \end{aligned}$$

We rewrite ξ_i $(i = 1, \cdots, r)$ in the form ξ_{jm} $(j = 1, \cdots, r_m)$ to denote the coordinates of the origins of the boundaries ending in the mth region. Then

$$\begin{aligned} &P\left(\mu_{i\,i+1} = r_{i\,i+1}, \ldots, \mu_{ik} = r_{ik} \mid n_i = r, \xi_{jm} \begin{pmatrix} j = 1, \ldots, r_m, \\ m = i+1, \ldots, k \end{pmatrix}\right) \\ &= \frac{r!}{r_{i\,i+1}! \ldots r_{ik}!\,(r - r_{i\,i+1} - \ldots - r_{ik})!} \prod_{j=1}^{r_{i\,i+1}} \Big[F(b_{i+1} - \xi_{j\,i+1}) - \end{aligned}$$

$$-F(b_{i+2}-\xi_{j\,i+1})\Big]\prod_{j=1}^{r_{i\,i+2}}[F(b_{i+2}-\xi_{j\,i+2})-F(b_{i+3}-\xi_{j\,i+2})\Big]\times\cdots$$

$$\cdots\times\prod_{j=1}^{r_{i\,k-1}}[F(b_{k-1}-\xi_{j\,k-1})-F(b_k-\xi_{j\,k-1})]\prod_{j=1}^{r_{ik}}F(b_k-\xi_{jk})$$

$$\times\prod_{j=1}^{r-r_{i\,i+1}-\cdots-r_{ik}}[1-F(b_{i+1}-\xi_{ji})].$$

Integrating with respect to ξ_{jm}, we obtain

$$P(\mu_{i\,i+1}=r_{i\,i+1},\ \ldots,\ \mu_{ik}=r_{ik}\,|\,n_i=r)$$

$$=\frac{r!}{(b_{i+1}-b_i)^r}\prod_{l=i+1}^{k}\frac{(sB_{il})^{r_{il}}}{r_{il}!}\cdot\frac{[b_{i+1}-b_i-s(1-\varepsilon(b_{i+1}-b_i))]^{r-r_{i\,i+1}-\cdots-r_{ik}}}{(r-r_{i\,i+1}-\ \ldots\ -r_{ik})!},\qquad(11)$$

where

$$\left.\begin{array}{c}B_{il}=\varepsilon(b_l-b_{i+1})-\varepsilon(b_l-b_i)-\varepsilon(b_{l+1}-b_{i+1})+\varepsilon(b_{l+1}-b_i)\\ \text{for}\quad i=1,\ \ldots,\ k-1,\\ B_{ik}=\varepsilon(b_k-b_{i+1})-\varepsilon(b_k-b_i).\end{array}\right\}\qquad(12)$$

Summing (11) over r and taking account of the fact that

$$P(n_i=r)=e^{-\lambda(b_{i+1}-b_i)}\frac{(\lambda(b_{i+1}-b_i))^r}{r!},$$

we finally obtain for $i=1,\cdots,k-1$

$$P(\mu_{i\,i+1}=r_{i\,i+1},\ \ldots,\ \mu_{ik}=r_{ik})=e^{-\lambda s(1-\varepsilon(b_{i+1}-b_i))}\prod_{l=i+1}^{k}\frac{(\lambda sB_{il})^{r_{il}}}{r_{il}!}.\qquad(13)$$

We note that, for the joint distribution of the first chain, (10) coincides with (13) if we put $\epsilon(b_j-b_0)=0$, so that (13) is also valid for $i=0$.

Since the chains are independent, we have

$$P(\mu_{01}=r_{01},\ \ldots,\ \mu_{k-1\,k}=r_{k-1\,k})=\prod_{i=0}^{k-1}P(\mu_{i\,i+1}=r_{i\,i+1},\ \ldots,\ \mu_{ik}=r_{ik}).\qquad(14)$$

We have therefore calculated the distribution of $\mu_{01},\cdots,\mu_{k-1\,k}$ when the boundary origins have a Poisson flow.

In the case of other distributions of flows of boundary origins it is naturally harder to calculate the distributions of $\mu_{01},\cdots,\mu_{k-1\,k}$, but it turns out that, if this distribution depends on the parameter λ and we are seeking the asymptotic distribution of U as $\lambda\to\infty$, then all that is needed for a knowledge of the limiting characteristic function of U is the limiting characteristic of the vector $(N_1,\cdots,N_k)$; and this is obviously easier to find. We shall meet this situation in Example 2. We assume further that we know the distribution of $\mu_{01},\cdots,\mu_{k-1\,k}$ and thereby also the distribution of $N_1,\cdots,N_k$. Then

$$\alpha_i = EN_i = \sum_{j=0}^{i-1}\sum_{l=i}^{k} \alpha_{jl}, \text{ where } \alpha_{jl} = E\mu_{jl} \ (i=1, \ldots, k),$$

$$\beta_{ii'} = EN_iN_{i'} = \sum_{j=0}^{i'-1}\sum_{l=i'}^{k}\sum_{g=0}^{i-1}\sum_{q=i}^{k} \beta_{gq,\,jl}, \text{ where } \beta_{gq,\,jl} = E\mu_{gq}\mu_{jl} \ (i',\ i=1, \ldots, k),$$

$$\gamma_{ii'} = E(N_i - EN_i)(N_{i'} - EN_{i'}) = \sum_{g=0}^{i-1}\sum_{q=i}^{k}\sum_{j=0}^{i'-1}\sum_{l=i'}^{k} \gamma_{gq,\,jl}, \qquad (15)$$

where

$$\gamma_{gq,\,jl} = E(\mu_{gq} - E\mu_{gq})(\mu_{jl} - E\mu_{jl}) \ (i,\ i'=1, \ldots k),$$

$$R_{ii'} = \frac{\gamma_{ii'}}{\sqrt{\gamma_{ii}}\sqrt{\gamma_{i'i'}}} \qquad (16)$$

is the coefficient of correlation between N_i and $N_{i'}$, and

$$\mathcal{F}(\theta_1, \ldots, \theta_k) = Ee^{\left(\theta_1 \frac{N_1-\alpha_1}{\sqrt{\gamma_{11}}} + \ldots + \theta_k \frac{N_k - \alpha_k}{\sqrt{\gamma_{kk}}}\right)}$$

is the characteristic function of the normalized random vector

$$\left(\frac{N_1-\alpha_1}{\sqrt{\gamma_{11}}}, \ldots, \frac{N_k-\alpha_k}{\sqrt{\gamma_{kk}}}\right).$$

We now consider the vector

$$(y(t_1, b_1), \ldots, y(t_k, b_k)) = \Big(Af\Big(t_1 - \frac{2}{v}\sqrt{b_1^2+z^2}\Big),$$
$$\ldots, Af\Big(t_k - \frac{2}{v}\sqrt{b_k^2+z^2}\Big)\Big).$$

Let

$$a_i = Ey(t_i, b_i) \ (i=1, \ldots, k),$$

$d_{ij} = Ey(t_i, b_i)\,y(t_j, b_j)$ $(i,\ j=1, \ldots, k)$, $c_{ij} = \operatorname{cov}(y(t_i, b_i)\,y(t_j, b_j))$ $= d_{ij} - a_ia_j$ $(i,\ j=1, \ldots, k)$, $\Psi(\theta_1, \ldots, \theta_k) = Ee^{i(\theta_1 y(t_1, b_1) + \ldots + \theta_k y(t_k b_k))}$.

Then the moments of the vector $\mathrm{U} = (U_1, \cdots, U_k)$ are expressed in terms of the moments α_{ij}, α_i, β_{ij}, γ_{ij}, a_i, d_{ij} and c_{ij} in the following way:

$$EU_i = \alpha_i a_i \quad (i=1, \ldots, k),$$

$$EU_iU_j = c_{ij}\left(\sum_{l=\max(i,\,j)}^{k}\ \sum_{m=0}^{\min(i,\,j)-1} \alpha_{ml}\right) + a_ia_j\beta_{ij},$$

and, in particular, for $i = j$:

$$EU_i^2 = c_{ii}\alpha_i + a_i^2\beta_{ii}.$$

The elements of the variance matrix of U are

$$\varkappa_{ij} = c_{ij}\left(\sum_{l=\max(i,\,j)}^{k} \sum_{m=0}^{\min(i,\,j)-1} \alpha_{ml}\right) + a_i a_j \gamma_{ij} \tag{17}$$
$$(i,\ j=1,\ \ldots,\ k),$$

and, in particular, the variance of U_i $(i = 1, \cdots, k)$ is given by

$$\varkappa_{ii} = c_{ii}\alpha_i + a_i^2 \gamma_{ii}. \tag{18}$$

The coefficient of correlation between U_i and U_j is

$$\rho_{ij} = \frac{\varkappa_{ij}}{\sqrt{\varkappa_{ii}}\sqrt{\varkappa_{jj}}}. \tag{19}$$

In place of U we introduce the normalized vector $w = (w_1, \cdots, w_k)$, where $w_i = (U_i - EU_i)/\sqrt{\kappa_{ii}}$. The characteristic function of w is

$$\varphi(\theta_1, \ldots, \theta_k) = E \exp\{i(w_1\theta_1 + \ldots + w_k\theta_k)\}$$
$$= e^{-i\sum_{l=1}^{k} \frac{a_l \alpha_l \theta_l}{\sqrt{\varkappa_{ll}}}} \sum_{r_{01}} \cdots \sum_{r_{k-1\,k}} P(\mu_{01} = r_{01}, \ldots, \mu_{k-1\,k} = r_{k-1\,k})$$
$$\times \left[\Psi\left(\frac{\theta_1}{\sqrt{\varkappa_{11}}}, \ldots, \frac{\theta_k}{\sqrt{\varkappa_{kk}}}\right)\right]^{r_{0k}} \left[\Psi\left(\frac{\theta_1}{\sqrt{\varkappa_{11}}}, \ldots, \frac{\theta_{k-1}}{\sqrt{\varkappa_{k-1\,k-1}}}, 0\right)\right]^{r_{0k-1}} \cdots$$
$$\cdots \left[\Psi\left(\frac{\theta_1}{\sqrt{\varkappa_{11}}}, 0, \ldots, 0\right)\right]^{r_{01}} \left[\Psi\left(0, \frac{\theta_2}{\sqrt{\varkappa_{22}}}, \ldots, \frac{\theta_k}{\sqrt{\varkappa_{kk}}}\right)\right]^{r_{1k}}$$
$$\times \left[\Psi\left(0, \frac{\theta_2}{\sqrt{\varkappa_{22}}}, \ldots, \frac{\theta_{k-1}}{\sqrt{\varkappa_{k-1\,k-1}}}, 0\right]^{r_{1\,k-1}} \cdots \left[\Psi\left(0, \frac{\theta_2}{\sqrt{\varkappa_{22}}}, 0, \ldots, 0\right)\right]^{r_{12}}$$
$$\cdots \times \left[\Psi\left(0, \ldots, 0, \frac{\theta_k}{\sqrt{\varkappa_{kk}}}\right)\right]^{r_{k-1k}}. \tag{20}$$

We note that $\Psi(\theta_1, \cdots, \theta_k)$ denotes the characteristic function of the vector $(y(t_1, b_1), \cdots, y(t_k, b_k))$, none of whose components vanish, which means that the boundary giving a wave of the form $(y(t_1, t_1), \cdots, y(t_k, b_k))$ intersects all the lines $x = b_i$ $(i = 1, \cdots, k)$. If, however, the boundary intersects only the lines $x = b_m, \cdots, x = b_r$ $(1 \leq m \leq r \leq k)$, then the characteristic function of the vector $(0, \cdots, 0, y(t_m, b_m), \cdots, y(t_r, b_r), 0, \cdots, 0)$ is written

$$\Psi(0, \ldots, 0, \theta_m, \ldots, \theta_r, 0, \ldots, 0).$$

Example 1 (continuation). For a simple flow of boundary origins we obtain, using (14) and (13),

$$\varphi(\theta_1, \ldots, \theta_k) = e^{-i\sum_{l=1}^{k}\frac{a_l\alpha_l\theta_l}{\sqrt{\varkappa_{ll}}}} \prod_{j=0}^{k-1} e^{-\lambda s(1-\varepsilon(b_{j+1}-b_j))}$$

$$\times \prod_{l=j+1}^{k} e^{\lambda s B_{jl}\Psi\left(0, \ldots, 0, \frac{\theta_{j+1}}{\sqrt{\varkappa_{j+1\,j+1}}}, \ldots, \frac{\theta_l}{\sqrt{\varkappa_{ll}}}, 0, \ldots, 0\right)}.$$

In the distribution of $(\mu_{0\,1}, \cdots, \mu_{k-1\,k})$ depends on the parameter λ, then, as in the univariate case, we can seek an asymptotic distribution of U as $\lambda \to \infty$, generalizing Robbins' theorem in the following way.

Theorem 1$'$. *If as* $\lambda \to \infty$

1) $\kappa_{ii} = \operatorname{var} U_i \to \infty$ $(i = 1, \cdots, k)$,

2) $\operatorname{var}\mu_{ij} = \gamma_{ij,\,ij} = o(\sigma^2)$, *where* $\sigma = \min_i\{\kappa_{ii}\}$ $(i = 1, \cdots, k)$, *then*

$$\varphi(\theta_1, \ldots, \theta_k) = \mathscr{F}(\delta_1\theta_1, \ldots, \delta_k\theta_k)\, e^{-\frac{1}{2}\sum_{i=1}^{k}\sum_{j=1}^{k}(\rho_{ij}-R_{ij}\delta_i\delta_j)\theta_i\theta_j} + o(1), \tag{21}$$

where $\phi(\theta_1, \cdots, \theta_k)$ *is the characteristic function of the vector*

$$w = \left(\frac{U_1 - EU_1}{\sqrt{\varkappa_{11}}}, \ldots, \frac{U_k - EU_k}{\sqrt{\varkappa_{kk}}}\right),$$

$$\delta_i = \frac{a_i\sqrt{\gamma_{ii}}}{\sqrt{\varkappa_{ii}}} \quad (i = 1, \ldots, k), \quad 0 \leqslant \delta_i \leqslant 1. \tag{22}$$

Proof. We rewrite (20) in the form

$$\varphi(\theta_1, \ldots, \theta_k) = \sum_{r_{01}} \cdots \sum_{r_{k-1\,k}} P(\mu_{01} = r_{01}, \ldots, \mu_{k-1\,k} = r_{k-1\,k})\, e^{i\sum_{l=1}^{k}\frac{(N_l\,\alpha_l)a_l\theta_l}{\sqrt{\varkappa_{ll}}}}$$

$$\times \left\{\prod_{j=1}^{k}\prod_{l=j}^{k}\left[e^{-\frac{ia_j\theta_j}{\sqrt{\varkappa_{jj}}} - \ldots - \frac{ia_l\theta_l}{\sqrt{\varkappa_{ll}}}}\Psi\left(0, \ldots, 0, \frac{\theta_j}{\sqrt{\varkappa_{jj}}}, \ldots, \frac{\theta_l}{\sqrt{\varkappa_{ll}}}, 0, \ldots, 0\right)\right]^{r_{j-1\,l}}\right\}.$$

We introduce the function $\phi_1(\theta_1, \cdots, \theta_k)$ which differs from $\phi(\theta_1, \cdots, \theta_k)$ in that the power to which the square brackets are raised is changed from $r_{j-1\,l}$ to $\alpha_{j-1\,l}$.

Then

$$\varphi_1(\theta_1, \ldots, \theta_k) = \mathscr{F}(\delta_1, \theta_1, \ldots, \delta_k\theta_k)$$

$$\times \left\{\prod_{j=1}^{k}\prod_{l=j}^{k}\left[e^{-\frac{ia_j\theta_j}{\sqrt{\varkappa_{jj}}} - \ldots - \frac{ia_l\theta_l}{\sqrt{\varkappa_{ll}}}}\Psi\left(0, \ldots, 0, \frac{\theta_j}{\sqrt{\varkappa_{jj}}}, \ldots, \frac{\theta_l}{\sqrt{\varkappa_{ll}}}, \ldots, 0\right)\right]^{\alpha_{j-1\,l}}\right\}. \tag{23}$$

First we prove that as $\lambda \to \infty$

$$\varphi_1(\theta_1, \ldots, \theta_k) = \mathscr{F}(\delta_1\theta_1, \ldots, \delta_k\theta_k)\, e^{-\frac{1}{2}\sum_{i=1}^{k}\sum_{j=1}^{k}(\rho_{ij}-R_{ij}\delta_i\delta_j)\theta_i\theta_j}(1+o(1)).$$

We expand the characteristic function

$$\Psi\left(0, \ldots, 0, \frac{\theta_j}{\sqrt{\varkappa_{jj}}}, \ldots, \frac{\theta_l}{\sqrt{\varkappa_{ll}}}, 0, \ldots, 0\right) \quad (j=1, \ldots, k;\ j \leqslant l \leqslant k)$$

in the neighborhood of $\theta_j/\sqrt{\kappa_{jj}} = 0, \cdots, \theta_l/\sqrt{\kappa_{ll}} = 0$ as a series (see [6]):

$$\Psi\left(0, \ldots, 0, \frac{\theta_j}{\sqrt{\varkappa_{jj}}}, \ldots, \frac{\theta_l}{\sqrt{\varkappa_{ll}}}, 0, \ldots, 0\right) = e^{\frac{i\theta_j a_j}{\sqrt{\varkappa_{jj}}}+\ldots+i\frac{a_l\theta_l}{\sqrt{\varkappa_{ll}}}}$$

$$\times\left(1-\frac{1}{2}\sum_{\nu=j}^{l}\sum_{\mu=j}^{l}c_{\nu\mu}\frac{\theta_\nu\theta_\mu}{\sqrt{\varkappa_{\nu\nu}}\sqrt{\varkappa_{\mu\mu}}}+o\left(\sum_{\nu=j}^{l}\sum_{\mu=j}^{l}\frac{c_{\mu\nu}\theta_\mu\theta_\nu}{\sqrt{\varkappa_{\mu\mu}}\sqrt{\varkappa_{\nu\nu}}}\right)\right) \tag{24}$$

or

$$\Psi\left(0, \ldots, 0, \frac{\theta_j}{\sqrt{\varkappa_{jj}}}, \ldots, \frac{\theta_l}{\sqrt{\varkappa_{ll}}}, 0, \ldots, 0\right) = e^{i\sum_{\nu=j}^{l}\frac{\theta_\nu a_\nu}{\sqrt{\varkappa_{\nu\nu}}}}$$

$$\times e^{-\frac{1}{2}\sum_{\nu=j}^{l}\sum_{\mu=j}^{l}c_{\nu\mu}\frac{\theta_\nu\theta_\mu}{\sqrt{\varkappa_{\nu\nu}}\sqrt{\varkappa_{\mu\mu}}}}\left(1+o\left(\sum_{\nu=j}^{l}\sum_{\mu=j}^{l}\frac{c_{\mu\nu}\theta_\mu\theta_\nu}{\sqrt{\varkappa_{\mu\mu}}\sqrt{\varkappa_{\nu\nu}}}\right)\right). \tag{25}$$

Substituting (25) into (23), we obtain

$$\varphi_1(\theta_1, \ldots, \theta_k) = \mathscr{F}(\theta_1\delta_1, \ldots, \delta_k\theta_k)\left\{\prod_{j=1}^{k}\prod_{l=j}^{k} e^{-\frac{\alpha_{j-1\,l}}{2}\sum_{\nu=j}^{l}\sum_{\mu=j}^{l}c_{\mu\nu}\frac{\theta_\nu\theta_\mu}{\sqrt{\varkappa_{\nu\nu}}\sqrt{\varkappa_{\mu\mu}}}}\right.$$

$$\left.\times\left(1+o\left(\alpha_{j-1\,l}\sum_{\mu=j}^{l}\sum_{\nu=j}^{l}c_{\mu\nu}\frac{\theta_\nu\theta_\mu}{\sqrt{\varkappa_{\nu\nu}}\sqrt{\varkappa_{\mu\mu}}}\right)\right)\right\}$$

$$= \mathscr{F}(\delta_1\theta_1, \ldots, \delta_k\theta_k)\left\{e^{-\frac{1}{2}\sum_{j=1}^{k}\sum_{l=j}^{k}\sum_{\nu=1}^{l}\sum_{\mu=j}^{l}c_{\mu\nu}\frac{\theta_\nu\theta_\mu}{\sqrt{\varkappa_{\mu\mu}}\sqrt{\varkappa_{\nu\nu}}}\alpha_{j-1\,l}}\right.$$

$$\left.\times\left(1+o\left(\sum_{j=1}^{k}\sum_{l=j}^{k}\sum_{\nu=j}^{l}\sum_{\mu=j}^{l}c_{\mu\nu}\alpha_{j-1\,l}\frac{\theta_\nu\theta_\mu}{\sqrt{\varkappa_{\mu\mu}}\sqrt{\varkappa_{\nu\nu}}}\right)\right)\right\}.$$

Since the quadratic forms $\Sigma_{\nu=j}^{l}\Sigma_{\mu=j}^{l} c_{\mu\nu}\theta_\nu\theta_\mu$ are nonnegative definite (for $C_{\mu\nu}$ are the elements of the variance matrix) and $\alpha_{j-1\,l}\geq 0$, the index is also a nonnegative definite quadratic form. Changing the order of summation, we obtain

$$\varphi_1(\theta_1, \ldots, \theta_k) = \mathscr{F}(\delta_1\theta_1, \ldots, \delta_k\theta_k) \times \left\{ e^{-\frac{1}{2}\sum_{\nu=1}^{k}\sum_{\mu=1}^{k}\frac{\theta_\mu\theta_\nu c_{\mu\nu}}{\sqrt{\varkappa_{\mu\mu}}\sqrt{\varkappa_{\nu\nu}}}\left(\sum_{l=\max(\nu,\mu)}^{k}\sum_{j=1}^{\min(\nu,\mu)}\alpha_{j-1\,l}\right)} \right.$$

$$\left. \times\left(1 + o\left(\sum_{\nu=1}^{k}\sum_{\mu=1}^{k}\frac{\theta_\mu\theta_\nu c_{\mu\nu}}{\sqrt{\varkappa_{\mu\mu}}\sqrt{\varkappa_{\nu\nu}}}\left(\sum_{l=\max(\mu,\nu)}^{k}\sum_{j=1}^{\min(\mu,\nu)}\alpha_{j-1\,l}\right)\right)\right)\right\}.$$

But, in accordance with (16), (17), (19) and (22), we have

$$\frac{c_{\mu\nu}}{\sqrt{\varkappa_{\mu\mu}}\sqrt{\varkappa_{\nu\nu}}}\left(\sum_{l=\max(\nu,\mu)}^{k}\sum_{j=1}^{\min(\mu,\nu)}\alpha_{j-1\,l}\right) = \rho_{\mu\nu} - R_{\mu\nu}\delta_\mu\delta_\nu,$$

and therefore

$$\varphi_1(\theta_1, \ldots, \theta_k) = \mathscr{F}(\delta_1\theta_1, \ldots, \delta_k\theta_k)\, e^{-\frac{1}{2}\sum_{\mu=1}^{k}\sum_{\nu=1}^{k}(\rho_{\mu\nu}-R_{\mu\nu}\delta_\mu\delta_\nu)\theta_\nu\theta_\mu}(1 + o(1)).$$

This expansion is valid in the neighborhood of the point $\theta_1/\sqrt{\kappa_{11}} = 0, \ldots, \theta_k/\sqrt{\kappa_{kk}} = 0$. By condition 1) of Theorem 1′ we have $\kappa_{ii} \to \infty$ as $\lambda \to \infty$; that is, this expansion is valid as $\lambda \to \infty$ for any finite $\theta_1, \ldots, \theta_k$.

We prove further that for any $\varepsilon > 0$ there exists λ_0 such that

$$\lambda > \lambda_0 \quad |\varphi(\theta_1, \ldots, \theta_k) - \varphi_1(\theta_1, \ldots, \theta_k)| < \varepsilon.$$

This will complete the proof of the theorem.

We denote the domain of variation of $r_{01}, \ldots, r_{k-1\,k}$ by $\mathfrak{D}$. Then

$$|\varphi(\theta_1, \ldots, \theta_k) - \varphi_1(\theta_1, \ldots, \theta_k)| \leqslant \sum_{\mathfrak{D}} P(\mu_{01} = r_{01}, \ldots, \mu_{k-1\,k} = r_{k-1\,k})$$

$$\times\left| e^{i\sum_{l=1}^{k}\frac{(N_l-\alpha_l)\alpha_l\theta_l}{\sqrt{\varkappa_{ll}}}}\right| \left| \prod_{j=1}^{k}\prod_{l=j}^{k}\left[e^{-\frac{i\alpha_j\theta_j}{\sqrt{\varkappa_{jj}}}-\cdots-\frac{i\alpha_l\theta_l}{\sqrt{\varkappa_{ll}}}}\Psi\left(0, \ldots, 0, \frac{\theta_j}{\sqrt{\varkappa_{jj}}}, \ldots\right.\right.\right.$$

$$\left.\left. \ldots, \frac{\theta_l}{\sqrt{\varkappa_{ll}}}, 0, \ldots, 0\right)\right]^{r_{j-1\,l}} - \prod_{j=1}^{k}\prod_{l=j}^{k}\left[e^{-\frac{i\alpha_j\theta_j}{\sqrt{\varkappa_{jj}}}-\cdots-\frac{i\alpha_l\theta_l}{\sqrt{\varkappa_{ll}}}} \times\right.$$

$$\times\Psi\left(0,\ldots,0,\frac{\theta_j}{\sqrt{\varkappa_{jj}}},\ldots,\frac{\theta_l}{\sqrt{\varkappa_{ll}}},0,\ldots,0\right)\Bigg]^{\alpha_{j-1\,l}}\Bigg|\leqslant\sum_{\mathfrak{D}}P(\mu_{01}=r_{01},\ldots$$

$$\ldots,\mu_{k-1\,k}=r_{k-1\,k})\prod_{j=1}^{k}\prod_{l=j}^{k}\Bigg|\Bigg[e^{-\frac{ia_j\theta_j}{\sqrt{\varkappa_{jj}}}-\ldots-\frac{ia_l\theta_l}{\sqrt{\varkappa_{ll}}}}$$

$$\times\Psi\left(0,\ldots,0,\frac{\theta_j}{\sqrt{\varkappa_{jj}}},\ldots,\frac{\theta_l}{\sqrt{\varkappa_{ll}}},0,\ldots,0\right)\Bigg]^{r_{j-1\,l}-\alpha_{j-1\,l}}-1\Bigg|$$

$$=\sum_{\mathfrak{D}^*}P(\mu_{01}=r_{01},\ldots,\mu_{k-1\,k}=r_{k-1\,k})$$

$$\times\prod_{j=1}^{k}\prod_{l=j}^{k}\Bigg|\Bigg[e^{-i\sum_{\nu=j}^{l}\frac{a_\nu\theta_\nu}{\sqrt{\varkappa_{\nu\nu}}}}\Psi\left(0,\ldots,0,\frac{\theta_j}{\sqrt{\varkappa_{jj}}},\ldots,\frac{\theta_l}{\sqrt{\varkappa_{ll}}},0,\ldots,0\right)\Bigg]^{r_{j-1\,l}-\alpha_{j-1\,l}}-1\Bigg|$$

$$+\sum_{\mathfrak{D}'}P(\mu_{01}=r_{01},\ldots,\mu_{k-1\,k}=r_{k-1\,k})$$

$$\times\prod_{j=1}^{k}\prod_{l=j}^{k}\Bigg|\Bigg[e^{-i\sum_{\nu=j}^{l}\frac{a_\nu\theta_\nu}{\sqrt{\varkappa_{\nu\nu}}}}\Psi\left(0,\ldots,0,\frac{\theta_j}{\sqrt{\varkappa_{jj}}},\ldots,\frac{\theta_l}{\sqrt{\varkappa_{ll}}},0,\ldots,0\right)\Bigg]^{r_{j-1\,l}-\alpha_{j-1\,l}}-1\Bigg|,$$

where the domain $\mathfrak{D}^*$ is given by

$$\left\{\left|\frac{r_{01}-\alpha_{01}}{\sigma}\right|<d,\ldots,\left|\frac{r_{k-1\,k}-\alpha_{k-1\,k}}{\sigma}\right|<d\right\},$$

d is a number and $\mathfrak{D}'$ is the complement of $\mathfrak{D}^*$ with respect to $\mathfrak{D}$.

We estimate the sum over $\mathfrak{D}'$ given by

$$\sum_{\mathfrak{D}'}P(\mu_{01}=r_{01},\ldots,\mu_{k-1\,k}=r_{k-1\,k})$$

$$\times\prod_{j=1}^{k}\prod_{l=j}^{k}\Bigg|\Bigg[e^{-i\sum_{\nu=l}^{j}\frac{\theta_\nu a_\nu}{\sqrt{\varkappa_{\nu\nu}}}}\Psi\left(0,\ldots,0,\frac{\theta_j}{\sqrt{\varkappa_{jj}}},\ldots,\frac{\theta_l}{\sqrt{\varkappa_{ll}}},0,\ldots,0\right)\Bigg]^{r_{j-1\,l}-\alpha_{j-1\,l}}-1\Bigg|$$

$$\leqslant 2^{\sum_{j=1}^{k}\sum_{l=j}^{k}1}\sum_{\mathfrak{D}'}P(\mu_{01}=r_{01},\ldots,\mu_{k-1\,k}=r_{k-1\,k})$$

$$\leqslant 2^{\sum_{j=1}^{k}\sum_{l=j}^{k}1}\sum_{j=1}^{k}\sum_{l=j}^{k}P\left(\left|\frac{\mu_{j-1\,l}-\alpha_{j-1\,l}}{\sigma}\right|>d\right),$$

but $E((\mu_{ij}-\alpha_{ij})/\sigma)=0$ and by condition 2) of Theorem 1′ $E((\mu_{ij}-\alpha_{ij})/\sigma)^2=\gamma_{ij,ij}/\sigma^2\longrightarrow 0$ as $\lambda\longrightarrow\infty$; that is, $(\mu_{ij}-\alpha_{ij})/\sigma$ converges in probability to zero as $\lambda\longrightarrow\infty$. Thus for any $\epsilon>0$ and d we can find λ_0 such that for $\lambda>\lambda_0$

$$2\sum_{j=1}^{k}\sum_{l=j}^{k}{}^{1}\sum_{j=1}^{k}\sum_{l=j}^{k} P\left(\left|\frac{\mu_{j-1\,k}-\alpha_{j-1\,k}}{\sigma}\right|>d\right)<\frac{\varepsilon}{2}.$$

Let us estimate the sum over $\mathfrak{D}^*$. For sufficiently large λ, as before, $\Psi(0,\cdots,0,\theta_j/\sqrt{\kappa_{jj}},\cdots,\theta_l/\sqrt{\kappa_{ll}},0,\cdots,0)$ can be expanded as a series in accordance with (25). From now on we denote $|(r_{j-1,l}-\alpha_{j-1\,l}/\sigma|$ by $p_{j-1,l}$, and then

$$\sum_{\mathfrak{D}^*}P(\mu_{01}=r_{01},\ldots,\mu_{k-1\,k}=r_{k-1\,k})$$

$$\times\prod_{j=1}^{k}\prod_{l=j}^{k}\left|\left[e^{-i\sum\limits_{\nu=l}^{j}\frac{\theta_\nu\alpha_\nu}{\sqrt{\varkappa_{\nu\nu}}}}\Psi\left(0,\ldots,0,\frac{\theta_j}{\sqrt{\varkappa_{jj}}},\ldots,\frac{\theta_l}{\sqrt{\varkappa_{ll}}},0,\ldots,0\right)\right]^{r_{j-1\,l}-\alpha_{j-1\,l}}-1\right|$$

$$=\sum_{\mathfrak{D}^*}P(\mu_{01}=r_{01},\ldots,\mu_{k-1\,k}=r_{k-1\,k})$$

$$\times\prod_{j=1}^{k}\prod_{l=j}^{k}\left|\left[e^{-\frac{1}{2}\sum\limits_{\nu=j}^{l}\sum\limits_{\mu=j}^{l}c_{\nu\mu}\frac{\theta_\nu\theta_\mu}{\sqrt{\varkappa_{\nu\nu}}\sqrt{\varkappa_{\mu\mu}}}}\left(1+o\left(\sum_{\nu=j}^{l}\sum_{\mu=j}^{l}\frac{c_{\nu\mu}\theta_\nu\theta_\mu}{\sqrt{\varkappa_{\nu\nu}}\sqrt{\varkappa_{\mu\mu}}}\right)\right)\right]^{\sigma p_{j-1\,l}}-1\right|$$

$$\leqslant\prod_{j=1}^{k}\prod_{l=j}^{k}\operatorname*{Max}_{|p_{j-1\,l}|<d}\left|e^{-\frac{\sigma p_{j-1\,l}}{2}\sum\limits_{\nu=j}^{l}\sum\limits_{\mu=j}^{l}c_{\mu\nu}\frac{\theta_\nu\theta_\mu}{\sqrt{\varkappa_{\nu\nu}}\sqrt{\varkappa_{\mu\mu}}}}-1\right|+o(1)$$

Since $C_{\mu\nu}$, θ_ν, θ_μ and $\sqrt{\sigma}/\sqrt{\kappa_{\mu\mu}}$ are bounded, d (up to now arbitrary) can be chosen so small, and λ so large, that this expression is less that $\epsilon/2$. The proof of the theorem is therefore complete.

We note further that (21) can be rewritten in the matrix form

$$\varphi(\theta')=\mathcal{F}(\boldsymbol{\delta}'\theta)\,e^{-\frac{1}{2}\theta'\Sigma\theta}+o(1),\tag{26}$$

where

$$\boldsymbol{\delta}=\begin{pmatrix}\delta_1\\ \vdots\\ \delta_k\end{pmatrix},\quad \theta=\begin{pmatrix}\theta_1\\ \vdots\\ \theta_k\end{pmatrix},\quad \Sigma=\|\rho_{\mu\nu}-R_{\mu\nu}\delta_\mu\delta_\nu\|.$$

Corollary. *If the vector* $((N_1-\alpha_1)/\sqrt{\gamma_{11}},\cdots,(N_k-\alpha_k)/\sqrt{\gamma_{kk}})$ *is assumed to have a limiting distribution function* $K(\mathbf{x}')$, $\mathbf{x}=\begin{bmatrix}x_1\\ \vdots\\ x_k\end{bmatrix}$, *so that the limit*

$$\lim_{\lambda\to\infty}\mathcal{F}(\theta') = \mathcal{F}_1(\theta')$$

exists, and if the limits

$$\mathfrak{d}^* = \lim_{\lambda\to\infty}\mathfrak{d}, \quad \Sigma^* = \lim_{\lambda\to\infty}\Sigma$$

also exist, then (26) *may be put in the form*

$$\lim_{\lambda\to\infty}\varphi(\theta') = \mathcal{F}_1(\mathfrak{d}^{*\prime}\theta)\, e^{-\frac{1}{2}\theta'\Sigma^*\theta},$$

and the limiting distribution function of the vector

$$\left(\frac{U_1 - \alpha_1}{\sqrt{\varkappa_{11}}}, \ldots, \frac{U_k - \alpha_k}{\sqrt{\varkappa_{kk}}}\right)$$

is equal to

$$H(\mathbf{x}') = \mathcal{K}((\tilde{\mathfrak{d}}^*)'\mathbf{x}) * H_0(\mathbf{x}),$$

where $H_0(\mathbf{x}')$ *is a multivariate normal distribution with parameters* $(0, \Sigma^*)$ *and*

$$\tilde{\mathfrak{d}}^* = \begin{pmatrix} \frac{1}{\delta_1} \\ \vdots \\ \frac{1}{\delta_k} \end{pmatrix}.$$

Example 1 (continuation). We prove that in this case the distribution of the vector $(N_1, \cdots, N_k)$ is asymptotically normal and conditions 1) and 2) of Theorem $1'$ are satisfied. Then, as we see from (21), the distribution of the vector $\mathbf{U} = (U_1, \cdots, U_k)$ is also asymptotically normal as $\lambda \to \infty$. Using (13), we can put the characteristic function of $(N_1, \cdots, N_k)$ into the following form:

$$\mathcal{F}^*(\theta_1, \ldots, \theta_k) = Ee^{i\sum_{\nu=1}^{k}\theta_\nu N_\nu} = \sum_{r_{01}} \cdots \sum_{r_{k-1\,k}} \prod_{j=0}^{k-1} P(\mu_{j\,j+1} = r_{j\,j+1}, \ldots, \mu_{jk} = r_{jk})$$

$$\times\, e^{i\sum_{j=0}^{k-1}\sum_{l=j+1}^{k} r_{jl}\left(\sum_{\nu=j+1}^{l}\theta_\nu\right)} = e^{\sum_{j=0}^{k-1}\left\{-\lambda s(1-\varepsilon(b_{j+1}-b_j)) + \lambda s\left[\sum_{l=j+1}^{k} e^{i\left(\sum_{\nu=j+1}^{l}\theta_\nu\right)} B_{jl}\right]\right\}}.$$

From this we easily calculate

$$\alpha_\mu = EN_\mu = i\,\frac{\partial \ln \mathcal{F}^*(\theta_1\ldots\theta_k)}{\partial\theta_\mu}\bigg|_{\substack{\theta_1=0\\ \cdots\\ \theta_k=0}} = \lambda s \quad (\mu = 1, \ldots, k),$$

$$\gamma_{\mu\mu} = \operatorname{var} N_\mu = - \frac{\partial^2 \ln \mathcal{F}^* (\theta_1 \ldots \theta_k)}{\partial \theta_\mu^2} \Bigg|_{\substack{\theta_1=0 \\ \cdots \\ \theta_k=0}} = \lambda s \quad (\mu = 1, \ldots, k),$$

$$\gamma_{\mu\mu'} = \operatorname{cov} N_\mu N_{\mu'} = - \frac{\partial^2 \ln \mathcal{F}^* (\theta_1 \ldots \theta_k)}{\partial \theta_\mu \partial \theta_{\mu'}} \Bigg|_{\substack{\theta_1=0 \\ \cdots \\ \theta_k=0}} = \lambda s \varepsilon (b_{\max(\mu, \mu')} - b_{\min(\mu, \mu')}),$$

$$R_{\mu\mu'} = \varepsilon (b_{\max(\mu, \mu')} - b_{\min(\mu, \mu')}). \tag{27}$$

It is easily verified that conditions 1) and 2) of Theorem 1′ are in fact satisfied. We now prove that the characteristic function of the normalized vector $((N_1 - \alpha_1)/\sqrt{\gamma_{11}}, \cdots, (N_k - \alpha_k)/\sqrt{\gamma_{kk}})$ tends to the characteristic function of the normal distribution. For,

$$\mathcal{F}(\theta_1, \ldots, \theta_k) = E e^{i \sum_{\nu=1}^{k} \theta_\nu \frac{N_\nu - \alpha_\nu}{\sqrt{\gamma_{\nu\nu}}}}$$

$$= e^{-i \sum_{\nu=1}^{k} \theta_\nu \sqrt{\lambda s} - \sum_{j=0}^{k-1} \lambda s (1 - \varepsilon (b_{j+1} - b_j)) + \lambda s \sum_{j=0}^{k-1} \sum_{l=j+1}^{k} e^{i \sum_{\nu=j+1}^{l} \frac{\theta_\nu}{\sqrt{\lambda s}}} B_{jl}}.$$

For any fixed θ_ν $(\nu = 1, \cdots, k)$ and sufficiently large λ we can expand

$$e^{i \sum_{\nu=j+1}^{l} \frac{\theta_\nu}{\sqrt{\lambda s}}}$$

as a series. Then

$$\mathcal{F}(\theta_1, \ldots, \theta_k) = e^{-i \sum_{\nu=1}^{k} \theta_\nu \sqrt{\lambda s} - \sum_{j=0}^{k-1} \lambda s (1 - \varepsilon (b_{j+1} - b_j))}$$

$$\times e^{\lambda s \left(\sum_{j=0}^{k-1} \sum_{l=j+1}^{k} B_{jl} \left(1 + \frac{i}{\sqrt{\lambda s}} \sum_{\nu=j+1}^{l} \theta_\nu - \frac{1}{2 \lambda s} \sum_{\nu=j+1}^{l} \sum_{\mu=j+1}^{l} \theta_\nu \theta_\mu + o\left(\frac{1}{\lambda}\right) \right) \right)}. \tag{28}$$

But, using (12), we easily obtain that

$$\sum_{j=0}^{k-1} \sum_{l=j+1}^{k} B_{jl} = \sum_{j=0}^{k-1} (1 - \varepsilon (b_{j+1} - b_j)),$$

$$\sum_{j=0}^{k-1} \sum_{l=j+1}^{k} \sum_{\nu=j+1}^{l} \theta_\nu B_{jl} = \sum_{\nu=1}^{k} \theta_\nu \left(\sum_{j=0}^{\nu-1} \sum_{l=\nu}^{k} B_{jl} \right) = \sum_{\nu=1}^{k} \theta_\nu,$$

$$\sum_{j=0}^{k-1} \sum_{l=j+1}^{k} \sum_{\nu=j+1}^{l} \sum_{\mu=j+1}^{l} \theta_\nu \theta_\mu B_{jl} = \sum_{\mu=1}^{k} \sum_{\nu=1}^{k} \theta_\nu \theta_\mu$$

$$\times \left(\sum_{l=\max(\mu, \nu)}^{k} \sum_{j=0}^{\min(\mu, \nu)-1} B_{jl} \right) = \sum_{\mu=1}^{k} \sum_{\nu=1}^{k} \theta_\nu \theta_\mu R_{\mu\nu}.$$

On substituting the expressions obtained into (28), we see that the distribution of $(N_1, \cdots, N_k)$ is asymptotically normal as $\lambda \to \infty$.

Example 2. Suppose that the assumptions regarding boundary depths and lengths be the same as in Example 1 but that the boundary origins form a stationary nonordinary flow without aftereffect.[1] As A. Ja. Hinčin has shown (see [2], pp. 35–42), the scheme of occurrences of boundary origins can in this case be represented in the following way. Boundary origins occur in "exciting" moments. "Exciting" moments form a Poisson flow with parameter λ. (We note that in this case λ is not equal to the intensity of the whole flow.) The number of boundary origins entering into any "exciting" moment is defined as a random variable which assumes a positive integral value k with probability p_k, so that the generating function of the random variable is

$$C(x)=\sum_k p_k x^k, \quad C(1)=1, \quad C'(1)=\sum_k kp_k.$$

In the absence of an aftereffect the characteristic function of the vectors $(N_1, \cdots, N_k)$ can be written in the form

$$\mathcal{F}^*(\theta_1, \ldots, \theta_k)=Ee^{i\sum_{\nu=1}^{k}\theta_\nu N_\nu}=\prod_{j=0}^{k-1} E\exp\left\{i\sum_{l=j+1}^{k}\mu_{jl}\left(\sum_{\nu=j+1}^{l}\theta_\nu\right)\right\}. \tag{29}$$

Since the distribution of N_1 in the first section is unknown, we proceed as follows.

We consider the section $x = b_0$ to the left of the first section $x = b_1$ and assume that the boundary origins occur in the interval (b_0, ∞) and that $N(b_0) = 0$. Also we transfer the origin to $x = b_0$. The zero region is now $[b_0 = 0, b_1)$ and the first section differs in no way from the others.

Let us consider

$$J_j=E\exp\left\{i\sum_{l=j+1}^{k}\mu_{jl}\left(\sum_{\nu=j+1}^{l}\theta_\nu\right)\right\}.$$

We denote by $(r, n_i, \xi_i, n_{il};\ i = 1, \cdots, r;\ l = j+1, \cdots, k)$ the following event: in the jth interval there are r "exciting" moments, the ith exciting moment is contained within the interval $(\xi_i, \xi_i + d\xi_i)$ in which there are n_i boundary origins, and n_{il} of these n_i boundaries end in the lth interval $(l = j+1, \cdots, k)$. Taking account of the fact that the flow of "exciting" moments is Poisson and that therefore $\xi_1, \cdots, \xi_r$ are uniformly distributed in the interval $[b_j, b_{j+1})$, we see that the probability of this event is given by

1) A nonordinary flow enables a medium with shears to be studied.

$$P(r,\ n_i,\ \xi_i,\ n_{il};\ i=1,\ \ldots,\ r;\ l=j+1,\ \ldots,\ k)$$
$$= e^{-\lambda(b_{j+1}-b_j)} \frac{(\lambda(b_{j+1}-b_j))^r}{r!} p_{n_1}\cdots p_{n_r} \frac{d\xi_1\ldots d\xi_r}{(b_{j+1}-b_j)^r}$$
$$\times \frac{n_1!}{n_{1j+1}!\ldots n_{1k}!\,(n_1-n_{1j+1}-\ldots-n_{1k})!}\times\cdots$$

$$\cdots\times\frac{n_r!}{n_{rj+1}!\ldots n_{rk}!\,(n_r-n_{rj+1}-\ldots-n_{rk})!}[F(b_k-\xi_1)]^{n_{1k}}\ldots[F(b_k-\xi_r)]^{n_{rk}}$$
$$\times[F(b_{k-1}-\xi_1)-F(b_k-\xi_1)]^{n_{1k-1}}\ldots[F(b_{k-1}-\xi_r)-F(b_k-\xi_r)]^{n_{rk-1}}\times\cdots$$
$$\cdots\times[F(b_{j+1}-\xi_1)-F(b_{j+2}-\xi_1)]^{n_{1j+1}}\ldots[F(b_{j+1}-\xi_r)-F(b_{j+2}-\xi_r)]^{n_{rj+1}}$$
$$\times[1-F(b_{j+1}-\xi_1)]^{n_1-n_{1j+1}-\ldots-n_{1k}}\ldots[1-F(b_{j+1}-\xi_r)]^{n_r-n_{rj+1}-\ldots-n_{rk}},$$

and under these conditions the random variable μ_{jl} assumes the value $\Sigma_{k=1}^{r} n_{kl}$. Summing over n_{il} $(i=1,\cdots,r;\ l=j+1,\cdots,k)$ and using the expressions found for calculating J_j, we obtain

$$J_j=\sum_{r=0}^{\infty} e^{-\lambda(b_{j+1}-b_j)}\frac{\lambda^r}{r!}\sum_{n_1}\cdots\sum_{n_r} p_{n_1}\cdots p_{n_r}\prod_{i=1}^{r}\int_{b_j}^{b_{j+1}} d\xi_i$$
$$\times\{e^{i(\theta_{j+1}+\ldots+\theta_k)}F(b_k-\xi_i)+e^{i(\theta_{j+1}+\ldots+\theta_{k-1})}[F(b_{k-1}-\xi_i)-F(b_k-\xi_i)]+\cdots$$
$$\cdots+e^{i\theta_{j+1}}[F(b_{j+1}-\xi_i)-F(b_{j+2}-\xi_i)]+1-F(b_{j+1}-\xi_i)\}^{n_i}.$$

Further, on noting that the modulus of the expression in curly brackets is less than or equal to 1 and remembering that $C(x)=\Sigma_k\, p_k x^k$, we obtain

$$J_j=\exp\Big\{\lambda\int_0^{b_{j+1}-b_j}[C(e^{i(\theta_{j+1}+\ldots+\theta_k)}F(b_k-b_{j+1}+v)$$
$$+e^{i(\theta_{j+1}+\ldots+\theta_{k-1})}(F(b_{k-1}+b_{j+1}+v)-F(b_k-b_{j+1}+v))+\cdots$$
$$\cdots+e^{i\theta_{j+1}}(F(v)-F(b_{j+2}-b_{j+1}+v))+1-F(v))-1]\,dv\Big\}.$$

On substituting the expression obtained in (29), we obtain[1)]

$$\mathcal{F}^*(\theta_1,\ \ldots,\ \theta_k)=\prod_{j=0}^{k-1}\exp\Big\{\lambda\int_0^{b_{j+1}-b_j}[C(e^{i(\theta_{j+1}+\ldots+\theta_k)}F(b_k-b_{j+1}+v)$$
$$+e^{i(\theta_{j+1}+\ldots+\theta_{k-1})}(F(b_{k-1}-b_{j+1}+v)-F(b_k-b_{j+1}+v))+\cdots$$
$$\cdots+e^{i\theta_{j+1}}(F(v)-F(b_{j+2}-b_{j+1}+v))+1-F(v))-1]\,dv\Big\}.$$

1) For $k=2$ this formula is a correction of the basic result in the paper by G. P. Klimov [7]. The formula he finds for the generating function of the number of conversations

In order to discard the artifically introduced initial conditions $N(b_0 = 0) = 0$ we let $b_1 \to \infty$; then

$$\mathcal{F}^*(\theta_1, \ldots, \theta_k) = \exp\Big\{\lambda \int_0^\infty [C(e^{i(\theta_1 + \ldots + \theta_k)} F(b_k - b_1 + v) + \ldots$$
$$\ldots + e^{i\theta_1}(F(v) - F(b_2 - b_1 + v)) + 1 - F(v)) - 1]\, dv$$
$$+ \sum_{j=1}^{k-1} \lambda \int_0^{b_{j+1} - b_j} [C(e^{i(\theta_{j+1} + \ldots + \theta_k)} F(b_k - b_{j+1} + v) + \ldots + e^{i\theta_{j+1}}$$
$$\times (F(v) - F(b_{j+2} - b_{j+1} + v)) + 1 - F(v)) - 1]\, dv\Big\}. \qquad (30)$$

The mathematical expectations and elements of the covariance matrix are as before easily calculated: using (30), we obtain

$$\alpha_\mu = i \frac{\partial \ln \mathcal{F}^*(\theta_1, \ldots, \theta_k)}{\partial \theta_\mu}\Big|_{\substack{\theta_1 = 0 \\ \vdots \\ \theta_k = 0}} = \lambda s C'(1) \quad (\mu = 1, \ldots, k),$$

$$(\mu = 1, \ldots, k)$$

$$\gamma_{\mu\mu'} = -\frac{\partial^2 \ln \mathcal{F}^*(\theta_1, \ldots, \theta_k)}{\partial \theta_\mu \partial \theta'_\mu}\Big|_{\substack{\theta_1 = 0 \\ \vdots \\ \theta_k = 0}} = \lambda C'(1) \int_{b_{\max(\mu, \mu')} - b_{\min(\mu, \mu')}}^\infty F(u)\, du$$
$$+ \lambda C''(1) \int_0^\infty F(b_{\max(\mu, \mu')} - b_{\min(\mu, \mu')} + v) F(v)\, dv$$
$$(\mu, \mu' = 1, \ldots, k).$$

In particular, the variance is equal to $\gamma_{\mu\mu} = \lambda B$, where $B = sC'(1) + C''(1) \int_0^\infty F^2(u)\, du$.

Using $\mu_{\mu\mu'}$, we can find the elements of the correlation matrix

$$R_{\mu\mu'} = \frac{C'(1) \int_{b_{\max(\mu, \mu')} - b_{\min(\mu, \mu')}}^\infty F(u)\, du + C''(1) \int_0^\infty F(b_{\max(\mu, \mu')} - b_{\min(\mu, \mu')} + v) F(v)\, dv}{B}.$$
$$(\mu, \mu' = 1, \ldots, k)$$

in progress at times t_0 and $t_0 + t$ with a nonordinary Poisson flow of calls, an infinite number of instruments and an arbitrary distribution of serving time, is incorrect. In fact from this formula two different expressions can be derived for the generating function of the number of conversations in progress at time $t_0 + t$. The error in the final result arises from the fact that step 1° of the proof is erroneous because the formula it contains is true for an ordinary Poisson flow only. There is a similar error in Mejzler's paper [8].

It is easy to show that conditions 1) and 2) of Theorem 1′ are satisfied.

We now prove that the distribution of $(N_1, \cdots, N_k)$ is asymptotically normal; it will then follow from Theorem 1′ that the distribution of $(U_1, \cdots, U_k)$ is also asymptotically normal as $\lambda \to \infty$. Let $\mathcal{F}(\theta_1, \cdots, \theta_k)$ be the characteristic function of the normalized vector $((N_1 - \alpha_1)/\sqrt{\gamma_{11}}, \cdots, (N_k - \alpha_k)/\sqrt{\gamma_{kk}})$.

We have

$$\ln \mathcal{F}(\theta_1, \ldots, \theta_k) = -i \sum_{\nu=1}^{k} \frac{\theta_\nu \sqrt{\lambda}\, sC'(1)}{\sqrt{B}} + \lambda \int_0^\infty \Bigg[C\Bigg(e^{i \sum_{\nu=1}^{k} \frac{\theta_\nu}{\sqrt{\lambda B}}} F(b_k - b_1 + v) + \ldots$$

$$\ldots + 1 - F(v) \Bigg) - 1 \Bigg] dv + \lambda \sum_{j=1}^{k-1} \int_0^{b_{j+1} - b_j} \Bigg[C\Bigg(e^{i \sum_{\nu=j+1}^{k} \frac{\theta_\nu}{\sqrt{\lambda B}}}$$

$$\times F(b_k - b_{j+1} + v) + \ldots + 1 - F(v) \Bigg) - 1 \Bigg] dv. \tag{31}$$

Expanding

$$C\Bigg(e^{i \sum_{\nu=j+1}^{k} \frac{\theta_\nu}{\sqrt{\lambda B}}} F(b_k - b_{j+1} + v) + \ldots + e^{i \frac{\theta_{j+1}}{\sqrt{\lambda B}}} (F(v) - F(b_{j+2} - b_{j+1} + v))$$

$$+ 1 - F(v) \Bigg) \quad (j = 0, \ldots, k)$$

as a series in the neighborhood $\theta'_\nu = \theta_\nu / \sqrt{\lambda B} = 0$ $(\nu = j+1, \cdots, k)$, we obtain

$$C\Bigg(e^{i \sum_{\nu=j+1}^{k} \frac{\theta_\nu}{\sqrt{\lambda B}}} F(b_k - b_{j+1} + v) + \ldots + 1 - F(v) \Bigg)$$

$$= 1 + \sum_{\mu=j+1}^{k} \frac{\partial C}{\partial \theta'_\mu} \Bigg|_{\substack{\theta'_\nu = 0, \\ \nu = j+1, \ldots, k}} \frac{\theta_\mu}{\sqrt{\lambda B}} + \frac{1}{2} \sum_{\mu=j+1}^{k} \sum_{\mu'=j+1}^{k} \frac{\partial^2 C}{\partial \theta'_\mu \partial \theta'_{\mu'}} \Bigg|_{\substack{\theta'_\nu = 0, \\ \nu = j+1, \ldots, k}}$$

$$\times \frac{\theta_\mu \theta_{\mu'}}{\lambda B} + \frac{1}{3!} \sum_{\mu=j+1}^{k} \sum_{\mu'=j+1}^{k} \sum_{\mu''=j+1}^{k} \frac{\partial^3 C}{\partial \theta'_\mu \partial \theta'_{\mu'} \partial \theta'_{\mu''}} \Bigg|_{\substack{0 < \theta''_\nu < \theta'_\nu, \\ \nu = j+1, \ldots, k}} \frac{\theta_\mu \theta_{\mu'} \theta_{\mu''}}{(\lambda B)^{3/2}}.$$

On substituting the expressions obtained into (31) and assuming that $\int_0^\infty F^3(t)\, dt < \infty$, we obtain

$$\ln \mathcal{F}^*(\theta_1, \ldots, \theta_k) = -\frac{1}{2} \sum_{\mu=1}^{k} \sum_{\mu'=1}^{k} R_{\mu\mu'} \theta_\mu \theta_{\mu'} + o\left(\frac{1}{\sqrt{\lambda}}\right).$$

This proves that $(N_1, \cdots, N_k)$ is asymptotically normal as $\lambda \to \infty$.

Let us now compare the results of Examples 1 and 2. In the second example the expectation is obtained by multiplying λs (the expectation in the first case) by $C'(1)$, the expectation of the number of boundaries in the "exciting" moment.

In the second example we add the term $\lambda C''(1)\int_0^\infty F^2(v)\,dv$ to the variance obtained in the first example. It is also easy to show that the univariate distribution of N_j $(j = 1, \cdots, k)$ in the first example is Poisson with parameter λs, while it follows from (30) that in the second example the univariate characteristic function of N_j $(j = 1, \cdots, k)$ has the form

$$\mathcal{F}^*(\theta) = \exp\left\{\lambda \int_0^\infty [C(e^{i\theta}F(u) + 1 - F(u)) - 1]\,du,\right.$$

which is no longer the characteristic function of a Poisson distribution.

A comparison of the results of Examples 1 and 2 therefore shows that disturbing the ordinariness of a simple flow of boundary origins complicates the process from which the sum wave is derived; but both processes have the same asymptotic behavior.

BIBLIOGRAPHY

[1] L. A. Zolotuhina, K. P. Latyšev and V. N. Čugueva, *Stochastic models of a stratified medium and probabilistic properties of reflected waves propagated in these media*, Trudy Mat. Inst. Steklov. 95 (1968), 42–97 = Proc. Steklov Inst. Math. 95 (1968), 47–113.

[2] A. Ja. Hinčin, *Studies on mathematical queueing theory* Fizmatgiz, Moscow, 1963. (Russian) MR 30 #3509.

[3] L. Takács, *On a coincidence problem concerning telephone traffic*, Acta Math. Acad. Sci. Hungar. 9 (1958), 45–81. MR 20 #2804.

[4] Herbert Robbins, *The asymptotic distribution of the sum of a random number of random variables*, Bull. Amer. Math. Soc. 54 (1948), 1151–1161. MR MR 10, 385.

[5] G. Orazov, *Limit theorems for a random number of random terms*, Izv. Akad. Nauk Turkmen. SSR Ser. Fiz.-Tehn. Him. Geol. Nauk 1964, no. 1, 24–29. (Russian) MR 28 #5456.

[6] Harald Cramér, *Mathematical methods of statistics*, Princeton Math. Series, vol. 9, Princeton Univ. Press, Princeton, N. J., 1946; Russian transl., IL, Moscow, 1948. MR 8, 39.

[7] G. P. Klimov, *The determination of transition probabilities of the number of telephone conversations*, Computing Methods and Programming, Izdat. Moskov. Gos. Univ., Moscow, 1962. (Russian)

[8] D. Mejzler, *A note on Erlang's formulas*, Israel J. Math. 3 (1965), 157–162. MR 33 #8046.

Trudy Mat. Inst. Steklov.
95 (1968)

Proc. Steklov Inst. Math.
95 (1968)

STOCHASTIC MODELS OF A STRATIFIED MEDIUM AND PROBABILISTIC PROPERTIES OF REFLECTED WAVES PROPAGATED IN THESE MEDIA

UDC 519.217

L. A. ZOLOTUHINA, K. P. LATYŠEV AND V. N. ČUGUEVA

Abstract. The problem of interpreting seismic data by the method of reflected waves, using probability theory and mathematical statistics, is investigated. A stratified medium is taken to be the realization of a random process and several stochastic models of plane-parallel media are constructed. The sum wave obtained in these models is a stochastic process, and much of the paper is devoted to a study of its properties. Exact and asymptotic characteristics of the process are obtained for once-reflected waves. In the case of multiply-reflected waves asymptotic properties are found when the intensity of the discontinuities of the medium increases without limit.

Bibliography 9 items. 5 figures.

§1. Introduction

One of the most important parts of geophysical prospecting is seismic prospecting with methods based on the study of the propagation of elastic waves in the medium under study which have been excited artificially by explosions or shocks. Elastic waves propagated from a source, as with light or sound waves, pass into the interior of the earth's crust, undergoing refraction and reflection, and partially return to the earth's surface where the vibrations they cause are recorded by special instruments in the form of seismograms. Knowing certain characteristics together with arrival time of the wave, the amplitude, distance to the source, and so on, we can determine the depth of those discontinuities of the medium which cause the refraction or reflection of the wave. This information is used in mineral prospecting.

Let us consider the following simple scheme. Suppose that a source of elastic vibrations is situated at the point S of the earth's surface and a seismic detector is located at a point Q at a distance $2b$ from S. We take the vertical plane section of the earth's crust through S and Q to give a plane half-space bounded by the straight line through S and Q. We shall study elastic waves arriving at Q from the source S. If these are not discontinuities in the medium the detector Q will record only the wave propagated along the straight line SQ.

Knowing the distance SQ and the arrival time of the wave at the seismogram we can determine the velocity of propagation of the elastic wave.

If the medium is not homogeneous waves reflected from discontinuities will after a certain time reach the detector Q. Suppose, for example, that an extended boundary parallel to the earth's surface is situated at depth z. If it intersects the vertical line through the point dividing the segment SQ into two halves, then after time $2\sqrt{b^2 + z^2}/v$ a wave reflected from this boundary will be recorded at Q. [1] Knowing the arrival time of the reflected wave and the propagation velocity, we can determine the depth of the reflecting boundary. It is true that nothing can be said in this case about the extent of the boundary, except that at least it is greater than the length of the wave. [2] Extended boundaries are of the greatest interest in seismic prospecting, and in order to be able to trace the extent of the boundary on a line passing through the points SQ we require not one but several seismic detectors placed at points $Q_1, Q_2, \cdots$ $\cdots, Q_K$. In this case the problem of determining the depth of the boundary, its extent and other characteristics (that is, everything that in seismic prospecting is called interpretation of seismic data) presents no fundamental difficulty.

The problem of interpretation at once becomes more complex if the inhomogeneity of the medium is caused not by one boundary but by several. Then in addition to single waves we have double waves, multiple waves, refracted waves, diffracted waves, and so on.

In real geological media there is a dispersion of seismic waves, as a result of which the energy of the fundamental wave, say a single wave, is partly expended on the formation of dispersed elastic waves. This is a great degree further complicates the character of the behavior of the sum wave and makes the problem of interpreting seismic data extremely complex.

In seismic prospecting reflected waves are of two types: useful waves and wave-noise. A quite extensive class of waves come within the second category. The so-called microseisms have the most random character. By microseisms we mean vibrations of the soil, both those of natural origin and those caused by man's activity. Natural microseisms are caused by ocean breakers, the wind, rain, moles and other similar phenomena. Artificial microseisms are induced by construction equipment, railroad trains, automobiles, etc. To this category we may also ascribe noise arising in the recording instrument itself.

---

1) We assume that waves are propagated in the medium according to the rules of geometric optics. The arrival time of the wave $2\sqrt{b^2 + z^2}/v$ is then obtained by considering the isosceles triangle with base $SQ = 2b$ and height z; the velocity of wave propagation is assumed constant.

2) For in this case we have a reflection. For simplicity we shall assume further that the critical angle of reflection is 90°.

Apart from microseisms, noise is made up of waves generated by discontinuities in the medium of a different kind, multiple waves, diffracted waves and others.

It should be noted that noise of the second type, apart from microseisms, remains the same for repeated trials, whereas microseisms vary from trial to trial, and, as already mentioned, have an element of randomness. Since the formation of microseisms is random in character, we can consider the microseisms in each trial as the realization of a stochastic process and therefore use the methods of analyzing processes statistically to exclude microseisms from the sum wave. Microseisms, however, contribute only slightly to the energy balance of the sum wave, and therefore, in general, we do not attempt to isolate them.

We remarked earlier that extended boundaries are of the greatest interest in seismic prospecting because useful minerals are most often associated with an extended boundary. How the extent of the boundary is related to the location of useful minerals, what the extent of the boundary should be to distinguish it from a short boundary, and other similar questions of the structure of the medium, will not be considered here. It is important to distinguish clearly two concepts: a useful wave and wave-noise. We must note first of all that in seismic prospecting there is no single definition of a useful wave. If, say, we need to separate out single waves generated by an unknown extended boundary, then these waves are considered useful and all the others regarded as noise, even double waves obtained from the same boundary. Therefore as soon as the problem of interpretation is posed we know at once what wave or waves are to be regarded as useful.

If we succeed in separating out the useful wave from the sum wave the problem of interpretation is solved. But in order to do this we need to know the laws governing the behavior of the wave-noise.

An attempt is made in this paper to use a probability-theoretic approach to describe the properties of the medium and the reflected waves generated by it.

We first consider a model of a stratified medium from the point of view of stochastic processes and find the relation between this model and certain models of queueing theory.

We also study the behavior of once- and twice-reflected sum waves given in the form of a stochastic process. For double waves, for example, the process itself is the combination of a random number of strongly dependent random terms depending on certain parameters. Various cases tending towards normality are studied.

The results obtained enable various statistical problems to be posed and solved, either of testing hypotheses or of estimating parameters defining the process. In this case much depends on the problems immediately confronting the seismic prospector.

§2. Stochastic models of a medium and basic assumptions on wave propagation

The problem of seismic prospecting is the interpretation of the seismic data on which conclusions regarding the structure of the geological medium are based. Although the effect of the source of elastic vibrations and the recording system on the seismogram is known to the interpreter, or at least can be estimated in principle, it is impossible to exclude the influence of the geological medium because the structure of the medium is itself the object of study, and therefore, if we have the same source and the same system of recording elastic vibrations but different geological media, experimental results will differ significantly. The geological medium therefore plays a leading part in the composition of the seismogram.

According to the mechanism of their formation geological media can be divided into four large classes, one of which is the so-called class of sedimentary media. If we consider sedimentary media from the point of view of their construction we can single out the subclass of plane-parallel media. This subclass of a geological medium has served as the object for constructing the stochastic model of a stratified medium considered in this paper. We shall assume that the discontinuities in a plane-parallel medium are parallel not only to each other but also to the diurnal surface. It will be shown that the problem of studying plane-parallel media which are not parallel to the diurnal surface reduces to the problem of studying plane-parallel media which are parallel to the diurnal surface.

Let us take a vertical section of this medium and consider the section of the earth's crust obtained. It is a half-plane bounded above by the diurnal surface. Any discontinuity of the medium passing through this section leaves a trace on the half-plane in the form of a straight line or segment parallel to the earth's surface. In view of this the problem of interpretation will from now on be regarded as a problem of seismic prospecting for a half-plane representing a homogeneous medium containing discontinuities in the form of parallel lines and segments.

The existing classification of geological media from the point of view of the mechanism of their formation is no more than a probable hypothesis about the true character of their formation. The stochastic models of a plane-parallel medium suggested below are also hypotheses starting from the assumption that the medium is the realization of a stochastic process of the formation of a geological medium.

We use a rectangular coordinate system xoz with origin and axis ox on the earth's surface and axis oz directed vertically downwards. We shall keep to this coordinate system throughout the rest of the paper.

Let us consider the proposed models in order of complexity:

M1. Discontinuities in the medium are given in the form of infinite lines parallel to ox. The number of boundaries (discontinuities) is a random variable N with a countable number of states $\{0, 1, 2, \cdots\}$ and distribution law

$$P(N=n)=p_n, \quad n=0, 1, 2, \ldots,$$
$$\sum_{n=0}^{\infty} p_n=1.$$

Each boundary is characterized by its depth z. The depths of the boundaries are independent, identically distributed random variables with distribution function $G(z)$.

This model finds almost no application in practical work, but it is nonetheless very convenient in the study of some theoretical questions. In any case it can be considered as one of the variants of the simplest cases of plane-parallel geological media.

In what follows we shall need certain concepts of the theory of stochastic processes.

The theory of stochastic processes is defined, generally speaking, as the "dynamic" part of probability theory in which questions are studied relating to the set of random variables from the point of view of their interdependence and limiting behavior. In formal terms the stochastic process S_x is the family of random variables $\{S_x, x \in \mathfrak{X}\}$ depending on the parameter x from the parametric set $\mathfrak{X}$ [1]. One of the ways of describing a stochastic process is to assign a joint distribution function for the set of random variables $S_{x_1}, S_{x_2}, \cdots, S_{x_n}$ for all integral n and points $x_1, x_2, \cdots, x_n$ from $\mathfrak{X}$.

Let us consider random variables given on the positive semiaxis. Suppose that for every $x > 0$ the random variable (process) S_x is the number of events occurring in the interval $(0, x]$ and that the following conditions hold:

A1. $S_0 = 0$.

A2. The process S_x has stationary, independent increments.

A3. For any $x > 0$ we have

$$0<P\{S_x>0\}<1.$$

A4. For any $x \geq 0$ we have

$$\lim_{h\to 0}\frac{P\{S_{x+h}-S_x\geqslant 2\}}{P\{S_{x+h}-S_x=1\}}=0.$$

The random process S_x with these properties is called a simple stream. It can be proved that the process $\{S_x, x\geq 0\}$ satisfying A1–A4 is a Poisson process with intensity $\lambda>0$ for which

$$P\{S_x=n\}=e^{-\lambda x}\frac{(\lambda x)^n}{n\,!} \quad (x\geqslant 0;\ n=0, 1, 2, \ldots). \tag{2.1}$$

Remarks. 1. The assumption that in the initial state $S_0=0$ is not essential.

2. Apart from being a condition of stationarity, A2 is simultaneously a condition of no aftereffect, which means that the probability of the occurrence of k events in the interval $(h, h+x]$ does not depend on the order of events up to the point $h>0$. In other words, the conditional probability of the occurrence of k events in $(h, h+x]$, calculated under any assumptions regarding the order of events up to the point h, is equal to the unconditional probability of the same event.

3. A4 is a condition of ordinariness of the process. Ordinariness expresses the practical impossibility that two or more events should coincide at the same point.

Let us consider the next model.

M2*. Suppose that each discontinuity of a plane-parallel medium is represented by a segment of constant length $s>0$ situated at depth z. In the coordinate system already introduced each discontinuity will be characterized by two random independent coordinates: the depth z and the coordinate ξ of the beginning of the segment (the left end of the discontinuity). As in M1 the depths z are independent, identically distributed random variables with distribution function $G(z)$. We project the beginning of each segment onto ox and calculate the number of points ξ_i along ox within the interval $(0, X]$, denoting this number by $N(X)$. It is obvious that $N(X)$ is a stochastic process, sometimes called a counter process.

We assume that conditions A1–A4 hold for $N(X)$; that is, $N(X)$ is a Poisson process. The probabilistic properties of the random variables ξ are then completely defined by the probabilistic properties of $N(X)$.

Conditions A1–A4 of course limit the applicability of the above models. But the question of whether these conditions hold in real media is more experimental than theoretical in character. Only an experiment can answer the question whether a real geological medium, considered as the realization of a random process, has the properties A1–A4. Nevertheless we shall always assume that conditions A1–A4 are satisfied. The ordinariness condition A4 signifies that there are no so-called shears in the model of the geological medium.

Finally, let us consider the last and most general model of a geological plane-parallel medium. The model $M2^*$ may be generalized by replacing discontinuities of constant length $s > 0$ by discontinuities whose extent is a random variable l with the same distribution function $F_l(x)$.

M2. Each discontinuity in the medium is described by three random variables: the coordinate ξ of the beginning of the discontinuity, the depth z with distribution function $G(z)$, and the extent l with distribution function $F_l(x)$ such that $E\{l\} < \infty$. We shall use the following notation:

$$\Phi(\alpha, \beta) = \int_\alpha^\beta (1 - F_l(u))\, du. \tag{2.2}$$

The probabilistic properties of the random variables ξ are described by those of the Poisson process with intensity $\lambda > 0$.

Remark. In a seismic experiment observations of reflected waves occupy a bounded time interval $(0, T)$. We can therefore easily determine the maximum depth H which can be reached by a wave before reflection and arrival at the seismic detector if we know T and the propagation velocity v of elastic vibrations in the medium. The region in which z is assigned is therefore bounded and the distribution function $G(z)$ is truncated at the point H. In some places we assume for simplicity that z is uniformly distributed in the interval $(0, H)$.

In conclusion we introduce some concepts and definitions which we shall need throughout the subsequent analysis. Let us consider the coordinate system xoz. We place the source of elastic vibrations at the point $a > 0$ on ox and the detectors at distances $2b_1, \cdots, 2b_K$ from a along ox, where $0 < b_1 < b_2 < \cdots < b_K$. This is an arrangement used in practice and will therefore be used in this paper. We shall consider only reflected waves, for which we make the following assumptions:

a) Reflection takes place according to the laws of geometric optics.

b) A reflected wave is described by a known function of time, called the signal function and defined by

$$Af(t-\tau) = \begin{cases} 0 & \text{for } t \leqslant \tau, \\ Af_c(t-\tau) & \text{for } t > \tau, \end{cases} \tag{2.3}$$

where $f(t - \tau)$ is a known function of time and the same for all discontinuities;

τ is the arrival time of the wave; [1] and A is the amplitude factor. The amplitude factors of different discontinuities are independent, identically distributed random variables with distribution function $F_A(x)$, $EA^2 < \infty$.

c) The wave propagation velocity v and the reflection factor q are constant.

d) z, l, A and $N(X)$ are mutually independent.

§3. Single waves and their probabilistic properties

The sum wave entering the seismic detector consists of a large number of once-reflected and multiply-reflected waves. The study of this sum wave is extremely complex and we therefore start with the simpler case of a sum wave formed by once-reflected waves, or, as they are still called, single waves. Let us consider the stochastic model M2 of the medium. We assume that there is only one detector, situated at a distance $2b$ $(b > 0)$ from the source of the elastic vibrations, and that the source is situated at the point $x = a$ in the coordinate system xoz. The random variable $N(a + 2b)$ is the number of discontinuities appearing in that part of the medium from the origin to the point $x = a + 2b$ with depths in the range $(0, H)$ (Figure 1).

We extend each discontinuity to infinity in both directions, thus obtaining the stochastic model M1 of the medium. In this model the sum wave at each point of the earth's surface, and in particular at the point $x = a + zb$, will be made up of single waves reflected from each boundary. Let us consider, for example, a boundary of depth z_k. At any time t the detector records a single wave, reflected from this discontinuity, of the form (see (2.3)) $qA_k f(t - 2\sqrt{b^2 + z_k^2}/v)$, where $\tau = 2\sqrt{b^2 + z_b^2}/v$ is found from geometric considerations and q is the reflection factor which is constant for all discontinuities.

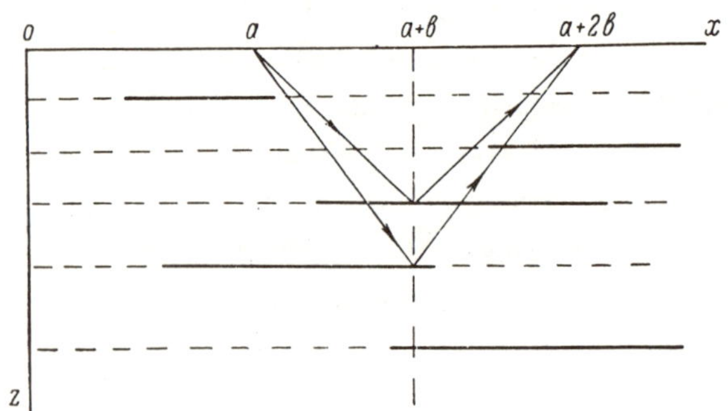

Figure 1. The arrangement of the "source-detector" system.

The sum wave then has the form

1) Throughout this paper the dependence of functions will be indicated only for those arguments of interest in the given context.

$$U_1(t,\ a,\ b,\ N)=\sum_{k=1}^{N} qA_k f\left(t-\frac{2}{v}\sqrt{b^2+z_k^2}\right), \tag{3.1}$$

where the random variable N is defined in M1. It is obvious that the random variable $U_1(t, a, b, N)$ is a process whose complete assignment requires knowledge of the multivariate distribution function (see §2).

Since the terms in the summation (3.1) are independent, we can use the method of characteristic functions to find the distribution function of $U_1(t, a, b, N)$. For convenience we write $U_1(t, a, b, N) = U_1$, remembering that the process (3.1) depends on the parameters t, a and b. The characteristic function of the random variable U_1 is then

$$\varphi_{U_1}(\theta)=E\{e^{i\theta U_1}\}$$

or

$$\varphi_{U_1}(\theta)=\sum_{n=0}^{\infty} p_n E\left\{e^{i\theta U_1}\big|_{N=n}\right\}. \tag{3.2}$$

From (3.1), since the terms are independent, we obtain

$$\begin{aligned} E\left\{e^{i\theta U_1}\big|_{N=n}\right\} &= E\left\{\exp iq\theta \sum_{k=1}^{n} A_k f(t-\tau_k)\right\} \\ &= \prod_{k=1}^{n} E\{\exp iq\theta A_k f(t-\tau_k)\}. \end{aligned} \tag{3.3}$$

Also,

$$E\{\exp iq\theta A_k f(t-\tau_k)\}=\int\int e^{iq\theta x f(t-\tau)} dF_A(x)\, dG(z), \tag{3.4}$$

where $\tau = 2\sqrt{b^2+z^2}/v$ and $F_A(x)$ and $G(z)$ are the distribution functions of the random variables A and z respectively. On substituting (3.4) into (3.3) and the result into (3.2), we obtain

$$\varphi_{U_1}(\theta)=\sum_{n=0}^{\infty} p_n \left\{\int\int e^{iq\theta x f(t-\tau)} dF_A(x)\, dG(z)\right\}^n.$$

If $p_n = e^{-\lambda}\lambda^n/n!$ $(n = 0, 1, \cdots)$, the univariate characteristic function $\phi_{U_1}(\theta)$ of the process U_1 takes the form

$$\varphi_{U_1}(\theta)=\exp \lambda \int\int \left(e^{iq\theta x f(t-\tau)}-1\right) dF_A(x)\, dG(z).$$

We see that the univariate characteristic function obtained does not depend on a. This was, in essence, determined by the choice of the model M1.

We turn now to the model M2. Since each discontinuity on the segment from $x = 0$ to $x = a + 2b$ (Figure 1) has its own coordinate for its beginning, and length l, the sum wave in this case obviously differs from the sum wave (3.1). In fact, the sum wave in M2 is composed only of waves reflected from those boundaries intersecting the vertical line which bisects the line joining the source and the detector (Figure 1). The sum wave of the model M2 will therefore comprise only parts of the sum wave (3.1). In order to isolate those discontinuities the reflection from which contributes to the sum wave, we introduce the function of two variables $W(x, y)$ given by

$$W(x, y) = \begin{cases} 1 & \text{for } 0 \leqslant x < y, \\ 0 & \text{otherwise}. \end{cases} \tag{3.5}$$

The sum wave $U_1(t, a, b, N)$ for M2 can then be put in the form

$$U_1 \equiv U_1(t, a, b, N) = q \sum_{j=1}^{N(a+2b)} A_j f(t - \tau_j) W(a + b - \xi_j, l_j), \tag{3.6}$$

where $\tau_j = 2\sqrt{b^2 + z_j^2}/v$.

Writing U_1 in the form (3.6) is no more than a convenient method for investigation. In a somewhat different form, without the use of the indicator function (3.5), this wave was studied in Zolotuhina's paper [2].

As in the case of the model M1, we use the method of characteristic functions to determine the probabilistic behavior of the process (3.6).

Let there be K detectors located at distances $2b_1, 2b_2, \cdots, 2b_K$ from the source ($b_1 < b_2 < \cdots < b_K$). We denote by $\phi_{U_1}(\theta; a)$ the univariate characteristic function of the process U_1, and by $\phi_{U_1^{(1)} U_1^{(2)} \cdots U_1^{(K)}}(\theta_1, \cdots, \theta_K; a)$ the K-dimensional characteristic function of the same process; that is,

$$\varphi_{U_1^{(1)} U_1^{(2)} \ldots U_1^{(K)}}(\theta_1, \ldots, \theta_K; a) = E\{\exp i[\theta_1 U_1^{(1)} + \ldots + \theta_K U_1^{(K)}]\},$$

where

$$U_1^{(\mu)} = \sum_{j=1}^{N(a+2b_\mu)} q A_j f(t - \tau_j^{(\mu)}) W(a + b_\mu - \xi_j, l_j) \tag{3.7}$$
$$(\mu = 1, \ldots, K)$$

is the sum wave arriving at the detector situated at distance $2b_\mu$ from the source.

Theorem 3.1. *Let $U_1(t, a, b, N)$ be the stochastic process defined by* (3.6). *Then*:

1) *For any t and b and real θ,*

$$\varphi_{U_1}(\theta;\ a) = \exp \lambda\Phi(0,\ a+b) \int\int (e^{qi\theta x f(t-\tau)} - 1)\, dF_A(x)\, dG(z)$$

and in the case $E\{l\} = S < \infty$

$$\lim_{a\to\infty} \varphi_{U_1}(\theta; a) = \varphi_{U_1}(\theta) = \exp \lambda s \int\int (\exp\{i\theta q x f(t-\tau)\} - 1)\, dF_A(x)\, dG(z). \quad (3.8)$$

2) *For any integer* K, *any* t, $b_1 < b_2 < \cdots < b_K$ *and real numbers* $\theta_1, \theta_2, \cdots, \theta_K$,

$$\varphi_{U_1^{(1)}\ldots U_1^{(K)}}(\theta_1, \ldots, \theta_K;\ a)$$
$$= \exp \lambda \sum_{m=1}^{K} \sum_{r=0}^{K-m} (a+2b_K)\, P_{m,r}(a)\, E\left\{e^{iqA(\theta_m f_m + \cdots + \theta_{m+r} f_{m+r})} - 1\right\}, \quad (3.9)$$

where $P_{mr}(a)$ *are probabilities defined by* (3.17) *and* (3.20)–(3.23), *and*

$$\lim_{a\to\infty} \varphi_{U_1^{(1)}\ldots U_1^{(K)}}(\theta_1, \ldots, \theta_K;\ a) \equiv \varphi_{U_1^{(1)}\ldots U_1^{(K)}}(\theta_1, \ldots, \theta_K)$$
$$= \exp \lambda \sum_{m=1}^{K} \sum_{r=0}^{K-m} P_{mr} E\left\{e^{iAq(\theta_m f_m + \cdots + \theta_{m+r} f_{m+r})} - 1\right\}, \quad (3.10)$$

where

$$P_{mr} = \lim_{a\to\infty} P_{mr}(a)\,(a+2b_K).$$

3) *If* $E\{A^2 f^2(t-\tau)\} < \infty$ *then the first and second moments of* U_1 *are finite. We have*

$$\lim_{a\to\infty} E\{U_1\} = \lambda s q \int\int x f(t-\tau)\, dF_A(x)\, dG(z), \quad (3.11)$$

$$\lim_{a\to\infty} \operatorname{var}\{U_1\} = \lambda s q^2 \int\int x^2 f^2(t-\tau)\, dF_A(x)\, dG(z), \quad (3.12)$$

$$\lim_{a\to\infty} \operatorname{cov}(U_1(t,\ a,\ b_\mu,\ N),\ U_1(t,\ a,\ b_\nu,\ N))$$
$$= \lambda q^2 C_{\mu\nu} \int\int x^2 f(t-\tau^{(\mu)})\, f(t-\tau^{(\nu)})\, dF_A(x)\, dG(z) \quad (3.13)$$
$$(\mu,\ \nu = 1,\ \ldots,\ K),$$

where

$$C_{\mu\nu} = s - \Phi(0,\ |b_\mu - b_\nu|).$$

Before proving the theorem we use a lemma which was formulated as a remark in Doob's book [1]. A proof can be found, under more general conditions, in Prékopa's paper [3].

Lemma 3.1. *Suppose that* $\{N(x);\ x \geq 0\}$ *is a Poisson process with intensity* λ *and that* n *events occur in the interval* $(x_0,\ x_0 + y)$ $(y > 0)$. *Then the moments of occurrence of these events* $\xi_1, \cdots, \xi_n$ *are independent random variables uni-*

formly distributed in the interval $(x_0, x_0 + y)$.

Proof of Theorem 3.1. It is obvious that (3.7) can be put into the form

$$U_1^{(\mu)} = \sum_{j=1}^{N(a+2b_K)} qA_j f(t - \tau_j^{(\mu)}) W(a + b_\mu - \xi_j, l_j) \quad (\mu = 1, 2, \ldots, K).$$

Let us consider the random variable

$$V = \theta_1 U_1^{(1)} + \ldots + \theta_K U_1^{(K)} = q \sum_{j=1}^{N(a+2b_K)} h(\xi_j, l_j, z_j, A),$$

where

$$h(\xi, l, z, A) = \theta_1 Af(t - \tau^{(1)}) W(a + b_1 - \xi, l) + \ldots$$
$$\ldots + \theta_K Af(t - \tau^{(K)}) W(a + b_K - \xi, l),$$
$$\tau^{(\mu)} = \frac{2}{v}\sqrt{b_\mu^2 + z^2} \quad (\mu = 1, 2, \ldots, K).$$

Given that $N(0) = 0$, we have the following chain of equations:

$$E\{\exp i(\theta_1 U_1^{(1)} + \ldots + \theta_K U_1^{(K)})\} = E\{e^{iV}\}$$
$$= \sum_{n=0}^{\infty} E\{e^{iV} \mid N(a + 2b_K) = n\} P\{N(a + 2b_K) = n\}.$$

We calculate

$$E\{e^{iV} \mid N(a + 2b_K) = n\}.$$

Since the random variables z, l, A are independent and identically distributed for each j, we have

$$E\{\exp iV \mid N(a + 2b_K) = n\} = E\left\{\exp i \sum_{j=1}^{n} qh(\xi_j, l_j, z_j, A_j)\right\}$$
$$= \prod_{j=1}^{n} E\{\exp iqh(\xi_j, l_j, z_j, A_j)\} = (E\{\exp iqh(\xi, l, z, A)\})^n.$$

Since

$$P\{N(a + 2b_K) = n\} = \frac{e^{-\lambda(a+2b_K)}[\lambda(a + 2b_K)]^n}{n!}$$

for $n = 0, 1, \cdots$, we have

$$E\{e^{iV}\} = \sum_{n=0}^{\infty} e^{-\lambda(a+2b_K)} \frac{[\lambda(a + 2b_K)]^n}{n!} (E\{\exp iqh(\xi, l, z, A)\})^n$$
$$= \exp(-\lambda(a + 2b_K)[1 - E\{e^{iqh(\xi, l, z, A)}\}]). \tag{3.14}$$

The random variable $h(\xi, l, z, A)$ will be composed of a different number of terms depending on the values assumed by the indicator functions $W(x, y)$.

We use the notation

$$W_m = W(a+b_m-\xi, l) \quad (m=1, \ldots, K),$$

and $P_{mr}(a)$ is the probability that $W_m = \cdots = W_{m+r} = 1$ and also $W_1 = \cdots$ $\cdots = W_{m-1} = W_{m+r+1} = \cdots = W_K = 0$.

Obviously $P_{mr}(a)$ is the probability that there are no terms with index numbers $1, 2, \cdots, m-1, m+r+1, \cdots, K$ in the sum $h(\xi, l, z, A)$.

We now calculate $P_{mr}(a)$. Let $m=1$, $r=K-1$. Then it follows from the definition of $W(x, y)$ that $P_{1,K-1}(a)$ is the probability, first, that the beginning of the discontinuity ξ falls in the interval $(0, a+b_1]$ and, second, that the length l of this discontinuity is greater than $a+b_K-\xi$. In other words,

$$P_{1K-1}(a) = P\{l > a+b_K-\xi;\ 0<\xi<a+b_1\}.$$

By Lemma 3.1 and the fact that l is independent of ξ, we obtain

$$P_{1K-1}(a) = \frac{1}{a+2b_K}\int_0^{a+b_1} P\{l > a+b_K-v\}\,dv.$$

Since $F_l(x) = P\{l<x\}$ we have, on making the change of variables $a+b_K-v=u$,

$$P_{1K-1}(a) = \frac{1}{a+2b_K}\int_{b_K-b_1}^{a+b_K}[1-F_l(u)]\,du = \frac{1}{a+2b_K}\Phi(b_K-b_1, a+b_K). \quad (3.15)$$

Therefore, from the condition $E\{l\}=s$, we have

$$s = \int_0^\infty [1-F_l(u)]\,du = \Phi(0, b_K-b_1) + \Phi(b_K-b_1, a+b_K) + \Phi(a+b_K, \infty).$$

Hence

$$\Phi(b_K-b_1, a+b_K) = s - \Phi(0, b_K-b_1) - \Phi(a+b_K, \infty). \quad (3.16)$$

Substituting (3.16) into (3.15), we obtain

$$P_{1K-1}(a) = \frac{1}{a+2b_K}(s-\Phi(0, b_K-b_1)-\Phi(a+b_K, \infty)). \quad (3.17)$$

Let $m=1$ and $r=0, 1, \cdots, K-2$. Then $P_{1r}(a)$ is the probability, first, that the beginning of the discontinuity ξ falls in $(0, a+b_1)$ and, second, that the length of this discontinuity l falls in $(a+b_{r+1}-\xi,\ a+b_{r+2}-\xi)$. In other words,

$$P_{1r}(a)=P\{a+b_{r+1}-\xi<l<a+b_{r+2}-\xi;\ 0<\xi<a+b_1\}.$$

By Lemma 3.1 and the fact that l is independent of ξ, we have

$$P_{1r}(a)=\frac{1}{a+2b_K}\int_0^{a+b_1} P\{a+b_{r+1}-v<l<a+b_{r+2}-v\}\,dv$$

$$=\frac{1}{a+2b_K}\Phi(b_{r+1}-b_1,\ a+b_{r+1})-\frac{1}{a+2b_K}\Phi(b_{r+2}-b_1,\ a+b_{r+2}). \quad (3.18)$$

Using, as before, the decomposition of $E\{l\}=s$ into its component terms, we express s in the form

$$\left.\begin{aligned} s&=\Phi(0,\ b_{r+1}-b_1)+\Phi(b_{r+1}-b_1,\ a+b_{r+1})+\Phi(a+b_{r+1},\ \infty),\\ s&=\Phi(0,\ b_{r+2}-b_1)+\Phi(b_{r+2}-b_1,\ a+b_{r+2})+\Phi(a+b_{r+2},\ \infty).\end{aligned}\right\} \quad (3.19)$$

On substituting the corresponding values of the integrals from (3.19) into (3.18), we obtain

$$P_{1r}(a)=\frac{1}{a+2b_K}(\Phi(0,\ b_{r+2}-b_1)-\Phi(0,\ b_{r+1}-b_1)$$
$$+\Phi(a+b_{r+2},\ \infty)-\Phi(a+b_{r+1},\ \infty)). \quad (3.20)$$

Now let $m=2,\cdots,K-1$ and $r=0,1,\cdots,K-1-m$. Then

$$P_{mr}(a)=P\{a+b_{m+r}-\xi<l<a+b_{m+r+1}-\xi;\ a+b_{m-1}<\xi<a+b_m\}.$$

By Lemma 3.1 and the fact that l is independent of ξ, proceeding as in the previous derivations, we finally obtain

$$P_{mr}(a)=\frac{1}{a+2b_K}(\Phi(b_{m+r}-b_m,\ b_{m+r}-b_{m-1})$$
$$-\Phi(b_{m+r+1}-b_m,\ b_{m+r+1}-b_{m-1})). \quad (3.21)$$

Let $m=2,\cdots,K-1$ and $r=K-m$. Then

$$P_{m,\,K-m}(a)=P\{l>a+b_K-\xi;\ \ a+b_{m-1}<\xi<a+b_m\}.$$

Therefore

$$P_{m,\,K-m}(a)=\frac{1}{a+2b_K}\Phi(b_K-b_m,\ b_K-b_{m-1}). \quad (3.22)$$

Finally, for $m=K$ and $r=0$,

$$P_{K0}(a) = \frac{1}{a + 2b_K} \Phi(0,\ b_K - b_{K-1}). \tag{3.23}$$

Using the above notation and formulas for $P_{mr}(a)$, we obtain

$$\begin{aligned} E\{e^{iqh(\xi,\, l,\, z,\, A)}\} &= e^{i0}\Big[1 - \sum_{m,\, r} P_{mr}(a)\Big] + \sum_{m,\, r} P_{mr}(a)\, E\{e^{iqA(\theta_m f_m + \dots + \theta_{m+r} f_{m+r})}\} \\ &= 1 + \sum_{m,\, r} P_{mr}(a)\, E\{e^{iqA(\theta_m f_m + \dots + \theta_{m+r} f_{m+r})} - 1\}, \end{aligned} \tag{3.24}$$

where

$$f_\mu = f\Big(t - \frac{2}{v}\sqrt{b_\mu^2 + z^2}\Big) \quad (\mu = 1,\ 2,\ \dots,\ K).$$

On substituting (3.24) into the formula (3.14) for the characteristic function, we obtain formula (3.9) of the theorem. In order to free ourselves from initial conditions we let a tend to infinity. Since a only appears in $P_{mr}(a)$, it is easy to see from (3.17) and (3.20)–(3.23) that the limits

$$\lim_{a \to \infty} (a + 2b_K)\, P_{mr}(a) = P_{mr}$$

exist, where

$$\begin{gathered} P_{1\,K-1} = s - \Phi(0,\ b_K - b_1), \\ P_{1r} = \Phi(0,\ b_{r+2} - b_1) - \Phi(0,\ b_{r+1} - b_1) \quad (r = 0,\ 1,\ \dots,\ K-2), \\ P_{mr} = \Phi(b_{m+r} - b_m,\ b_{m+r} - b_{m-1}) - \Phi(b_{m+r+1} - b_m,\ b_{m+r+1} - b_{m-1}) \\ (r = 0,\ 1,\ \dots,\ K - 1 - m; \quad m = 2,\ 3,\ \dots,\ K-1), \\ P_{m,\,K-m} = \Phi(b_K - b_m,\ b_K - b_{m-1}) \quad (m = 2,\ 3,\ \dots,\ K-1), \\ P_{K0} = \Phi(0,\ b_K - b_1). \end{gathered} \tag{3.25}$$

As a result, passing to the limit with respect to a in (3.9) and using (3.25), we obtain the characteristic function (3.10).

The following formulas relating the moments of the random variables $U_1^{(1)}, \dots, U_1^{(K)}$ to their characteristic function are known:

$$\begin{gathered} \lim_{a\to\infty} E\{U_1^{(\mu)}\} = -i\frac{\partial}{\partial\theta_\mu} \ln \varphi_{U_1^{(1)} \dots U_1^{(K)}}(0,\ \dots,\ 0), \\ \lim_{a\to\infty} \operatorname{var}\{U_1^{(\mu)}\} = -\frac{\partial^2}{\partial\theta_\mu^2} \ln \varphi_{U_1^{(1)} \dots U_1^{(K)}}(0,\ 0,\ \dots,\ 0), \\ \lim_{a\to\infty} \operatorname{cov}\{U_1^{(\mu)} U_1^{(\nu)}\} = -\frac{\partial^2}{\partial\theta_\mu \partial\theta_\nu} \ln \varphi_{U_1^{(1)} \dots U_1^{(K)}}(0,\ \dots,\ 0) \\ (\mu,\ \nu = 1,\ 2,\ \dots,\ K). \end{gathered} \tag{3.26}$$

Using (3.10), we obtain

$$\ln \varphi_{U_1^{(1)} \ldots U_1^{(K)}} (\theta_1, \ldots, \theta_K)$$
$$= \lambda \sum_{m=1}^{K} \sum_{r=0}^{K-m} P_{mr} E \{ e^{iAq(\theta_m f_m + \ldots + \theta_{m+r} f_{m+r})} - 1 \}. \quad (3.27)$$

If we differentiate (3.27) with respect to θ_μ, the nonzero terms will be those for which $m \leq \mu \leq m + r$. Therefore

$$\frac{\partial}{\partial \theta_\mu} \ln \varphi_{U_1^{(1)} \ldots U_1^{(K)}} (\theta_1, \ldots, \theta_\mu)$$
$$= \lambda i q \sum_{m=1}^{\mu} \sum_{r=\mu-m}^{K-m} P_{mr} E \{ A f_\mu e^{iqA(\theta_m f_m + \ldots + \theta_{m+r} f_{m+r})} \}. \quad (3.28)$$

From (3.26) we obtain

$$\lim_{a \to \infty} E \{ U_1^{(\mu)} \} = \lambda q E \{ A f_\mu \} C_\mu,$$

and from (3.25) it follows that

$$f_\mu = f\left(t - \frac{2}{v} \sqrt{b_\mu^2 + z^2}\right) \quad (\mu = 1, 2, \ldots, K).$$

Therefore

$$\lim_{a \to \infty} E \{ U_1^{(\mu)} \} = q \lambda s E \{ A \} \int f\left(t - \frac{2}{v} \sqrt{b_\mu^2 + z^2}\right) dG(z). \quad (3.29)$$

On differentiating (3.28) with respect to θ_ν the nonzero terms will obviously be those for which $m \leq \nu \leq m + r$. Therefore

$$\frac{\partial^2}{\partial \theta_\mu \partial \theta_\nu} \ln \varphi_{U_1^{(1)} \ldots U_1^{(K)}} (\theta_1, \ldots, \theta_K)$$
$$= -\lambda q^2 \sum_{m=1}^{\min(\mu, \nu)} \sum_{r=\max(\mu-m, \nu-m)}^{K-m} P_{mr} E \{ A^2 f_\mu f_\nu e^{iAq(\theta_m f_m + \ldots + \theta_{m+r} f_{m+r})} \}.$$

Putting $\mu = \nu$, we have

$$\frac{\partial^2}{\partial \theta_\mu^2} \ln \varphi_{U_1^{(1)} \ldots U_1^{(K)}} (\theta_1, \ldots, \theta_K)$$
$$= -\lambda q^2 \sum_{m=1}^{\mu} \sum_{r=\mu-m}^{K-m} P_{mr} E \left\{ A^2 f_\mu^2 e^{iAq \sum_{\beta=0}^{r} \theta_{m+\beta} f_{m+\beta}} \right\}$$

and, using (3.26), we obtain

$$\lim_{a \to \infty} \operatorname{var} \{ U_1^{(\mu)} \} = \lambda q^2 s E \{ A^2 \} \int f^2 \left(t - \frac{2}{v} \sqrt{b_\mu^2 + z^2}\right) dG(z). \quad (3.30)$$

In exactly the same way, we obtain

$$\lim_{a\to\infty} \operatorname{cov}\{U_1^{(\mu)}, U_1^{(\nu)}\}$$

$$= \lambda q^2 C_{\mu\nu} E\{A^2\} \int f\left(t - \frac{2}{v}\sqrt{b_\mu^2 + z^2}\right) f\left(t - \frac{2}{v}\sqrt{b_\nu^2 + z^2}\right) dG(z),$$

where

$$C_{\mu\nu} = \sum_{m=1}^{\min(\mu,\nu)} \sum_{r=\max(\mu-m,\,\nu-m)}^{K-m} P_{mr}.$$

It can be shown from (3.25) that

$$C_{\mu\nu} = s - \Phi(0, |b_\mu - b_\nu|) \quad (\mu, \nu = 1, \ldots, K). \tag{3.31}$$

For, let $\mu < \nu$. Then $C_{\mu\nu}$ can be rewritten in the form

$$C_{\mu\nu} = \sum_{m=1}^{\mu} \sum_{r=\nu-m}^{K-m} P_{mr} = \sum_{r=\nu-1}^{K-1} P_{1r} + \sum_{r=\nu-2}^{K-2} P_{2r} + \cdots + \sum_{r=\nu-\mu}^{K-\mu} P_{\mu r}.$$

From (3.25) we can prove that

$$\sum_{r=\nu-1}^{K-1} P_{1r} = s - \Phi(0, b_\nu - b_1),$$

$$\sum_{r=\nu-2}^{K-2} P_{2r} = \Phi(b_\nu - b_2, b_\nu - b_1),$$

$$\cdots\cdots\cdots\cdots\cdots$$

$$\sum_{r=\nu-\mu}^{K-\mu} P_{\mu r} = \Phi(b_\nu - b_\mu, b_\nu - b_{\mu-1}).$$

Thus, for $\mu < \nu$,

$$C_{\mu\nu} = s - \Phi(0, b_\nu - b_1) + \Phi(b_\nu - b_2, b_\nu - b_1) + \cdots$$
$$\cdots + \Phi(b_\nu - b_\mu, b_\nu - b_{\mu-1}) = s - \Phi(0, b_\nu - b_1) + \Phi(b_\nu - b_\mu, b_\nu - b_1).$$

On the other hand,

$$\Phi(0, b_\nu - b_1) = \Phi(0, b_\nu - b_\mu) + \Phi(b_\nu - b_\mu, b_\nu - b_1).$$

Thus, for $\mu < \nu$,

$$C_{\mu\nu} = s - \Phi(0, b_\nu - b_\mu).$$

In exactly the same way we can show that, for $\mu > \nu$,

$$C_{\mu\nu} = s - \Phi(0, b_\mu - b_\nu).$$

The validity of (3.31) is thus established.

Therefore

$$\lim_{a\to\infty} \operatorname{cov}\{U_1^{(\mu)}, U_1^{(\nu)}\}$$
$$= C_{\mu\nu}\lambda q^2 E\{A^2\} \int f\left(t - \frac{2}{v}\sqrt{b_\mu^2 + z^2}\right) f\left(t - \frac{2}{v}\sqrt{b_\nu^2 + z^2}\right) dG(z).$$

It follows from the definition of the signal function $f(t - \tau)$ that as $|b_\mu - b_\nu|$ increases the coefficients $C_{\mu\nu} \to 0$, and therefore the correlation coefficient

$$\rho_{\mu\nu} = \frac{\operatorname{cov}\{U_1^{(\mu)}, U_1^{(\nu)}\}}{\sqrt{\operatorname{var}\{U_1^{(\mu)}\} \operatorname{var}\{U_1^{(\nu)}\}}}$$

tends to zero.

Finally, if we put $K = 1$ in the characteristic function (3.10) we obtain the univariate characteristic function (3.8) of the process (3.6). This proves the theorem.

Remarks. 1. In deriving the formulas for $P_{mr}(a)$ we used the fact that since the random variable z has a continuous distribution there is zero probability that two discontinuities of the medium will have the same depth z.

2. The theorem remains valid if we consider the multivariate characteristic function with respect to t of the process (3.6). By using the method of constructing the characteristic function considered in the theorem we can obtain, for any integer M, any $t_1 < t_2 < \cdots < t_M$ and real $\theta_1, \theta_2, \cdots, \theta_M$, the multivariate characteristic function

$$\varphi_{U_1(t_1)\dots U_1(t_M)}(\theta_1, \dots, \theta_M) = \exp \lambda s E\{e^{iAq(\theta_1 f(t_1-\tau)+\dots+\theta_M f(t_M-\tau))} - 1\}, \quad (3.32)$$

where $\tau = 2\sqrt{b^2 + z^2/v}$.

On using the formula (3.26) for finding the covariance from the characteristic function (3.32), we obtain

$$\lim_{a\to\infty} \operatorname{cov}(U_1(t_1), U_1(t_2)) = \lambda s q E\{A^2\} \int_0^H f(t_1 - \tau) f(t_2 - \tau)\, dG(z).$$

It follows from the definition of the signal function $f(t - \tau)$ that for sufficiently large $|t_1 - t_2|$ the correlation coefficient

$$\rho_{t_1 t_2} = \frac{\operatorname{cov}(U_1(t_1), U_1(t_2))}{\sqrt{\operatorname{var}\{U_1(t_1)\} \operatorname{var}\{U_1(t_2)\}}}$$

will be sufficiently close to zero.

§4. The asymptotic behavior of the sum wave for a single reflection when the intensity of the discontinuities is increased

A knowledge of the multivariate characteristic function enables us to solve a very wide range of problems in the theory of stochastic processes. Moments of interest to an investigator, such as the expectation, variance and product moments, can be found. It is even possible, by using inversion formulas for characteristic functions, to obtain distribution functions. In most cases, however, the distribution functions obtained (for only a knowledge of the distribution functions can give exhaustive answers to all the questions arising in an investigation) have a very complex form, and this renders their use in practice very difficult. Hence for both practical and theoretical purposes we try to find asymptotic forms of the distribution laws, which may have a simpler form, or, equivalently, find the asymptotic form of the characteristic functions when a parameter (or several parameters) on which the process depends tends to a limit. It need not be thought, of course, that the investigator, as it were, adjusts his results to those already known, and it should be considered a success if the limiting distribution coincides with a relationship which is already known and convenient in practice.

The aim of this section is to study the asymptotic behavior of the characteristic function (3.10) as $\lambda \to \infty$. To begin with we give a short account of certain results from probability theory.

We say that the stochastic process $\{S(x, \alpha), x \in \mathfrak{X}\}$ with finite second moments and a probability distribution law which depends on a parameter α belonging to a parametric set $\mathfrak{A}$ is *asymptotically normal* as α tends to a stated limit if, for every finite set of indices $x_1, \cdots, x_n$ from $\mathfrak{X}$, the following conditions are satisfied: for the joint characteristic function of the random variables $S^*(x_1, \alpha), \cdots, S^*(x_n, \alpha)$, where

$$S^*(x, \alpha) = \frac{S(x, \alpha) - E\{S(x, \alpha)\}}{\sqrt{\operatorname{var}\{S(x, \alpha)\}}},$$

and for each set of real numbers $\theta_1, \cdots, \theta_n$ there exists a limit

$$\lim_{\alpha\to\alpha_0} \ln \varphi_{S^*(x_1, \alpha)\dots S^*(x_n, \alpha)}(\theta_1, \theta_2, \dots, \theta_n) = -\frac{1}{2}\sum_{\mu, \nu=1}^{n} \theta_\mu \theta_\nu R_{\mu\nu} \tag{4.1}$$

as α tends to a given limit, where

$$R_{\mu\nu} = \lim_{\alpha\to\alpha_0} R\{S(x_\mu, \alpha), S(x_\nu, \alpha)\} = \lim_{\alpha\to\alpha_0} \frac{\operatorname{cov}\{S(x_\mu, \alpha), S(x_\nu, \alpha)\}}{\sqrt{\operatorname{var}\{S(x_\mu, \alpha)\}\operatorname{var}\{S(x_\nu, \alpha)\}}}.$$

Also, if $\phi_{S_1,\dots,S_n}(\theta_1, \cdots, \theta_n)$ is the joint characteristic function of the random variables $S_1, \cdots, S_n$, then the joint characteristic function of the random variables $S_1^0, \cdots, S_n^0$, where $S_\mu^0 = \alpha_\mu S_\mu + \beta_\mu$ (the α_μ and β_μ are constants; $\mu = 1, \cdots, n$), is given in terms of $\phi_{S_1,\cdots,S_n}(\Theta_1, \cdots, \theta_n)$ by the formula

$$\varphi_{S_1^0\ldots S_n^0}(\theta_1, \ldots, \theta_n) = e^{i(\theta_1\beta_1+\ldots+\theta_n\beta_n)}\varphi_{S_1\ldots S_n}(\alpha_1\theta_1, \ldots, \alpha_n\theta_n). \tag{4.2}$$

We now apply the above to the process (3.6). In (4.2) we put, for $\mu = 1, \cdots, K$,

$$\beta_\mu = -\sqrt{\lambda s}\frac{E\{A\}E\{f_\mu\}}{\sqrt{E\{A^2\}E\{f_\mu^2\}}}, \quad \alpha_\mu = \frac{1}{q}(\lambda s E\{A^2\}E\{f_\mu^2\})^{-1/2}. \tag{4.3}$$

From (3.10) we obtain the logarithm of the characteristic function:

$$\ln \varphi_{U_1^{(1)}\ldots U_1^{(K)}}(\alpha_1\theta_1, \ldots, \alpha_K\theta_K)$$
$$= \lambda \sum_{m=1}^{K}\sum_{r=0}^{K-m} P_{mr} E\left\{e^{iAq\sum_{\mu=0}^{r}\alpha_{m+\mu}\theta_{m+\mu}f_{m+\mu}} - 1\right\}. \tag{4.4}$$

We note that $E\{\exp iAq\sum_{\mu=0}^r \alpha_{m+\mu}\theta_{m+\mu}f_{m+\mu}\}$ is the characteristic function of the vector $(Aqf_m, \cdots, Aqf_{m+r})$, and $\alpha_{m+\mu}\theta_{m+\mu}$ $(\mu = 1, \cdots, r)$ are arbitrary real numbers. Let $E\{A^2\} < \infty$ and $E\{f^2(t-\tau)\} < \infty$. We expand this characteristic function in the neighborhood of the point $\alpha_m\theta_m = 0, \cdots, \alpha_{m+r}\theta_{m+r} = 0$:

$$E\left(\exp iAq\sum_{\mu=0}^{r}\alpha_{m+\mu}\theta_{m+\mu}f_{m+\mu}\right) = \exp\left[i\sum_{\mu=0}^{r}\alpha_{m+\mu}\theta_{m+\mu}E\{Aqf_{m+\mu}\}\right]$$
$$\times\left(1 - \frac{1}{2}\sum_{j,k=0}^{r}\operatorname{cov}(Aqf_{m+j}, Aqf_{m+k})\alpha_{m+j}\alpha_{m+k}\theta_{m+j}\theta_{m+k}\right.$$
$$\left. + o\left(\sum_{j=1}^{r}\alpha_{m+j}^2\theta_{m+j}^2\right)\right). \tag{4.5}$$

We see that $\alpha_{m+\mu}\theta_{m+\mu} \to 0$ as $\lambda \to \infty$ for any finite $\theta_{m+\mu}$ $(\mu = 1, \cdots, r)$. Hence this expansion is valid as $\lambda \to \infty$ for any finite $\theta_{m+\mu}$ $(\mu = 1, \cdots, r)$.

We expand $\exp i\sum_{\mu=0}^r \alpha_{m+\mu}\theta_{m+\mu}E\{Aqf_{m+\mu}\}$ as a power series in $i\sum_{\mu=0}^r \alpha_{m+\mu}\theta_{m+\mu}E\{Aqf_{m+\mu}\}$. Since $i\sum_{\mu=0}^r \alpha_{m+\mu}\theta_{m+\mu}E\{Aqf_{m+\mu}\} \to 0$ as $\lambda \to \infty$ for any finite $\theta_{m+\mu}$, we have

$$\exp i \sum_{\mu=0}^{r} \alpha_{m+\mu}\theta_{m+\mu}E\{Aqf_{m+\mu}\}$$

$$= 1 + i \sum_{\mu=0} \alpha_{m+\mu}\theta_{m+\mu}E\{Aqf_{m+\mu}\} - \frac{1}{2} \sum_{j,\,k=0}^{r} \alpha_{m+j}\alpha_{m+k}\theta_{m+j}\theta_{m+k}E\{Aqf_{m+j}\}$$

$$\times E\{Aqf_{m+k}\} + o\left(\sum_{j,\,k=0}^{r} \alpha_{m+j}\alpha_{m+k}\theta_{m+j}\theta_{m+k}E\{Aqf_{m+j}\}E\{Aqf_{m+k}\} \right).$$

On substituting this expansion into (4.5) and then the result into (4.4), we obtain

$$\ln \varphi_{U_1^{(1)} \dots U_1^{(K)}}(\alpha_1\theta_1, \ \dots, \ \alpha_K\theta_K)$$

$$= \lambda \sum_{m=1}^{K} \sum_{r=0}^{K-m} P_{mr} \left(i \sum_{\mu=0}^{r} \alpha_{m+\mu}\theta_{m+\mu}E\{Aqf_{m+\mu}\} \right.$$

$$\left. - \frac{1}{2} \sum_{j,\,k=0}^{r} \alpha_{m+j}\alpha_{m+k}\theta_{m+j}\theta_{m+k}E\{A^2q^2f_{m+j}f_{m+k}\} + o\left\{ \left(\sum_{j=0}^{r} \theta_{m+j}\alpha_{m+j} \right)^2 \right\} \right).$$

We introduce the normalized random variables $U_{10}^{(1)}, \cdots, U_{10}^{(K)}$ given by

$$U_{10}^{(\mu)} = \frac{U_1^{(\mu)} - E\{U_1^{(\mu)}\}}{\sqrt{\operatorname{var}\{U_1^{(\mu)}\}}} = \alpha_\mu U_1^{(\mu)} + \beta_\mu,$$

where α_μ and β_μ are calculated as in (4.3).

By (4.2)

$$\ln \varphi_{U_{10}^{(1)} \dots U_{10}^{(K)}}(\theta_1, \ \dots, \ \theta_K) = -i \sum_{\mu=1}^{K} \theta_\mu\beta_\mu$$

$$+ i\lambda \sum_{m=1}^{K} \sum_{r=0}^{K-m} \sum_{\mu=0}^{r} P_{mr}\alpha_{m+\mu}\theta_{m+\mu}E\{Aqf_{m+\mu}\}$$

$$- \frac{1}{2}\lambda \sum_{m=1}^{K} \sum_{r=0}^{K-m} \sum_{j,\,k=0}^{r} P_{mr}\alpha_{m+j}\alpha_{m+k}\theta_{m+j}\theta_{m+k}E\{A^2q^2f_{m+j}f_{m+k}\} + o(1).$$

It can be shown that

$$\sum_{m=1}^{K} \sum_{r=0}^{K-m} \sum_{\mu=0}^{r} P_{mr}\alpha_{m+\mu}\theta_{m+\mu}E\{Aqf_{m+\mu}\} = \sum_{\nu=1}^{K} C_\nu\alpha_\nu\theta_\nu E\{Aqf_\nu\},$$

where

$$C_\nu = \sum_{m-1}^{\nu} \sum_{r=\nu-m}^{K-m} P_{mr} = s.$$

Also,

$$\sum_{m=1}^{K}\sum_{r=0}^{K-m}\sum_{j,\,k=0}^{r} P_{mr}\alpha_{m+j}\alpha_{m+k}\theta_{m+j}\theta_{m+k}E\{A^2q^2f_{m+j}f_{m+k}\}$$
$$=\sum_{\mu,\,\nu=1}^{K} C_{\mu\nu}\alpha_{\mu}\alpha_{\nu}\theta_{\mu}\theta_{\nu}E\{A^2q^2f_{\nu}f_{\mu}\},$$

where

$$C_{\mu\nu}=\sum_{\alpha=1}^{\mu}\sum_{\beta=0}^{K-\nu} P_{\alpha\ \nu-\alpha+\beta}.$$

Hence

$$\ln\varphi_{U_{10}^{(1)}\ldots U_{10}^{(K)}}(\theta_1,\ldots,\theta_K)=-\frac{1}{2}\lambda q^2E\{A^2\}\sum_{\mu,\,\nu=1}^{K} C_{\mu\nu}\alpha_{\mu}\alpha_{\nu}\theta_{\mu}\theta_{\nu}E\{f_{\mu}f_{\nu}\}+o(1)$$
$$=-\frac{1}{2}\sum_{\nu,\,\mu=1}^{K}\frac{C_{\mu\nu}}{s}\theta_{\mu}\theta_{\nu}\frac{E\{f_{\nu}f_{\mu}\}}{\sqrt{E\{f_{\mu}^2\}E\{f_{\mu}^2\}}}+o(1).$$

It follows from (3.31) that $C_{\mu\nu}=s-\int_0^{|b_\mu-b_\nu|}[1-F_l(u)]du$, and therefore, from the formulas for variance (3.12) and covariance (3.13), we find that

$$R_{\mu\nu}=\frac{C_{\mu\nu}}{s}\frac{E\{f_{\mu}f_{\nu}\}}{\sqrt{E\{f_{\nu}^2\}E\{f_{\mu}^2\}}}$$

is an element of the correlation matrix of the process.

If we now let λ tend to infinity we find that condition (4.1) is satisfied for the characteristic function $\phi_{U_{10}^{(1)}\cdots U_{10}^{(K)}}(\theta_1,\cdots,\theta_K)$.

We have therefore proved the following theorem.

Theorem 4.1. *Let the second moments* $E\{A^2\}$ *and* $E\{f^2(t-\tau)\}$ *exist and be finite. Then as* $\lambda\to\infty$ *the stochastic process defined by* (3.6) *is asymptotically normal in the sense that, for any* $b_1,\cdots,b_K$, K *and real* $\theta_1,\cdots,\theta_K$, *there exists a limit of the characteristic function given by*

$$\lim_{\lambda\to\infty}\ln\varphi_{U_{10}^{(1)}\ldots U_{10}^{(K)}}(\theta_1,\ldots,\theta_K)=-\frac{1}{2}\sum_{\mu,\,\nu=1}^{K}\theta_{\mu}\theta_{\nu}R_{\mu\nu},$$

where

$$U_{10}=\frac{U_1-E\{U_1\}}{\sqrt{\operatorname{var}\{U_1\}}};\qquad R_{\mu\nu}=\frac{E\{f_{\mu}f_{\nu}\}}{\sqrt{E\{f_{\mu}^2\}E\{f_{\nu}^2\}}}\cdot\frac{C_{\mu\nu}}{s}.$$

Similar results can be obtained for the characteristic function (3.32) also. It is seen from the theorem that the mean value of the sum wave tends to infinity as the intensity λ increases without limit. From a physical standpoint this is obviously not justifiable, and it can be avoided by imposing limitations on certain quantities defining the sum wave. One of the possible restrictions, imposed on the amplitude factor, is given below, and the probabilistic behavior of the sum wave is discussed.

Suppose that the amplitudes of the reflected wave are such that $E\{A\} = 0$ and $E\{A^2\} = A_0^2$. Then it follows from (3.11) and (3.12) that

$$E\{U_1^{(\mu)}\} = 0,$$

$$\operatorname{var}\{U_{11}^{(\mu)}\} = \lambda sqA_0^2 \int f\left(t - \frac{2}{v}\sqrt{b_\mu^2 + z^2}\right) dG(z).$$

Also, let $A_0 \to 0$ as $\lambda \to \infty$ in such a way that $\lambda A_0^2 \to \sigma^2 < \infty$. In this case it is easy to see that

$$\operatorname{var}\{U_1^{(\mu)}\} \to sq\sigma^2 \int f\left(t - \frac{2}{v}\sqrt{b_\mu^2 + z^2}\right) dG(z).$$

Then it follows from Theorem 4.1 that the asymptotic distribution of the sum wave is normal with zero mean and finite variance. From the physical point of view the condition $\lambda A_0^2 \to \sigma^2$ as $\lambda \to \infty$ means a decrease in the contribution to the sum wave from a single reflected wave when the intensity λ of discontinuities in the medium is increased.

In conclusion, we consider one case of calculating the absorption of energy of a ray passing through a seismic medium. It is a question of the so-called geometric divergence of a wave. In this case the amplitude of the reflected wave depends on the path followed by the wave from source to detector and is given by

$$A' = \frac{A}{r},$$

where $r = 2\sqrt{b^2 + z^2}$ is the path followed by the wave from source to detector on reflection from a discontinuity at random depth z, and A is a random amplitude with properties as before.

With this approach, all the results obtained for the sum wave remain valid. In fact, the stochastic process (3.6) can be written in the form

$$U_1(a + 2b) = \sum_{j=1}^{N(a+2b)} qA_j \frac{f(t - \tau_j)}{r_j} W(a + b - \xi_j, l_j). \tag{4.6}$$

Since A is independent, $f(t - \tau)/r$ can be regarded as a new signal function. The process (4.6) therefore differs in no way from (3.6), and all investigations carried out for the process (3.6) can be repeated for (4.6).

Generally speaking, a study of the way in which the behavior of the sum wave is governed by the form of the amplitude of reflected waves belongs more to the domain of seismic prospecting, and naturally depends on the problem confronting the investigator. For example, the amplitude of vibrations in the study of a boundary layer differs from the amplitude in the study of waves reflected from distant interfaces. A study of all these problems lies outside the scope of the present paper; but we will consider below (§6) one further case of calculating the energy absorption of reflected waves. We shall study the behavior of the sum wave in calculating the number of interfaces which the wave overcomes before reaching the detector.

§5. Single waves in plane-parallel media when the discontinuities are situated obliquely to the diurnal surface

In the seismic prospecting of plane-parallel media, in addition to the cases already considered there are obviously cases in practice when plane-parallel discontinuities are inclined to the diurnal surface. In this section we study the behavior of the sum wave composed of single waves reflected from plane-parallel discontinuities situated at an angle to the diurnal surface, the angle being measured in a clockwise sense.

Let $0 < \alpha < \pi/2$. We assume, as before, that the source and detector of the elastic vibrations are situated at the points $(a, 0)$ and $(a + 2b, 0)$ respectively in the xoz coordinate system (Figure 2).

To calculate the single waves in this scheme we must find the locus of

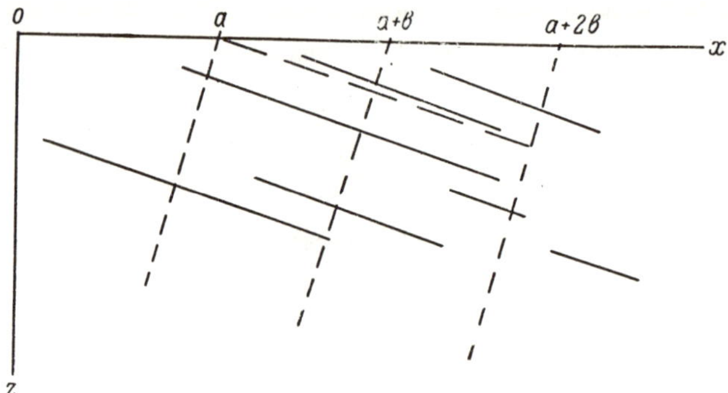

Figure 2. The arrangement of the discontinuities.

points of reflection in the xoz plane. In previous sections this locus was the vertical line passing through the midpoint of the line joining the source and the detector.

As mentioned earlier, an analogy can be traced between this seismic prospecting problem and certain problems of queueing theory as, for example, between the number of discontinuities intersecting the line $x = a + b$ in the xoz plane and the number of engaged lines, servers, and so on, at a given time in queueing theory. In the problems of seismic prospecting the x-coordinate plays the part of time in queueing theory. In problems of queueing theory it makes no difference, generally speaking, how the "calls" are arranged in space, but in problems of seismic prospecting the location of each discontinuity (the analog of a call in queueing theory) at a certain random depth is of importance. However, the fact that the points of reflection (for a single reflection) can be located only on the line $x = a + b$ (the analog of a given moment of time in queueing theory) in principle enables the study of the stochastic process U_1 to be reduced to known problems of queueing theory with certain natural modifications.

It is obvious that consideration of discontinuities inclined to the diurnal surface leads to a study of single waves whose points of reflection lie on a curve $x = x(z)$ which is not coincident with the line $x = a + b$. But in this case we cannot use the known results of queueing theory because the concept of a given moment of time has no meaning. Instead of the moment of time in queueing theory, we have a function of time. The absence of such a function in queueing theory can only be explained by the fact that problems similar to those in this paper have not yet been been considered in queueing theory.

Let us therefore formulate the problem of seismic prospecting. We consider a seismic medium whose structure is described by the stochastic model M2 except that all discontinuities are inclined at the same angle $\alpha\,(0 < \alpha < \pi/2)$ to the diurnal surface. The source and detector of the seismic vibrations are situated in the coordinate system xoz on the diurnal surface. The problem consists in determining the behavior of the sum wave as a stochastic process. All the requirements previously placed on the signal function $Af(t - \tau)$ and on the amplitudes of vibrations and other quantities, of course, remain in force.

First of all, to find the locus of points of reflection we take a new rectangular coordinate system $x'o'z'$ with origin at the source, x'-axis in the direction of the discontinuities and z'-axis perpendicular to the boundaries of the discontinuities (Figure 3).

Let us consider any point (x_0', z_0') in the new coordinate system. This point is a point of reflection if and only if

$$x_0' = \frac{bz_0' \cos \alpha}{z_0' + b \sin \alpha}. \tag{5.1}$$

This relation is easily found by using the rule of geometric optics that the angle of incidence is equal to the angle of reflection and considering the similar

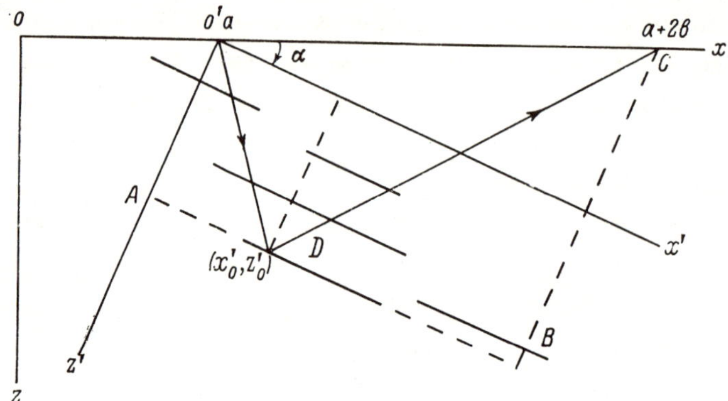

Figure 3. The path of a reflected wave.

triangles $Ao'D$ and CBD. The locus of points of reflection proceeds from the origin of the coordinate system $x'o'z'$ and is asymptotic to the line $x' = b\cos\alpha$ as $z' \to \infty$.

The problem could be solved by transferring to the coordinate system xoz, but we shall use a different approach. We return to the earlier problem of a plane-parallel medium with discontinuities parallel to the diurnal surface. We assume that the source of elastic vibrations is situated at depth h and that the detector is located on the diurnal surface at the point $(a + 2b, 0)$ in the coordinate system xoz (Figure 4).

Let us find the locus of points of reflection of single waves. The simple calculations referred to earlier lead to a relationship similar to (5.1), namely

$$x = 2b\frac{z-h}{2z-h} + a. \qquad (5.2)$$

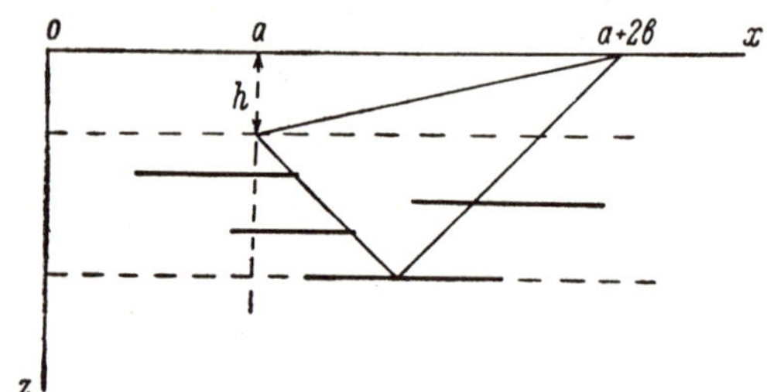

Figure 4. The position of the source and path of the wave.

The curve (5.2) is defined for $z \geq h$ and is asymptotic to $x = a + b$ as $z \to \infty$. If we express (5.1) in the coordinate system xoz, we obtain (5.2). Therefore the problem of investigating the sum wave when the discontinuities are inclined to the diurnal surface reduces to the problem in which the discontinuities are parallel to the diurnal surface and the source of vibrations is situated at a certain depth. It is obvious then that discontinuities at a depth less than h will be of no interest to us.

In expressing the sum wave in the form of a stochastic process we use the auxiliary function $W(x, y)$ defined in (3.5). Suppose that a Poisson process $N(a+2b)$ of intensity λ is given in the interval $(0, a+2b)$. Then

$$U_1(b+2b) = \sum_{j=1}^{N(a+2b)} qA_j f(t-\tau_j) W(x_j - \xi_j, l_j), \tag{5.3}$$

where the amplitude A and signal function $f(t-\tau)$ have the properties stated in §2. The arrival time τ is determined by

$$\tau = \frac{1}{v}\sqrt{4b^2 + (2z-h)^2},$$

while the random variable x, as a function of z, is found from (5.2).

If we compare the stochastic process (5.3) with (3.6) we see that the difference consists in the substitution of the constant term $(a+b)$ in the first argument of W by a random term of the form (5.2). This fact, however, does not alter the procedure for finding the characteristic functions of (5.3), and the derivation follows the same lines as in the case of (3.6).

In order not to overburden this account with tedious calculations we shall find only the univariate characteristic function of the process (5.3). Since the random variables z are independent, we have

$$\varphi_{U_1}(\theta, a) = E\{e^{i\theta U_1}\} = \sum_{n=0}^{\infty} P\{N(a+2b) = n\} E\{e^{i\theta U_1} \mid N = n\}. \tag{5.4}$$

As in the proof of Theorem 3.1, we have

$$E\{e^{i\theta U_1} \mid N(a+2b) = n\}$$

$$= \left[1 + \frac{1}{a+2b}\int\left\{(e^{iq\theta A(t-\tau)} - 1)\int_0^{x(z)}(1-F_l(u))\,du\right\} dF_A G(z)\right]^n, \tag{5.5}$$

where the function $x(z)$ is given by (5.2).

On substituting (5.5) into (5.4), we obtain

$$\varphi_{U_1}(\theta, a) = \exp\lambda\int\left\{(e^{iq\theta A f(t-\tau)} - 1)\int_0^{x(z)}(1-F_l(u))\,du\right\} dF_A G(z)$$

and, passing to the limit as $a \to \infty$, we have

$$\lim_{a\to\infty} \varphi_{U_1}(\theta,\ a) = \varphi_{U_1}(\theta) = \exp \lambda s \int_h^H (e^{iq\theta A f(t-\tau)} - 1)\, dF_A G(z). \qquad (5.6)$$

Formula (5.6) almost completely coincides with the formula for the univariate characteristic function of the process (3.6). The difference is that the domain of integration of the distribution function $G(z)$ is changed and the arrival time τ depends on the depth h. On account of this the results of §§3 and 4 automatically carry over.

Each value of α $(0 < \alpha < \pi/2)$ obviously defines h uniquely, and vice-versa.

In the case $\pi/2 < \alpha < \pi$ the problem is solved analogously. This can be seen by carrying out all the calculations after interchanging the positions of the source and the detector.

§6. Probabilistic properties of the sum wave taking account of the refractive index

Let us consider the model M2 of the medium. In contrast to previous sections we investigate here the behavior of single sum waves, taking into account the number of discontinuities which the wave crosses before and after reflection prior to being recorded by the detector.

We shall assume that the reflection factor q and the refractive index κ are constant for all the discontinuities and that the wave does not change its direction as it passes through the discontinuities.

Suppose that in the interval $(0,\ a + 2b]$ there are N boundaries at depths $z^{(1)}, \cdots, z^{(N)}$. We compose the variational series $z_1 < \cdots < z_N$. If the ith boundary is situated at depth z_i then the x-coordinate y_i of the point of reflection of the wave from this boundary is $y_i = a + b$. The equations of the lines through $(a,\ 0)$, $(a + b,\ z_i)$ and through $(a + b,\ z_i)$, $(a + 2b,\ 0)$ are, respectively,

$$y_i^{(1)} = b\frac{z}{z_i} + a, \qquad (6.1)$$

$$y_i^{(2)} = -b\frac{z}{z_i} + 2b + a. \qquad (6.2)$$

By means of the transformation $z = Hx$, where $x \in [0,\ 1]$, we have $y_i^{(1)} = bx/x_i + a$ and $y_i^{(2)} = -bx/x_i + 2b + a$. The sum wave at the detector at time t then has the form

$$U_{1\varkappa}(t,\ a,\ b,\ N)=\sum_{i=1}^{N} A_i \varkappa^{\sum_{m=1}^{i-1} W(y_{im}^{(1)}-\xi_m,\ l_m)+\sum_{m=1}^{i-1} W(y_{im}^{(2)}-\xi_m,\ l_m)}$$

$$\times qW(a+b-\xi_i,\ l_i) f\left(t-\frac{2}{v}\sqrt{b^2+H^2x_i^2}\right), \qquad (6.3)$$

where the random variables $y_{im}^{(1)}$ and $y_{im}^{(2)}$ are the x-coordinates of the points of intersection of the lines (6.1) and (6.2) with the mth discontinuity.

We write

$$u_i=\varkappa^{\sum_{m=1}^{i-1}[W(y_{im}^{(1)}-\xi_m,\ l_m)+W(y_{im}^{(2)}-\xi_m,\ l_m)]}$$

$$\times qA_i f\left(t-\frac{2}{v}\sqrt{b^2+H^2x_i^2}\right) W(a+b-\xi_i,\ l_i). \qquad (6.4)$$

The sum wave then takes the form

$$U_{1\varkappa}(t,\ a,\ b,\ N)=U_{1\varkappa}=\sum_{i=1}^{N} u_i.$$

We need to investigate the probabilistic properties of the random process $U_{1K}(t,\ a,\ b,\ N)$.

Let us calculate the expectation of the process (6.3). We have

$$E\{U_{1\varkappa}(t,\ a,\ b,\ N)\}=\sum_{n=0}^{\infty} P(N=n)\, E\{U_{1\varkappa}\mid N=n\}.$$

But

$$E\{U_{1\varkappa}\mid N=n\}=E\left\{\sum_{i=1}^{n} u_i\right\}=\sum_{i=1}^{n} E\{u_i\}.$$

It is known [4] that if the random variable X has density $f(x)$ and distribution function $F(x)$, then the joint distribution of the terms $X_{k_1}, X_{k_1+k_2}, \ldots, X_{k_1+\cdots+k_s}$ of the variational series $X_1<X_2<\cdots<X_n$ of an independent sample of size n has density

$$f(x_{k_1},\ x_{k_1+k_2},\ \ldots,\ x_{k_1+\ldots+k_s})=C_{k_1,\ k_1+k_2,\ \ldots,\ k_1+\ldots+k_s}(n)\,[F(x_{k_1})]^{k_1-1}$$

$$\times [F(x_{k_1+k_2})-F(x_{k_1})]^{k_2-1}\cdots[1-F(x_{k_1+\ldots+k_s})]^{n-k_1-\ldots-k_s}$$

$$\times f(x_{k_1})\cdots f(x_{k_1+\ldots+k_s}), \qquad (6.5)$$

where

$$C_{k_1,\ k_1+k_2,\ \ldots,\ k_1+\ldots+k_s}(n)=\frac{\Gamma(n+1)}{\Gamma(k_1)\cdots\Gamma(k_s)\,\Gamma(n-k_1-\ldots-k_s+1)}.$$

By Lemma 3.1 and the independence of l we have

$$E\{u_i\} = qE\{A_i\} \int_{0<x_1<\ldots<x_i<1} \cdots \int \prod_{m=1}^{i-1} I(x_m, x_i) f\left(t - \frac{2}{v}\sqrt{b^2 + H^2x_i^2}\right) \times E\{W(a+b-\xi_i, l_i)\} g(x_1, \ldots, x_i)\, dx_1 \ldots dx_i,$$

where

$$I(x_m, x_i) = E\left\{\varkappa^{W(y_{im}^{(1)}-\xi_m, l_m)+W(y_{im}^{(2)}-\xi_m, l_m)} \,\middle|\, X\right\}, \qquad (6.6)$$

and $g(x_1, \cdots, x_i)$ is the joint density of the ordered random variables $x_1 < x_2 < \cdots < x_i$ obtained from (6.5) with $F(x) \equiv G(x) = x$.

Since $W(a + b - \xi_i, l_i)$ does not depend on the variables of integration $x_1, \cdots, x_{i-1}$, we have

$$E\{u_i\} = \frac{qE\{A_i\}}{a+2b} \Phi(0, a+b) \int_{0<x_1<\ldots<x_{i-1}<x_i<1} \cdots \int \prod_{m=1}^{i-1} I(x_m, x_i) \times f\left(t - \frac{2}{v}\sqrt{b^2 + H^2x_i^2}\right) g(x_1, \ldots, x_i)\, dx_1 \ldots dx_i.$$

Let us find the expectation (6.6). By the definition of $W(x, y)$ the sum $W(y_{im}^{(1)} - \xi_m, l_m) + W(y_{im}^{(2)} - \xi_m, l_m)$ can obviously equal 0, 1 or 2 with corresponding probabilities. We use the notation

$$P\{W(y_{im}^{(1)} - \xi_m, l_m) = \alpha,\ W(y_{im}^{(2)} - \xi_m, l_m) = \beta\} = P_{\alpha\beta}(x_m, x_i),$$

where α and β are either 0 or 1. Then

$$\left.\begin{aligned} P\{W(y_{im}^{(1)} - \xi_m, l_m) + W(y_{im}^{(2)} - \xi_m, l_m) = 0\} &= P_{00}(x_m, x_i), \\ P\{W(y_{im}^{(1)} - \xi_m, l_m) + W(y_{im}^{(2)} - \xi_m, l_m) = 1\} &= \\ = P_{10}(x_m, x_i) + P_{01}(x_m, x_i)&, \\ P\{W(y_{im}^{(1)} - \xi_m, l_m) + W(y_{im}^{(2)} - \xi_m, l_m) = 2\} &= P_{11}(x_m, x_i). \end{aligned}\right\} \qquad (6.7)$$

It is obvious that $P_{00}(x_m, x_1) = 1 - P_{10}(x_m, x_i) - P_{01}(x_m, x_i) - P_{11}(x_m, x_i)$. Therefore

$$I(x_m, x_i) = 1 - (1 - \varkappa)[P_{10}(x_m, x_i) + P_{01}(x_m, x_i)] - (1 - \varkappa^2) P_{11}(x_m, x_i).$$

Let us calculate the values of the probabilities (6.7). By Lemma 3.1 we have

$$P_{10}(x_m, x_i) = \frac{1}{a+2b} \int_0^{y_{im}^{(1)}} P\{y_{im}^{(1)} - v < l < y_{im}^{(2)} - v\}\, dv$$

$$= \frac{1}{a+2b} [\Phi(0, y_{im}^{(1)}) - \Phi(y_{im}^{(2)} - y_{im}^{(1)}, y_{im}^{(2)})],$$

$$P_{01}(x_m, x_i) = \frac{1}{a+2b} \int_{y_{im}^{(1)}}^{y_{im}^{(2)}} P\{l > y_{im}^{(2)} - v\}\, dv = \frac{1}{a+2b} \Phi(0, y_{im}^{(2)} - y_{im}^{(1)}),$$

$$P_{11}(x_m, x_i) = \frac{1}{a+2b} \int_0^{y_{im}^{(1)}} P\{l > y_{im}^{(2)} - v\}\, dv$$

$$= \frac{1}{a+2b} [s - \Phi(0, y_{im}^{(2)} - y_{im}^{(1)}) - \Phi(y_{im}^{(2)}, \infty)].$$

Then

$$I(x_m, x_i) = 1 - (1-\varkappa) \frac{1}{a+2b} [\Phi(0, y_{im}^{(1)}) + 2\Phi(0, y_{im}^{(2)} - y_{im}^{(1)})$$

$$+ \Phi(y_{im}^{(2)}, \infty) - s] - \frac{(1-\varkappa^2)}{a+2b} [s - \Phi(0, y_{im}^{(2)} - y_{im}^{(1)}) - \Phi(y_{im}^{(2)}, \infty)]$$

$$= 1 - \frac{1-\varkappa}{a+2b} [s\varkappa + \Phi(0, y_{im}^{(1)}) + (1-\varkappa)\Phi(0, y_{im}^{(2)} - y_{im}^{(1)}) - \varkappa\Phi(y_{im}^{(2)}, \infty)].$$

We consider the following integral with variables $x_1, \cdots, x_{i-1}$:

$$\int_{0<x_1<\ldots<x_{i-1}<x_i} \cdots \int \prod_{m=1}^{i-1} I(x_m, x_i)\, g(x_1, \ldots, x_{i-1}, x_i)\, dx_1 \ldots dx_{i-1}.$$

Since $I(x_m, x_i)$ and $g(x_1, \cdots, x_{i-1}, x_i)$ are symmetric functions of $x_1, \cdots, x_{i-1}$, we have

$$\int_{0<x_1<\ldots<x_{i-1}<x_i} \cdots \int \prod_{m=1}^{i-1} I(x_m, x_i)\, g(x_1, \ldots, x_i)\, dx_1 \ldots dx_{i-1}$$

$$= \frac{1}{(i-1)!} \left[\int_0^{x_i} I(x, x_i)\, dx \right]^{i-1} \frac{\Gamma(n+1)}{\Gamma(n-i+1)} (1-x_i)^{n-i}.$$

Therefore

$$\sum_{i=1}^{n} E\{u_i\} = \frac{qE\{A\}}{a+2b} \Phi(0, a+b) \sum_{i=1}^{n} \int_0^1 \frac{\Gamma(n+1)}{\Gamma(i)\Gamma(n-i+1)} (1-x_i)^{n-i}$$

$$\times \left[\int_0^{x_i} I(x, x_i)\, dx \right]^{i-1} f\left(t - \frac{2}{v}\sqrt{b^2 + H^2 x_i^2}\right) dx_i. \tag{6.8}$$

An interchange of integration and summation gives

$$\sum_{i=1}^{n} \frac{\Gamma(n+1)}{\Gamma(i)\,\Gamma(n-i+1)}(1-x_i)^{n-i}\left(\int_0^{x_i} I(x,\ x_i)\,dx\right)^{i-1}$$
$$= n\left(1-x_i+\int_0^{x_i} I(x,\ x_i)\,dx\right)^{n-1}. \tag{6.9}$$

On substituting (6.9) into (6.8), we obtain

$$\sum_{i=1}^{n} E\{u_i\} = \frac{nqE\{A\}\,\Phi(0,\ a+b)}{a+2b}\int_0^1\left(1-y+\int_0^y I(x,\ y)\,dx\right)^{n-1}$$
$$\times f\left(t-\frac{2}{v}\sqrt{b^2+H^2y^2}\right)dy.$$

Therefore

$$E\{U_{1\varkappa}\} = \sum_{n=0}^{\infty} e^{-\lambda(a+2b)}\frac{[\lambda(a+2b)]^n}{n\,!}\cdot\frac{nqE\{A\}\,\Phi(0,\ a+b)}{a+2b}$$
$$\times\int_0^1\left(1-y+\int_0^y I(x,\ y)\,dx\right)^{n-1} f\left(t-\frac{2}{v}\sqrt{b^2+H^2y^2}\right)dy.$$

An interchange of integration and summation gives

$$E\{U_{1\varkappa}\} = \lambda qE\{A\}\,\Phi(0,\ a+b)\int_0^1 f\left(t-\frac{2}{v}\sqrt{b^2+H^2y^2}\right)$$
$$\times\exp\left\{-\lambda\left[(a+2b)\,y-(a+2b)\int_0^y I(x,\ y)\,dx\right]\right\}dy.$$

Let us consider the expression $(a+2b)y-(a+2b)\int_0^y I(x,\ y)dx$. The change of variables $x/y=\nu$ reduces it to

$$(1-\varkappa)\,y\left[\varkappa s+\int_0^1 (\Phi(0,\ b\nu+a)+(1-\varkappa)\,\Phi(0,\ 2b(1-\nu))\right.$$
$$\left.-\varkappa\Phi(-b\nu+2b+a,\ \infty))d\nu\right].$$

We write

$$B(a) = \varkappa s + \int_0^1 (\Phi(0,\ b\nu + a) + (1-\varkappa)\Phi(0,\ 2b(1-\nu))$$
$$- \varkappa\Phi(-b\nu + 2b + a,\ \infty))\, d\nu.$$

Then

$$E\{U_{1\varkappa}(t,\ a,\ b,\ N)\} = \lambda q E\{A\}\Phi(0,\ a+b)\int_0^1 e^{-\lambda y(1-\varkappa)B(a)} f\left(t - \frac{2}{v}\sqrt{b^2 + H^2y^2}\right) dy.$$

It is easy to see that the limit

$$\lim_{a\to\infty} B(a) = B = s(1+\varkappa) + (1-\varkappa)\int_0^1 \Phi(0,\ 2b(1-\nu))\, d\nu \qquad (6.10)$$

exists. Therefore the expectation of the stationary process $U_{1K}(t, b, N)$ is given by

$$E\{U_{1\varkappa}(t,\ b,\ N)\} = \lambda s q E\{A\}\int_0^1 e^{-\lambda(1-\varkappa)By} f\left(t - \frac{2}{v}\sqrt{b^2 + H^2y^2}\right) dy. \qquad (6.11)$$

The variance of the process $U_{1K}(t, a, b, N)$ is calculated from the formula

$$\operatorname{var}\{U_{1\varkappa}\} = E\{U_{1\varkappa}^2\} - (E\{U_{1\varkappa}\})^2.$$

By the theorem of total probability,

$$E\{U_{1\varkappa}^2\} = \sum_{n=0}^{\infty} P(N=n)\, E(U_{1\varkappa}^2 \mid N=n\}. \qquad (6.12)$$

Let us consider U_{1K}^2 with $N(a+2b) = n$:

$$U_{1\varkappa}^2 = \sum_{i=1}^{n} u_i^2 + 2\sum_{i=2}^{n}\sum_{j=1}^{i-1} u_i u_j.$$

By (6.4) we have

$$u_i u_j = q^2 A_i A_j \varkappa^{\sum_{m=1}^{j-1}(h_{jm}+h_{im}) + \sum_{m=j+1}^{i-1} h_{im} + h_{ij}} f\left(t - \frac{2}{v}\sqrt{b^2 + H^2x_i^2}\right)$$
$$\times f\left(t - \frac{2}{v}\sqrt{b^2 + H^2x_j^2}\right) W(a+b-\xi_j,\ l_j)\, W(a+b-\xi_i,\ l_i),$$

where $h_{\alpha m} = W(y_{\alpha m}^{(1)} - \xi_m, l_m) + W(y_{\alpha m}^{(2)} - \xi_m, l_m)$.

By Lemma 3.1 and the independence of A, l and ξ we have

$$E\{u_i u_j\} = q^2 (E\{A\})^2 \int\limits_{0<x_1<\ldots<x_i<1} \cdots \int \prod_{m=1}^{j-1} \mathcal{J}(x_m, x_j, x_i) \prod_{m=j+1}^{i-1} I(x_m, x_i)$$
$$\times \frac{1}{a+2b} \Phi(0, a+b) E\{\varkappa^{h_{ij}} W(a+b-\xi_j, l_j) \mid (X)\}$$
$$\times f\left(t - \frac{2}{v}\sqrt{b^2 + H^2 x_i^2}\right) f\left(t - \frac{2}{v}\sqrt{b^2 + H^2 x_j^2}\right) g(x_1, \ldots, x_i)\, dx_1 \ldots dx_i, \quad (6.13)$$

where $I(x_m, x_i)$ is given by (6.6) and

$$\mathcal{J}(x_m, x_j, x_i) = E\{\varkappa^{h_{jm}+h_{im}} \mid (X)\} \quad (m \neq j). \qquad (6.14)$$

Let us find the expectation (6.14). By the definition of $W(x, y)$ it is obvious that $h_{jm} + h_{im}$ can equal 0, 1, 2, 3 or 4 with corresponding probabilities.

We use the notation

$$P\{W(y_{im}^{(1)} - \xi_m, l_m) = \alpha_1,\ W(y_{jm}^{(1)} - \xi_m, l_m) = \alpha_2,\ W(y_{jm}^{(2)} - \xi_m, l_m) = \alpha_3,$$
$$W(y_{im}^{(2)} - \xi_m, l_m) = \alpha_4\} = P_{\alpha_1\alpha_2\alpha_3\alpha_4}(x_m, x_j, x_i),$$

where the value of α_ν is either 0 or 1. Then

$$\begin{gathered}
P\{h_{jm} + h_{im} = 0\} = P_{0000}(x_m, x_j, x_i), \\
P_1(x_m, x_j, x_i) = P\{h_{jm} + h_{im} = 1\} = P_{1000}(x_m, x_j, x_i) \\
+ P_{0100}(x_m, x_j, x_i) + P_{0010}(x_m, x_j, x_i) + P_{0001}(x_m, x_j, x_i), \\
P_2(x_m, x_j, x_i) = P\{h_{jm} + h_{im} = 2\} = P_{1100}(x_m, x_j, x_i) \\
+ P_{0110}(x_m, x_j, x_i) + P_{0011}(x_m, x_j, x_i), \\
P_3(x_m, x_j, x_i) = P\{h_{jm} + h_{im} = 3\} \\
= P_{1110}(x_m, x_j, x_i) + P_{0111}(x_m, x_j, x_i), \\
P_4(x_m, x_j, x_i) = P\{h_{jm} + h_{im} = 4\} = P_{1111}(x_m, x_j, x_i).
\end{gathered} \qquad (6.15)$$

It is obvious that

$$P_{0000}(x_m, x_j, x_i) = 1 - P\{h_{jm} + h_{im} > 0\}.$$

Therefore

$$\mathcal{J}(x_m, x_j, x_i) = 1 - P\{h_{jm} + h_{im} > 0\} + \sum_{\nu=1}^{4} \varkappa^\nu P_\nu(x_m, x_j, x_i)$$
$$= 1 - \sum_{\nu=1}^{4} (1 - \varkappa^\nu) P_\nu(x_m, x_j, x_i).$$

Let us calculate the values of the probabilities (6.15). By Lemma 3.1 and formulas (6.1) and (6.2) we have

$$P_{1000}(x_m, x_j, x_i)=\frac{1}{a+2b}\int_0^{y_{im}^{(1)}} P\{y_{im}^{(1)}-v<l<y_{jm}^{(1)}-v\}\,dv$$

$$=\frac{1}{a+2b}\left(\Phi\left(0,\ b\frac{x_m}{x_i}+a\right)+\Phi\left(0,\ b\frac{x_m}{x_jx_i}(x_i-x_j)\right)+\Phi\left(b\frac{x_m}{x_j}+a,\ \infty\right)-s\right),$$

$$P_{0100}(x_m, x_j\ x_i)=\frac{1}{a+2b}\int_{y_{im}^{(1)}}^{y_{jm}^{(1)}} P\{y_{jm}^{(1)}-v<l<y_{jm}^{(2)}-v\}\,dv$$

$$=\frac{1}{a+2b}\left(\Phi\left(0,\ b\frac{x_m}{x_j}+a\right)-\Phi\left(2b\left(1-\frac{x_m}{x_j}\right),\ 2b-b\frac{x_m}{x_jx_i}(x_i+x_j)\right)\right),$$

$$P_{0010}(x_m, x_j, x_i)=\frac{1}{a+2b}\int_{y_{jm}^{(1)}}^{y_{jm}^{(2)}} P(y_{jm}^{(2)}-v<l<y_{im}^{(2)}-v\}\,dv$$

$$=\frac{1}{a+2b}\left(\Phi\left(0,\ 2b\left(1-\frac{x_m}{x_j}\right)\right)-\Phi\left(b\frac{x_m}{x_ix_j}(x_i-x_j),\ 2b-b\frac{x_m}{x_ix_j}(x_i+x_j)\right)\right),$$

$$P_{0001}(x_m, x_j, x_i)=\frac{1}{a+2b}\int_{y_{jm}^{(2)}}^{y_{im}^{(2)}} P\{l>y_{im}^{(2)}-v\}\,dv=\frac{1}{a+2b}\Phi\left(0,\ b\frac{x_m}{x_ix_j}(x_i-x_j)\right),$$

$$P_{1100}(x_m, x_j, x_i)=\frac{1}{a+2b}\int_0^{y_{im}^{(1)}} P\{y_{jm}^{(1)}-v<l<y_{jm}^{(2)}-v\}\,dv$$

$$=\frac{1}{a+2b}\left(\Phi\left(0,\ 2b-b\frac{x_m}{x_ix_j}(x_i+x_j)\right)-\Phi\left(0,\ b\frac{x_m}{x_ix_j}(x_i-x_j)\right)\right.$$
$$\left.+\Phi\left(-b\frac{x_m}{x_j}+2b+a,\ \infty\right)-\Phi\left(b\frac{x_m}{x_j}+a,\ \infty\right)\right),$$

$$P_{0110}(x_m, x_j, x_i)=\frac{1}{a+2b}\int_{y_{im}^{(1)}}^{y_{jm}^{(1)}} P\{y_{jm}^{(2)}-v<l<y_{im}^{(2)}-v\}\,dv$$

$$=\frac{1}{a+2b}\left(\Phi\left(0,\ 2b-b\frac{x_m}{x_ix_j}(x_i+x_j)\right)-\Phi\left(0,\ 2b\left(1-\frac{x_m}{x_j}\right)\right)\right.$$
$$\left.+\Phi\left(2b\left(1-\frac{x_m}{x_j}\right),\ \infty\right)\right),$$

$$P_{0011}(x_m, x_j, x_i) = \frac{1}{a+2b} \int_{y_{jm}^{(1)}}^{y_{jm}^{(2)}} P\{l > y_{im}^{(2)} - v\}\, dv.$$

$$= \frac{1}{a+2b}\left(s - \Phi\left(0,\ b\frac{x_m}{x_i x_j}(x_i - x_j)\right) - \Phi\left(2b - b\frac{x_m}{x_i x_j}(x_i + x_j),\ \infty\right)\right),$$

$$P_{1110}(x_m, x_j, x_i) = \frac{1}{a+2b} \int_0^{y_{im}^{(1)}} P\{y_{jm}^{(2)} - v < l < y_{im}^{(2)} - v\}\, dv$$

$$= \frac{1}{a+2b}\left(\Phi\left(2b - b\frac{x_m}{x_i x_j}(x_i + x_j),\ -b\frac{x_m}{x_j} + 2b + a\right)\right.$$

$$\left. - \Phi\left(2b\left(1 - \frac{x_m}{x_i}\right),\ -b\frac{x_m}{x_i} + 2b + a\right)\right),$$

$$P_{0111}(x_m, x_j, x_i) = \frac{1}{a+2b} \int_{y_{im}^{(1)}}^{y_{jm}^{(1)}} P\{l > y_{im}^{(2)} - v\}\, dv$$

$$= \frac{1}{a+2b}\Phi\left(2b - b\frac{x_m}{x_i x_j}(x_i + x_j),\ 2b\left(1 - \frac{x_m}{x_i}\right)\right),$$

$$P_{1111}(x_m, x_j, x_i) \frac{1}{a+2b} \int_0^{y_{im}^{(1)}} P\{l > y_{im}^{(2)} - v\}\, dv$$

$$= \frac{1}{a+2b}\Phi\left(2b\left(1 - \frac{x_m}{x_i}\right),\ -b\frac{x_m}{x_i} + 2b + a\right).$$

Let us consider the integral with respect to $x_1, \cdots, x_{j-1}, x_{j+1}, \cdots, x_{i-1}$:

$$\underset{\substack{0 < x_1 < \cdots < x_{j-1} < x_j < \\ < x_{j+1} < \cdots < x_{i-1} < x_i}}{\int \cdots \int} \prod_{m=1}^{j-1} \mathcal{J}(x_m, x_j, x_i) \prod_{m=j+1}^{i-1} I(x_m, x_i)\, g(x_1, \ldots, x_i)$$

$$\times\, dx_1 \ldots dx_{j-1} dx_{j+1} \ldots dx_{i-1}.$$

Since $\Pi_{m=1}^{j-1}\mathcal{J}(x_m, x_j, x_i)$ and $\Pi_{m=j+1}^{i-1} I(x_m, x_i)$ are symmetric functions of $x_1, \cdots, x_{j-1}$ and $x_{j+1}, \cdots, x_{i-1}$, respectively, by the definition of $g(x_1, \cdots, x_i)$ we have

$$\int\limits_{\substack{0<x_1<\dots<x_{j-1}<x_j<\\<x_{j+1}<\dots<x_{i-1}<x_i}}\cdots\int \prod_{m=1}^{j-1} \mathcal{J}(x_m, x_j, x_i) \prod_{m=j+1}^{i-1} I(x_m, x_i)\, g(x_1, \dots, x_i)$$

$$\times dx_1 \dots dx_{j-1} dx_{j+1} \dots dx_{i-1} = \frac{n!}{(j-1)!\,(i-j-1)!\,(n-i)!}$$

$$\times \left[\int_0^{x_j} \mathcal{J}(x, x_j, x_i)\, dx\right]^{j-1} \left[\int_{x_j}^{x_i} I(y, x_i)\, dy\right]^{i-j-1} (1-x_i)^{n-i}. \qquad (6.16)$$

Let us find $E\{\kappa^{h_{ij}} W(a+b-\xi_j, l_j)|(X)\}$. The random variable $\kappa^{h_{ij}} W(a+b-\xi_j, l_i)$ can take the values 0, 1, κ and κ^2 with probabilities P_0, $P_1(x_j, x_i)$, $P_\kappa(x_j, x_i)$ and $P_{\kappa^2}(x_j, x_i)$, respectively, where

$$P_1(x_j, x_i) = P\{\kappa^{h_{ij}} W(a+b-\xi_j, l_j) = 1\}$$

$$= \frac{1}{a+2b}\left[\Phi\left(0,\ b\left(1-\frac{x_j}{x_i}\right)\right) - \Phi\left(b\left(1-\frac{x_j}{x_i}\right),\ 2b\left(1-\frac{x_i}{x_j}\right)\right)\right],$$

$$P_\kappa(x_j, x_i) = P\{\kappa^{h_{ij}} W(a+b-\xi_j, l_j) = \kappa\}$$

$$= \frac{1}{a+2b}\left[\Phi\left(0,\ 2b\left(1-\frac{x_j}{x_i}\right)\right) - \Phi\left(0,\ b\left(1-\frac{x_i}{x_j}\right)\right)\right.$$

$$\left. + \Phi\left(-b\frac{x_j}{x_i} + 2b + a,\ \infty\right) - \Phi(a+b,\ \infty)\right],$$

$$P_{\kappa^2}(x_j, x_i) = P\{\kappa^{h_{ij}} W(a+b-\xi_j, l_j) = \kappa^2\}$$

$$= \frac{1}{a+2b}\Phi\left(2b\left(1-\frac{x_j}{x_i}\right),\ -b\frac{x_j}{x_i} + 2b + a\right).$$

Then

$$E\{\kappa^{h_{ij}} W(a+b-\xi_j, l_j)\,|\,(X)\} = P_1(x_j, x_i) + \kappa P_\kappa(x_j, x_i) + \kappa^2 P_{\kappa^2}(x_j, x_i). \qquad (6.17)$$

On substituting (6.16) and (6.17) into (6.13), we obtain

$$E\{u_i u_j\} = (qE\{A\})^2 \frac{1}{a+2b}\Phi(0, a+b)\frac{n!}{(j-1)!\,(i-j-1)!\,(n-i)!}$$

$$\times \iint\limits_{0<x_1<x_2<1} (1-x_2)^{n-i} \left[\int_0^{x_1} \mathcal{J}(x, x_1, x_2)\, dx\right]^{j-1} \left[\int_{x_1}^{x_2} I(y, x_2)\, dy\right]^{i-j-1} \times$$

$$\times[P_1(x_1, x_2) + \varkappa P_\varkappa(x_1, x_2) + \varkappa^2 P_{\varkappa^2}(x_1, x_2)]$$
$$\times f\left(t - \frac{2}{v}\sqrt{b^2 + H^2x_1^2}\right) f\left(t - \frac{2}{v}\sqrt{b^2 + H^2x_2^2}\right) dx_1 dx_2. \qquad (6.18)$$

An interchange of integration and summatioh gives

$$\sum_{i=2}^{n}\sum_{j=1}^{i-1} E\{u_i u_j\} = (n)_2 \frac{(qE\{A\})^2}{a+2b}\Phi(0, a+b).$$
$$\times \iint\limits_{0<x_1<x_2<1}\left[1 - x_2 + \int_0^{x_1}\mathcal{J}(x, x_1, x_2)\,dx + \int_{x_1}^{x_2} I(y, x_2)\,dy\right]^{n-2}$$
$$\times[P_1(x_1, x_2) + \varkappa P_\varkappa(x_1, x_2) + \varkappa^2 P_{\varkappa^2}(x_1, x_2)]$$
$$\times f\left(t - \frac{2}{v}\sqrt{b^2 + H^2x_1^2}\right) f\left(t - \frac{2}{v}\sqrt{b^2 + H^2x_2^2}\right) dx_1 dx_2.$$

In calculating $E\{u_i^2\}$ we use known formulas obtained in the calculation of $E\{u_i\}$:

$$u_i^2 = q^2 A_i^2 W^2(a + b - \xi_i, l_i)$$
$$\times f^2\left(t - \frac{2}{v}\sqrt{b^2 + H^2x_i^2}\right)\varkappa^{2\sum_{m=1}^{i-1}[W(y_{im}^{(1)} - \xi_m, l_m) + W(y_{im}^{(2)} - \xi_m, l_m)]}$$

Then

$$E\{u_i^2\} = \frac{q^2E\{A^2\}}{a+2b}\Phi(0, a+b)$$
$$\times \int\limits_{0<x_1<\ldots<x_i<1}\!\!\cdots\!\int \prod_{m=1}^{i-1} E\left\{\varkappa^{2W(y_{im}^{(1)} - \xi_m, l_m) + 2W(y_{im}^{(2)} - \xi_m, l_m)}\,|\,(X)\right\}$$
$$\times f^2\left(t - \frac{2}{v}\sqrt{b^2 + H^2x_i^2}\right) g(x_1, \ldots, x_i)\,dx_1\ldots dx_i.$$

As before, we obtain

$$\prod_{m=1}^{i-1} E\left\{\varkappa^{2W(y_{im}^{(1)} - \xi_m, l_m) + 2W(y_{im}^{(2)} - \xi_m, l_m)}\,|\,(X)\right\}$$
$$= \prod_{m=1}^{i-1}\{1 - (1 - \varkappa^2)[P_{10}(x_m, x_i) + P_{01}(x_m, x_i)] - (1 - \varkappa^4)P_{11}(x_m, x_i)\}. \qquad (6.19)$$

Using the symmetry of the function (6.19) with respect to $x_1, \cdots, x_{i-1}$, we obtain

$$E\{u_i^2\} = \frac{q^2E\{A^2\}}{a+2b}\Phi(0, a+b)\int\limits_{0<x<1}\left[\int_0^x I^*(y, x)\,dy\right]^{i-1}\frac{\Gamma(n+1)}{\Gamma(i)\Gamma(n-i+1)}$$
$$\times(1-x)^{n-i} f^2\left(t - \frac{2}{v}\sqrt{b^2 + H^2x^2}\right) dx,$$

where

$$I^*(y, x) = 1 - (1 - \varkappa^2)[P_{01}(y, x) + P_{10}(y, x)] - (1 - \varkappa^4) P_{11}(y, x).$$

An interchange of integration and summation gives

$$\sum_{i=1}^{n} E\{u_i^2\} = n \frac{E\{A^2\} q^2}{a + 2b} \Phi(0, a + b) \int\limits_{0<x<1} \left[\int_0^x I^*(y, x)\, dy + 1 - x \right]^{n-1} \times f^2\left(t - \frac{2}{v}\sqrt{b^2 + H^2x^2}\right) dx. \tag{6.20}$$

Therefore, by (6.18) and (6.20),

$$\begin{aligned} E\{U_{1\varkappa}^2 |_{N=n}\} &= n \frac{E\{A^2\} q^2}{a + 2b} \Phi(0, a + b) \int_0^1 \left[\int_0^x I^*(y, x)\, dy + 1 - x \right]^{n-1} \\ &\quad \times f^2\left(t - \frac{2}{v}\sqrt{b^2 + H^2x^2}\right) dx \\ &+ 2n(n-1) \frac{(E\{A\} q)^2}{a + 2b} \Phi(0, a + b) \iint\limits_{0<x_1<x_2<1} \left[1 - x_2 + \int_0^{x_1} \mathcal{J}(x, x_1, x_2)\, dx \right. \\ &\quad \left. + \int_{x_1}^{x_2} I(y, x_2)\, dy \right]^{n-2} [P_1(x_1, x_2) + \varkappa P_\varkappa(x_1, x_2) + \varkappa^2 P_{\varkappa^2}(x_1, x_2)] \\ &\quad \times f\left(t - \frac{2}{v}\sqrt{b^2 + H^2x_1^2}\right) f\left(t - \frac{2}{v}\sqrt{b^2 + H^2x_2^2}\right) dx_1 dx_2. \end{aligned} \tag{6.21}$$

On substituting (6.21) into (6.12), we obtain, after simple calculations,

$$\begin{aligned} E\{U_{1\varkappa}^2\} &= q^2 E\{A^2\} \lambda \Phi(0, a + b) \int_0^1 \exp\left\{ \lambda(a + 2b) \right. \\ &\quad \left. \times \int_0^x [I^*(y, x) - 1]\, dy \right\} f^2\left(t - \frac{2}{v}\sqrt{b^2 + H^2x^2}\right) dx \\ &+ 2(\lambda q E\{A\})^2 \Phi(0, a + b) \iint\limits_{0<x_1<x_2<1} (a + 2b)[P_1(x_1, x_2) + \varkappa P_\varkappa(x_1, x_2) \\ &\quad + \varkappa^2 P_{\varkappa^2}(x_1, x_2)] f\left(t - \frac{2}{v}\sqrt{b^2 + H^2x_1^2}\right) f\left(t - \frac{2}{v}\sqrt{b^2 + H^2x_2^2}\right) \\ &\quad \times \exp\left\{ \lambda(a + 2b) \int_0^{x_1} [\mathcal{J}(x, x_1, x_2) - 1]\, dx \right. \\ &\quad \left. + \lambda(a + 2b) \int_{x_1}^{x_2} [I(y, x_2) - 1]\, dy \right\} dx_1 dx_2. \end{aligned} \tag{6.22}$$

It is easy to see from the formulas defining P_{10}, P_{01} and P_{11} that

$$\lim_{a\to\infty} \lambda (a+2b) \int_0^x [I^*(y, x)-1]\,dy = -(1-\varkappa^2)\,x \int_0^1 [2\Phi(0, 2b(1-v)) + (1+\varkappa^2)(s-\Phi(0, 2b(1-v)))]\,dv = -(1-\varkappa^2)\,xB_1, \tag{6.23}$$

where

$$B_1 = \int_0^1 [2\Phi(0, 2b(1-v)) + (1+\varkappa^2)(s-\Phi(0, 2b(1-v)))]\,dv > 0.$$

From the formulas defining $P_1(x_j, x_i)$, $P_\kappa(x_j, x_i)$ and $P_{\kappa^2}(x_j, x_i)$ we see that

$$\lim_{a\to\infty} (a+2b)[P_1(x_1, x_2) + \varkappa P_\varkappa(x_1, x_2) + \varkappa^2 P_{\varkappa^2}(x_1, x_2)] = G_1(x_1, x_2), \tag{6.24}$$

where

$$G_1(x_1, x_2) = \Phi\left(0, b\left(1-\frac{x_1}{x_2}\right)\right) - \Phi\left(b\left(1-\frac{x_1}{x_2}\right), 2b\left(1-\frac{x_1}{x_2}\right)\right) + 2\varkappa\Phi\left(b\left(1-\frac{x_1}{x_2}\right), 2b\left(1-\frac{x_1}{x_2}\right)\right) + \varkappa^2\left[s-\Phi\left(0, 2b\left(1-\frac{x_1}{x_2}\right)\right)\right]. \tag{6.25}$$

Also,

$$\lim_{a\to\infty} (a+2b)[\mathcal{J}(x, x_1, x_2)-1] = -G_2(x, x_1, x_2), \tag{6.26}$$

where

$$\begin{aligned} G_2(x, x_1, x_2) = (1-\varkappa)\Big[&3\Phi\left(0, b\frac{x}{x_1x_2}(x_2-x_1)\right) \\ &- \Phi\left(2b\left(1-\frac{x}{x_1}\right), 2b-b\frac{x}{x_1x_2}(x_1+x_2)\right) + \Phi\left(0, 2b\left(1-\frac{x}{x_1}\right)\right) \\ &- \Phi\left(b\frac{x}{x_1x_2}(x_2-x_1), 2b-b\frac{x}{x_1x_2}(x_1+x_2)\right)\Big] \\ &+ (1-\varkappa^2)\Big[\Phi\left(0, 2b-b\frac{x}{x_1x_2}(x_1+x_2)\right) - \Phi\left(0, b\frac{x}{x_1x_2}(x_2-x_1)\right) \\ &+ \Phi\left(2b\left(1-\frac{x}{x_1}\right), 2b-b\frac{x}{x_1x_2}(x_1+x_2)\right) - \Phi\left(2b-b\frac{x}{x_1x_2}(x_1+x_2),\right. \end{aligned}$$

$$2b\left(1-\frac{x}{x_2}\right)\right)+\Phi\left(b\frac{x}{x_1x_2}(x_2-x_1),\ 2b-b\frac{x}{x_1x_2}(x_1+x_2)\right)\Big]$$

$$+(1-\varkappa^3)\Big[\Phi\left(0,\ 2b\left(1-\frac{x}{x_2}\right)\right)-\Phi\left(0,\ 2b-b\frac{x}{x_1x_2}(x_1+x_2)\right)$$

$$+\Phi\left(2b-b\frac{x}{x_1x_2}(x_1+x_2),\ 2b\left(1-\frac{x}{x_2}\right)\right)\Big]$$

$$+(1-\varkappa^4)\Big[s-\Phi\left(0,\ 2b\left(1-\frac{x}{x_2}\right)\right)\Big]. \tag{6.27}$$

By the definition of $I(y,\ x_2)$

$$\lim_{a\to\infty}(a+2b)[I(y,\ x_2)-1]=-G_3(y,\ x_2), \tag{6.28}$$

where

$$G_3(y,\ x_2)=(1-\varkappa)\Big[s(1+\varkappa)+(1-\varkappa)\Phi\left(0,\ 2b\left(1-\frac{y}{x_2}\right)\right)\Big]. \tag{6.29}$$

On substituting (6.23), (6.24), (6.26) and (6.28) into (6.22), we obtain for U_{1K} as $a\to\infty$

$$\lim_{a\to\infty}E\{U_{1\varkappa}^2\}=\lambda sq^2E\{A^2\}\int_0^1\exp\{-\lambda(1-\varkappa^2)xB_1\}$$

$$\times f^2\left(t-\frac{2}{v}\sqrt{b^2+H^2x^2}\right)dx+2\lambda^2s(qE\{A\})^2\iint_{0<x_1<x_2<1}G_1(x_1,\ x_2)$$

$$\times f\left(t-\frac{2}{v}\sqrt{b^2+H^2x_1^2}\right)f\left(t-\frac{2}{v}\sqrt{b^2+H^2x_2^2}\right)$$

$$\times\exp\left\{-\lambda\left[\int_0^{x_1}G_2(x,\ x_1,\ x_2)\,dx+\int_{x_1}^{x_2}G_3(y,\ x_2)\,dy\right\}dx_1dx_2. \tag{6.30}$$

We see from (6.25) that if we put $x_1=ux_2$, then $G_1(x_1,\ x_2)$ is transformed into a function $\psi(u)$ with a single argument. From (6.27) it is easy to see that the substitution $x=\alpha ux_2$, $x_1=ux_2$ leads to the equation $G_2(x,\ x_1,\ x_2)=(1-\kappa)\psi_1(\alpha,\ u)$ and in exactly the same way the substitution $y=\beta x_2$ in (6.29) gives $G_3(y,\ \kappa_2)=(1-\kappa)\psi_2(\beta)$, where ψ_1 and ψ_2 are functions of the variables α, u and β respectively.

On carrying out these substitutions in the second integral of (6.30) and rewriting $x_2=x$, we obtain

$$\lim_{a\to\infty} E\{U^2_{1\varkappa}\} = \lambda s q^2 E\{A^2\} \int_0^1 e^{-\lambda(1-\varkappa^2)B_1 x} f^2\left(t - \frac{2}{v}\sqrt{b^2 + H^2x^2}\right)dx$$

$$+ 2\lambda^2 s (E\{A\}q)^2 \int_0^1\int_0^1 x f\left(t - \frac{2}{v}\sqrt{b^2 + H^2x^2}\right) f\left(t - \frac{2}{v}\sqrt{b^2 + H^2(ux)^2}\right)$$

$$\times \psi(u)\, e^{-\lambda(1-\varkappa)xB_2(u)}\, dx\, du,$$

where $B_2(u) = u\int_0^1 \psi_1(\alpha, u)d\alpha + \int_u^1 \psi_2(\beta)d\beta > 0$.

Therefore the variance of the stochastic process U_{1K}, which is stationary as $a \to \infty$, is given by

$$\operatorname{var}\{U_{1\varkappa}\} = q^2\lambda s E\{A^2\} \int_0^1 e^{-\lambda(1-\varkappa^2)B_1 x} f^2\left(t - \frac{2}{v}\sqrt{b^2 + H^2x^2}\right)dx$$

$$+ 2\lambda^2 s (qE\{A\})^2 \int_0^1\int_0^1 x f\left(t - \frac{2}{v}\sqrt{b^2 + H^2x^2}\right) f\left(t - \frac{2}{v}\sqrt{b^2 + H^2(ux)^2}\right)$$

$$\times \psi(u)\, e^{-\lambda(1-\varkappa)xB_2(u)} du\, dx - (q\lambda s E\{A\})^2$$

$$\times \left(\int_0^1 e^{-\lambda(1-\varkappa)xB} f\left(t - \frac{2}{v}\sqrt{b^2 + H^2x^2}\right)dx\right)^2. \qquad (6.31)$$

From these calculations we see that there are difficulties in obtaining simple characteristics of the sum wave U_{1K} such as the expectation and variance. Their investigation is also rendered difficult by the unwieldiness of (6.31), and we therefore proceed to simplify the problem.

We have calculated the exact number of boundaries crossed by a wave before it reaches the detector when the wave is reflected from a boundary at depth z. We assume that there are M_z such boundaries. The amplitude of the wave is then diminished by a factor κ^{M_z} in the course of a single reflection if the refractive index κ is the same for all discontinuities in the medium and M_z is random. We calculate by how many times on average the amplitude of a wave reflected from a boundary at depth z is diminished; that is, we calculate

$$E\{\varkappa^{M_z}\} = E\left\{\varkappa^{\sum_{m=1}^{N_z}\left(W(y^{(1)}_{zm} - \xi_m,\, l_m) + W(y^{(2)}_{zm} - \xi_m,\, l_m)\right)}\right\},$$

where $N_z = N_z(a + 2b)$ is the number of discontinuities occurring in the interval $(0, a + 2b)$ at depths up to z. It can be shown that N_z is a random variable distributed according to Poisson's law with parameter $\lambda z/H$. Therefore, by the theorem of total probability, we have

$$E\{\varkappa^{M_z}\} = \sum_{n=0}^{\infty} P_z(n) E\{\varkappa^{M_z} \mid N_z = n\}, \tag{6.32}$$

where

$$P_z(n) = \exp\left\{-\frac{\lambda(a+2b)z}{H}\right\} \frac{[\lambda z(a+2b)]^n}{n!\,H^n}.$$

We make the substitution $z = Hx$ and find the conditional expectation, given by

$$E\{\varkappa^{M_x} \mid N_x = n\} = E\left\{\varkappa^{\sum_{m=1}^{n} [W(y_{xm}^{(1)}-\xi_m,\, l_m) + W(y_{xm}^{(2)}-\xi_m,\, l_m)]}\right\},$$

where $y_{xm}^{(1)}$ and $y_{xm}^{(2)}$ are the coordinates of the intersection of the mth discontinuity with the lines joining the source and detector with the point of reflection of the discontinuity at depth x.

Using (6.5) we obtain

$$E\{\varkappa^{M_x} \mid N_x = n\} = \frac{n!}{x^n} \int_{0<x_1<\cdots<x_n<x} \cdots \int \prod_{m=1}^{n} I(x_m, x)\, dx_1 \ldots dx_n,$$

where $I(x_m, x) = 1 - (1-\kappa)[P_{10}(x_m, x) + P_{01}(x_m, x)] - (1-\kappa^2)P_{11}(x_m, x)$, and $P_{10}(x_m, x)$, $P_{01}(x_m, x)$ and $P_{11}(x_m, x)$ are calculated from (6.7) with $x_i = x$.

By the symmetry of $\Pi_{m=1}^{n} I(x_m, x)$ with respect to $x_1, \cdots, x_n$, we obtain

$$E\{\varkappa^{M_x} \mid N_x = n\} = \left[\frac{1}{x}\int_0^x I(y, x)\, dy\right]^n. \tag{6.33}$$

On substituting (6.33) into (6.32) and carrying out the summation, we obtain as $a \to \infty$,

$$E\{\varkappa^{M_x}\} = e^{-\lambda(1-\varkappa)xB}, \tag{6.34}$$

where B is given by (6.10).

We now return to the process U_{1K}. If we replace the random factor

$$\varkappa^{\sum_{m=1}^{i-1} (W(y_{im}^{(1)}-\xi_m,\, l_m) + W(y_{im}^{(2)}-\xi_m,\, l_m))}$$

in the system u_i (see (6.4)) by its expectation in accordance with (6.34), we obtain the new process

$$U_{1\varkappa}^{*}=\sum_{i=1}^{N} qA_{i}e^{-\lambda(1-\varkappa)Bx_i} f\left(t-\frac{2}{v}\sqrt{b^2+H^2x_i^2}\right)W(a+b-\xi_i,\ l_i), \quad (6.35)$$

whose terms

$$u_i^{*}=qA_{i}e^{-\lambda(1-\varkappa)Bx_i} f\left(t-\frac{2}{v}\sqrt{b^2+H^2x_i^2}\right)W(a+b-\xi_i,\ l_i)$$

are independent. If we put

$$\psi(t,\ \lambda,\ x_i)=e^{-\lambda(1-\varkappa)Bx_i} f\left(t-\frac{2}{v}\sqrt{b^2+H^2x_i^2}\right)$$

and consider $\psi(t, \lambda, x)$ as a new signal function, it becomes clear from previous analysis (§3) that a characteristic function can be obtained for U_{1K}^{*}, and then the probabilistic behavior of the sum wave (6.35) can be investigated.

Using the results of §3, we can show that the univariate characteristic function of U_{1K}^{*} in the stationary case is given by

$$\varphi_{U_{1\varkappa}^{*}}(\theta)=\exp\left\{\lambda s\int_0^1 \left(e^{i\theta q\xi\psi(t,\ \lambda,\ x)}-1\right)dx\,dF_A(\xi)\right\}.$$

From the characteristic function we can easily determine the expectation and variance of U_{1K}^{*}. By (3.29) and (3.30) we have

$$E\{U_{1\varkappa}^{*}\}=\lambda sqE\{A\}\int_0^1 \psi(t,\ \lambda,\ x)\,dx,$$

$$\operatorname{var}\{U_{1\varkappa}^{*}\}=\lambda sq^2E\{A^2\}\int_0^1 \psi^2(t,\ \lambda,\ x)\,dx.$$

It follows from these formulas that $E\{U_{1K}\}=E\{U_{1K}^{*}\}$ (see (6.11)).

The question naturally arises as to the validity of replacing the random variable κ^{M_x} by its expectation $E\{\kappa^{M_x}\}=e^{-\lambda(1-\kappa)Bx}$. In order to determine this we calculate the variance of κ^{M_x}:

$$\operatorname{var}\{\varkappa^{M_x}\}=E\{\varkappa^{2M_x}\}-e^{-2\lambda(1-\varkappa)Bx}.$$

By the use of methods developed in this section for calculating $E\{U_{1K}^2\}$ it can be shown that

$$E\{\varkappa^{2M_x}\}=e^{-\lambda(1-\varkappa^2)xB_1},$$

where $B_1=(1+\kappa^2)s+(1-\kappa^2)\int_0^1\Phi(0,\ 2b(1-y))dy$.

Therefore

$$\operatorname{var}\{\varkappa^{M}x\} = e^{-\lambda(1-\varkappa^2)xB_1} - e^{-2\lambda(1-\varkappa)Bx}.$$

We see from this formula that $\operatorname{var}\{\kappa^{M}x\} \to 0$ as $\lambda \to \infty$. Hence by Čebyšev's theorem $\kappa^{M}x - e^{-\lambda(1-\kappa)Bx}$ tends in probability to zero as $\lambda \to \infty$. Hence for sufficiently large λ the study of the probabilistic properties of $U_{1\kappa}$ can be replaced by a study of $U^*_{1\kappa}$.

§7. The sum wave taking account of double reflections and its probabilistic characteristics

Up to this point we have studied the stochastic behavior of the sum wave taking account of single waves only. As we have already noted in the Introduction, the representation of the sum wave by single waves is the first step in the study of the true character of the behavior of the sum wave. In this section we investigate the stochastic behavior of the sum wave taking account of double waves.

A double wave is understood to be one which undergoes three reflections from discontinuities in the medium. One of the reflecting boundaries may be the diurnal surface. It is clear from this definition that either two or three reflecting boundaries must play a part in the formation of a double wave, and the following four types of double waves can be identified (Figure 5).

1. A double wave formed by reflections from three boundaries at different

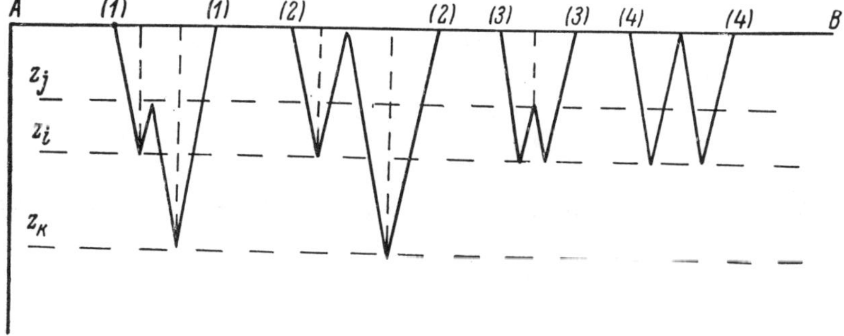

Figure 5. Types of double waves.

depths z_i, z_j, z_k.

2. A double wave formed by reflections from two boundaries at different depths z_i, z_j and the diurnal surface.

3. A double wave formed by reflections from two boundaries at different depths z_i and z_j.

4. A double wave formed by reflections from one boundary at depth z_i and the diurnal surface.

In the investigation of the sum wave we shall use the model M2 of the stochastic medium in the xoz coordinate system. The double wave can be shown to reach the detector when the coordinates of the points of reflection satisfy (7.1)–(7.4).[1)]

Type 1:

$$
\begin{aligned}
y_{i1}^{(1)} &= b\frac{z_i}{z_i - z_j + z_k} + a,\\
y_{j2}^{(1)} &= b\frac{2z_i - z_j}{z_i - z_j + z_k} + a,\\
y_{k3}^{(1)} &= b\frac{2z_i - 2z_j + z_k}{z_i - z_j + z_k} + a,
\end{aligned}
\tag{7.1}
$$

where $(y_{i1}^{(1)}, z_i)$, $(y_{j2}^{(1)}, z_j)$ and $(y_{k3}^{(1)}, z_k)$ are the coordinates of the points of reflection for

$$0 < z_i - z_j + z_k < H, \quad 0 < z_j < z_i \cdot z_k,$$

The upper index of y denotes the type of reflected wave, the first of the lower indices is the index number of the reflecting boundary and the second of the lower indices is the index number of the reflection.

Type 2. When $z_j = 0$, from (7.1) we obtain

$$
\begin{aligned}
y_{i1}^{(2)} &= b\frac{z_i}{z_i + z_k} + a,\\
y_{j2}^{(2)} &= b\frac{2z_i}{z_i + z_k} + a,\\
y_{k3}^{(2)} &= b\frac{2z_i + z_k}{z_i + z_k} + a,
\end{aligned}
\tag{7.2}
$$

where $(y_{i1}^{(2)}, z_i)$, $(y_{j2}^{(2)}, 0)$ and $(y_{k3}^{(2)}, z_k)$ are the coordinates of the points of reflection for $0 < z_i + z_k < H$.

Type 3. When $z_i = z_k$, from (7.1) we obtain

$$
\begin{aligned}
y_{i1}^{(3)} &= b\frac{z_i}{2z_i - z_j} + a,\\
y_{j2}^{(3)} &= b + a,\\
y_{i3}^{(3)} &= b\frac{3z_i - 2z_j}{2z_i - z_j} + a,
\end{aligned}
\tag{7.3}
$$

where $(y_{i1}^{(3)}, z_i)$, $(y_{j2}^{(3)}, z_j)$ and $(y_{i3}^{(3)}, z_i)$ are the coordinates of the points of reflection for $0 < 2z_i - z_j < H$, $0 < z_j < z_i$.

---

1) Formulas (7.1)–(7.4) are easily derived from geometric considerations similar to those of §5.

Type 4. When $z_j = 0$ and $z_i = z_k$, from (7.1) we obtain

$$y_{i1}^{(4)} = \frac{b}{2} + a,$$
$$y_{j2}^{(4)} = b + a, \qquad (7.4)$$
$$y_{i3}^{(4)} = \frac{3}{2} b + a,$$

where $(y_{i1}^{(4)}, z_i)$, $(y_{j2}^{(4)}, 0)$ and $(y_{i3}^{(4)}, z_i)$ are the coordinates of the points of reflection for $0 < 2z_i < H$.

If the boundaries with depths z_i, z_j and z_k were infinitely extended in both directions parallel to ox, a double wave would always arrive at the detector and the coordinates of the points of reflection would be given by (7.1)–(7.4). However, in the model M2 every reflecting boundary is characterized by an initial point ξ with respect to the x-axis, depth z and length l; that is, by the number triple (ξ, z, l). A double wave therefore arrives at the detector only if the following conditions corresponding to each type are satisfied:

for type 1

$$\left.\begin{aligned} \xi_i < y_{i1}^{(1)} < \xi_i + l_i, \\ \xi_j < y_{j2}^{(1)} < \xi_j + l_j, \\ \xi_k < y_{k3}^{(1)} < \xi_k + l_k; \end{aligned}\right\} \qquad (7.5)$$

for type 2

$$\left.\begin{aligned} \xi_i < y_{i1}^{(2)} < \xi_i + l_i, \\ \xi_k < y_{k3}^{(2)} < \xi_k + l_k; \end{aligned}\right\} \qquad (7.6)$$

for type 3

$$\left.\begin{aligned} \xi_i < y_{i1}^{(3)} < \xi_i + l_i, \\ \xi_j < y_{j2}^{(3)} < \xi_j + l_j, \\ \xi_i < y_{i3}^{(3)} < \xi_i + l_i; \end{aligned}\right\} \qquad (7.7)$$

for type 4

$$\left.\begin{aligned} \xi_i < \frac{b}{2} + a < \xi_i + l_i, \\ \xi_i < \frac{3}{2} b + a < \xi_i + l_i. \end{aligned}\right\} \qquad (7.8)$$

Suppose that (7.5) holds. Then the double wave arriving at the detector at time t is of the form

$$A_i A_j A_k q^3 f(t - \tau),$$

where q is the reflection factor; A_m $(m = 1, 2, \cdots)$ are independent identically distributed random amplitudes and $\tau = 2\sqrt{b^2 + (z_i - z_j + z_k)^2}/v$ is the delay time.

But (7.5) can be written in another form if we use the auxiliary function

$W(z, y)$ defined by (3.5). A double wave arriving at the detector at time t will then have the form

$$q^3A_iA_jA_kf(t-\tau)W(y_{i1}^{(1)}-\xi_i, l_i)W(y_{j2}^{(1)}-\xi_j, l_j)W(y_{k3}^{(1)}-\xi_k, l_k), \quad (7.9)$$

where the random variables $y_{i1}^{(1)}$, $y_{j2}^{(1)}$ and $y_{k3}^{(1)}$ are defined by (7.1).

From now on we denote the expression (7.9) by the random variable u_{ijk} dependent upon the parameters t, a and b. The indices ijk signify that the first, second and third reflections take place from discontinuities at depths z_i, z_j, z_k respectively.

Similarly, by using (7.6)–(7.8) we obtain the following random variables of the type (7.9):

$$u_{i0k}=q^3A_iA_kA_0f\left(t-\frac{2}{v}\sqrt{b^2+(z_i+z_k)^2}\right)W(y_{i1}^{(2)}-\xi_i, l_i)W(y_{k3}^{(2)}-\xi_k, l_k),$$

$$u_{iji}=q^3A_i^2A_jf\left(t-\frac{2}{v}\sqrt{b^2+(2z_i-z_j)^2}\right)W(y_{i1}^{(3)}-\xi_i, l_i)W(y_{i3}^{(3)}-\xi_i, l_i)$$
$$\times W(y_{j2}^{(3)}-\xi_j, l_j)\Big),$$

$$u_{i0i}=q^3A_i^2A_0f\left(t-\frac{2}{v}\sqrt{b^2+4z_i^2}\right)W\left(\frac{b}{2}+a-\xi_i, l_i\right)W\left(\frac{3}{2}b+a-\xi_i, l_i\right),$$

where A_0, the random amplitude of vibrations generated by reflection from the diurnal surface, is independent of and has the same distribution as A_m $(m=1, 2, \cdots)$.

In accordance with the stochastic model M2 the appearance of discontinuities in the interval $(0+2b)$ is governed by the Poisson process $N(a+2b)$ with intensity λ. Suppose that there are $N(a+2b)\equiv N$ discontinuities in $(0, a+2b)$, the depths being numbered in ascending order: $0<z_1<\cdots<z_N<H$. The sum wave, formed of double waves only, will then at time t have the form

$$U_2(t, a, b, N)=\sum_{i,j,k}^{N} u_{ijk}, \quad (7.10)$$

where the summation is taken over all i, j, k from 0 to N such that $j<i, k$.

The method of characteristic functions used in studying the behavior of the stochastic process (3.6) is not applicable to (7.10), because all the terms of the sum in (7.10) are strongly dependent random variables. In such cases the first step most frequently adopted is to calculate the simplest characteristics, such as the expectation and variance of the process. Let us calculate the expectation of (7.10).

As before, the expectation is calculated from the formula

$$E\{U_2(t,\ a,\ b,\ N)\} = \sum_{n=0}^{\infty} P\{N=n\}\, E\{U_2 \mid N=n\}, \tag{7.11}$$

where, as before, for brevity we write $U_2(t, a, b, N) \equiv U_2$, and the probabilities $P\{N = n\}$ $(n = 0, 1, 2, \cdots)$ are defined by (2.1) with $x = a + 2b$.

In calculating $E\{U_2 | N = n\}$ we decompose the sum wave (7.10) into terms corresponding to the four different types of reflected waves:

$$U_2|_{N=n} = \sum_{j=1}^{n-2} \sum_{\substack{i,\,k=1\\ i\neq k}}^{n} u_{ijk} + \sum_{\substack{i,\,k=1\\ i\neq k}}^{n} u_{i0k} + \sum_{j=1}^{n-1} \sum_{i=j+1}^{n} u_{iji} + \sum_{i=1}^{n} u_{i0i}. \tag{7.12}$$

Let us find the expectation of the first sum in (7.12). We have

$$E\left\{\sum_{j=1}^{n-2} \sum_{\substack{i,\,k=1\\ i\neq k}}^{n} u_{ijk}\right\} = \sum_{j=1}^{n-2} \sum_{i=j+1}^{n-1} \sum_{k=i+1}^{n} E\{u_{ijk}\} + \sum_{j=1}^{n-2} \sum_{k=j+1}^{n-1} \sum_{i=k+1}^{n} E\{u_{ijk}\}. \tag{7.13}$$

For $0 < j < i < k$ we have

$$E\{u_{ijk}\} = q^3 E\{A_i A_j A_k f_{ijk} W_{i1} W_{j2} W_{k3}\}, \tag{7.14}$$

where we have introduced notation for the random variables as follows:

$$\begin{aligned} f_{ijk} &= f\left(t - \frac{2}{v}\sqrt{b^2 + (z_i - z_j + z_k)^2}\right), \\ W_{i1} &= W(y_{i1}^{(1)} - \xi_i,\ l_i), \\ W_{j2} &= W(y_{j2}^{(1)} - \xi_j,\ l_j), \\ W_{k3} &= W(y_{k3}^{(1)} - \xi_k,\ l_k). \end{aligned} \tag{7.15}$$

Then

$$\begin{aligned} &E\{A_i A_j A_k f_{ijk} W_{i1} W_{j2} W_{k3}\} \\ &= E\{A_i A_j A_k f_{ijk} E(W_{i1} W_{j2} W_{k3} \mid (Z))\}, \end{aligned} \tag{7.16}$$

where $E\{W_{i1}W_{j2}W_{k3}|(z)\}$ is calculated for fixed z_i, z_j and z_k such that $0 < z_j < z_i < z_k < H$. By Lemma 3.1 and the independence of l_i, l_j and l_k we obtain

$$E\{W_{i1} W_{j2} W_{k3} \mid (Z)\} = \frac{1}{(a+2b)^3}\, \Phi(0,\ y_{i1}^{(1)})\, \Phi(0,\ y_{j2}^{(1)})\, \Phi(0,\ y_{k3}^{(1)}). \tag{7.17}$$

On substituting (7.17) into (7.16) and using the independence of A_i, A_j and A_k, we obtain

$$E\{u_{ijk}\}=\frac{(qE\{A\})^3}{(a+2b)^3}\iiint\limits_{(Z)} f_{ijk}\Phi(0,\ y_{i1}^{(1)})\,\Phi(0,\ y_{j2}^{(1)})\,\Phi(0,\ y_{k3}^{(1)})$$
$$\times g(z_j,\ z_i,\ z_k)\,dz_i dz_j dz_k, \tag{7.18}$$

where the domain of integration (z) is defined by

$$0<z_j<z_i<z_k<H,\ 0\leqslant z_i-z_j+z_k\leqslant H,$$

and $g(z_j, z_i, z_k)$ is the joint density of the order statistics $z_j<z_i<z_k$.

It follows from (6.5) that for $F(z)\equiv G(z)=z/H$ we have

$$g(z_j,\ z_i,\ z_k)=C_{jik}(n)\,z_j^{j-1}(z_i-z_j)^{i-j-1}(z_k-z_i)^{k-i-1}(H-z_k)^{n-k}H^{-n}.$$

Also, if we use the transformation $z=Hx$, where the random variable x is given on $(0, 1)$, formula (6.5) takes the final form

$$g(x_j,\ x_i,\ x_k)=C_{jik}(n)\,x_j^{j-1}(x_i-x_j)^{i-j-1}(x_k-x_i)^{k-i-1}(1-x_k)^{n-k}. \tag{7.19}$$

On substituting (7.19) into (7.18) and noting that the variables of integration do not depend on the indices of summation i, j and k, we obtain

$$E\{u_{ijk}\}=\frac{(qE\{A\})^3}{(a+2b)^3}C_{jik}(n)\iiint\limits_{\substack{0<x_1<x_2<x_3<1\\ 0\leqslant x_2-x_1+x_3\leqslant 1}} f\Big(t-\frac{2}{v}\sqrt{b^2+H^2(x_2-x_1+x_3)^2}\Big)$$
$$\times\Phi\Big(0,\ b\frac{x_2}{x_2-x_1+x_3}+a\Big)\Phi\Big(0,\ b\frac{2x_2-x_1}{x_2-x_1+x_3}+a\Big)\Phi\Big(0, b\frac{2x_2-2x_1+x_3}{x_2-x_1+x_3}+a\Big)$$
$$\times x_1^{j-1}(x_2-x_1)^{i-j-1}(x_3-x_2)^{k-i-1}(1-x_3)^{n-k}\,dx_1dx_2dx_3.$$

Therefore

$$\sum_{j=1}^{n-2}\sum_{i=j+1}^{n-1}\sum_{k=i+1}^{n}E\{u_{ijk}\}$$
$$=\Big(\frac{qE\{A\}}{a+2b}\Big)^3\sum_{j=1}^{n-2}\sum_{i=j+1}^{n-1}\sum_{k=i+1}^{n}C_{jik}(n)\iiint\limits_{\substack{0<x_1<x_2<x_3<1\\ 0\leqslant x_2-x_1+x_3\leqslant 1}} x_1^{j-1}(x_2-x_1)^{i-j-1}(x_3-x_2)^{k-i-1}$$
$$\times(1-x_3)^{n-k}f\Big(t-\frac{2}{v}\sqrt{b^2+H^2(x_2-x_1+x_3)^2}\Big)\Phi\Big(0,\ b\frac{x_2}{x_2-x_1+x_3}+a\Big)$$
$$\times\Phi\Big(0,\ b\frac{2x_2-x_1}{x_2-x_1+x_3}+a\Big)\Phi\Big(0,\ b\frac{2x_2-2x_1+x_3}{x_2-x_1+x_3}+a\Big)dx_1dx_2dx_3.$$

Interchanging the order of integration and summation and noting that

$$\sum_{j=1}^{n-2}\sum_{i=j+1}^{n-1}\sum_{k=i+1}^{n}C_{jik}(n)\,x_1^{j-1}(x_2-x_1)^{i-j-1}(x_3-x_2)^{k-i-1}(1-x_3)^{n-k}$$
$$=n(n-1)(n-2)\equiv(n)_3,$$

we obtain

$$\sum_{j=1}^{n-2}\sum_{i=j+1}^{n-1}\sum_{k=i+1}^{n} E\{u_{ijk}\}$$

$$=\left(\frac{qE\{A\}}{a+2b}\right)^3 (n)_3 \iiint\limits_{\substack{0<x_1<x_2<x_3<1\\ 0\leqslant x_2-x_1+x_3\leqslant 1}} f\left(t-\frac{2}{v}\sqrt{b^2+H^2(x_2-x_1+x_3)^2}\right)$$

$$\times\Phi\left(0,\ b\frac{x_2}{x_2-x_1+x_3}+a\right)\Phi\left(0,\ b\frac{2x_2-x_1}{x_2-x_1+x_3}+a\right)$$

$$\times\Phi\left(0,\ b\frac{2x_2-2x_1+x_3}{x_2-x_1+x_3}+a\right)dx_1dx_2dx_3. \tag{7.20}$$

By a similar calculation we find that the second sum in (7.13) coincides with (7.20) except for the domain of integration. If we unite the domains of integration $\{0<x_1<x_2<x_3<1\}$ and $\{0<x_1<x_3<x_2<1\}$, we obtain

$$E\left\{\sum_{j=1}^{n-2}\sum_{\substack{i,\,k=1\\ i\neq k}}^{n} u_{ijk}\right\}=(n)_3\left[\frac{qE\{A\}}{a+2b}\right]^3 I_3(a),$$

where

$$I_3(a)=\iiint\limits_{\substack{0<x_1<x_2,\ x_3<1\\ 0\leqslant x_2-x_1+x_3\leqslant 1}} f\left(t-\frac{2}{v}\sqrt{b^2+H^2(x_2-x_1+x_3)^2}\right)$$

$$\times\Phi\left(0,\ b\ \frac{x_2}{x_2-x_1+x_3}+a\right)\Phi\left(0,\ b\ \frac{2x_2-x_1}{x_2-x_1+x_3}+a\right)$$

$$\times\Phi\left(0,\ b\ \frac{2x_2-2x_1+x_3}{x_2-x_1+x_3}\right)dx_1dx_2dx_3.$$

The expectations of the remaining sums in (7.12) are calculated similarly, and we finally obtain

$$E\{U_2\,|\,N=n\}=(n)_3\left[\frac{qE\{A\}}{a+2b}\right]^3 I_3(a)+(n)_2\frac{q^3}{(a+2b)^2}$$

$$\times\left[(E\{A\})^3 I_2(a)+E\{A\}E\{A^2\}I_2^*(a)\right]+\frac{q^3E\{A\}E\{A^2\}}{a+2b}I_1(a), \tag{7.21}$$

where

$$I_2(a)=\iint\limits_{\substack{0<x_1,\ x_2<1\\ 0\leqslant x_1+x_2\leqslant 1}} f\left(t-\frac{2}{v}\sqrt{b^2+H^2(x_1+x_2)^2}\right)\Phi\left(0,\ b\ \frac{x_1}{x_1+x_2}+a\right)$$

$$\times\Phi\left(0,\ b\ \frac{2x_1+x_2}{x_1+x_2}+a\right)dx_1dx_2,$$

$$I_2^*(a) = \iint\limits_{\substack{0<x_1,\, x_2<1 \\ 0\leqslant 2x_2-x_1\leqslant 1}} f\left(t - \frac{2}{v}\sqrt{b^2 + H^2(2x_2 - x_1)^2}\right)\Phi(0,\ b+a)$$

$$\times\left[s - \Phi\left(0,\ 2b\,\frac{x_2 - x_1}{2x_2 - x_1}\right) - \Phi\left(b\,\frac{3x_2 - 2x_1}{2x_2 - x_1} + a,\ \infty\right)\right] dx_1 dx_2,$$

$$I_1(a) = \left[s - \Phi(0,\ b) - \Phi\left(\frac{3}{2}b + a,\ \infty\right)\right]\int\limits_0^{1/2} f\left(t - \frac{2}{v}\sqrt{b^2 + 4H^2x^2}\right) dx.$$

On substituting (7.21) into (7.11) and using (2.1) for the values of the possibilities $P\{N = n\}$, we obtain

$$E\{U_2(t,\ a,\ b,\ N)\} = (qE\{A\})^3 I_3(a)\lambda^3$$
$$+ q^3[(E\{A\})^3 I_2(a) + E\{A\} E\{A^2\} I_2^*(a)]\lambda^2$$
$$+ q^3E\{A\} E\{A^2\} I_1(a)\lambda. \tag{7.22}$$

But it is easy to see that as $a \to \infty$ the integrals $I_1(a)$, $I_2(a)$, $I_2^*(a)$ and $I_3(a)$ have limits given by

$$\lim_{a\to\infty} I_1(a) = [s - \Phi(0,\ b)]\int\limits_0^{1/2} f\left(t - \frac{2}{v}\sqrt{b^2 + 4H^2x^2}\right)dx = I_1,$$

$$\lim_{a\to\infty} I_2(a) = s^2 \iint\limits_{\substack{0<x_1,\, x_2<1 \\ 0\leqslant x_1+x_2\leqslant 1}} f\left(t - \frac{2}{v}\sqrt{b^2 + H^2(x_1 + x_2)^2}\right) dx_1 dx_2 = I_2,$$

$$\lim_{a\to\infty} I_2^*(a) = s \iint\limits_{\substack{0<x_1<x_2<1 \\ 0\leqslant 2x_2-x_1\leqslant 1}} f\left(t - \frac{2}{v}\sqrt{b^2 + H^2(2x_2 - x_1)^2}\right)$$

$$\times\left[s - \Phi\left(0,\ \frac{2b(x_2 - x_1)}{2x_2 - x_1}\right)\right] dx_1 dx_2 = I_2^*,$$

$$\lim_{a\to\infty} I_3(a) = s^3 \iiint\limits_{\substack{0<x_1<x_2,\, x_3<1 \\ 0\leqslant x_2-x_1+x_3\leqslant 1}} f\left(t - \frac{2}{v}\sqrt{b^2 + H^2(x_2 - x_1 + x_3)^2}\right) dx_1 dx_2 dx_3 = I_3.$$

The expectation of the process (7.10), taking account of single waves, will therefore have the form

$$\lim_{a\to\infty} E\{U_2(t,\ a,\ b,\ N) + U_1(t,\ a,\ b,\ N)\} = [E\{A\}qI_0 + E\{A^2\}q^3E\{A\}I_1]\lambda$$
$$+ q^3([E\{A\}]^3 I_2 + E\{A\}E\{A^2\}I_2^*)\lambda^2 + q^3E^3\{A\}I_3\lambda^3, \tag{7.23}$$

where by (3.11) $I_0 = s\int_0^1 f(t - 2\sqrt{b^2 + H^2x^2}/v)dx$; that is, the expectation of the sum wave is represented by a polynomial in λ of the third degree.

We now consider the variance of the sum wave $U_1(t, a, b, N) + U_2(t, a, b, N)$, which is calculated from

$$\operatorname{var}\{U_1(t, a, b, N) + U_2(t, a, b, N)\} = E\{U_1 + U_2\}^2 - E^2\{U_1 + U_2\}. \quad (7.24)$$

Let us consider the square of the sum $U_1 + U_2$. On expressing U_2 as the sum of the four types of double waves and denoting by u_i the single wave generated by the ith discontinuity, we obtain

$$\{U_1 + U_2\}^2 = \left\{\sum_{i=1}^{N} u_i + \sum_{i=1}^{N} u_{i0i} + \sum_{1\leqslant j<i\leqslant N} u_{iji} + \sum_{\substack{i,\,k=1\\ i\neq k}}^{N} u_{i0k} + \sum_{\substack{1\leqslant j<i,\ k\leqslant N\\ i\neq k}} u_{ijk}\right\}^2. \quad (7.25)$$

As before, we have

$$E\{U_1 + U_2\}^2 = \sum_{n=0}^{\infty} P\{N = n\} E\{[U_1 + U_2]^2 \mid N = n\}. \quad (7.26)$$

On squaring the expression within the curly brackets in (7.25) when $N = n$, we obtain the sum of fifteen different terms, each of which in turn is the sum of paired products of the random variables u.

In order not to overburden this account by very tedious calculations, let us consider just one of the terms, say

$$\Sigma u_{i_1 j_1 k_1} u_{i_2 j_2 k_2}, \quad (7.27)$$

and find its expectation.

We suppose that the sum (7.27) is taken over all i_1, j_1, k_1, i_2, j_2 and k_2 such that

$$1 \leqslant j_1 < i_1,\ k_1, \quad 1 \leqslant j_2 < i_2,\ k_2 \quad \text{for} \quad i_1 \neq k_1, \quad i_2 \neq k_2. \quad (7.28)$$

We see that the terms of (7.27) consist of the paired products of the two random variables u_{i_1, j_1, k_1} and u_{i_2, j_2, k_2}. Each random variable is characterized by a set of three indices satisfying (7.28). But the set (i_1, j_1, k_1) is independent of (i_2, j_2, k_2), and therefore relationships can exist between these number triples, imposing certain connections on u_{i_1, j_1, k_1} and u_{i_2, j_2, k_2}. For example, we have cases in (7.27) when not one of the numbers (i_1, j_1, k_1) is equal to any one of (i_2, j_2, k_2). The product $u_{i_1 j_1 k_1} u_{i_2 j_2 k_2}$ will then have six different indices. In addition, there are cases when only one of the numbers (i_1, j_1, k_1) is equal to one of (i_2, j_2, k_2). There will then be five different indices in the product $u_{i_1 j_1 k_1} u_{i_2 j_2 k_2}$, and the sum of this pair will be taken over all $i_1, j_1, k_1, i_2, j_2, k_2$

satisfying (7.28) such that one of the numbers (i_1, j_1, k_1) equals one of (i_2, j_2, k_2). There are also other variants when two or three of the numbers (i_1, j_1, k_1) coincide with two or three of (i_2, j_2, k_2) and the pair $u_{i_1j_1k_1}u_{i_2j_2k_2}$ will be summed over the indices satisfying (7.28) and these additional conditions. All the possible variants of the terms of (7.27) can be enumerated and the corresponding calculations carried out. But there are no fundamental differences among these variants when finding the expectation, and we shall therefore carry out the calculation for one case only.

Let us consider the number triples (i_1, j_1, k_1) and (i_2, j_2, k_2), and suppose that they satisfy (7.28), and that, in addition, $i_1 = j_2$. It can then be shown that the summation must be taken over the domains

$$\left.\begin{array}{ll}\{j_1<i_1<k_1,\ i_1<i_2<k_2\}, & \{j_1<k_1<i_1,\ i_1<i_2<k_2\},\\ \{j_1<i_1<k_1,\ i_1<k_2<i_2\}, & \{j_1<k_1<i_1,\ i_1<k_2<i_2\}.\end{array}\right\} \quad (7.29)$$

Let us take one of these domains, for example $\{j_1 < i_1 < k_1,\ i_1 < i_2 < k_2\}$, and denote it by S. We find the conditional expectation of $\Sigma_{(S)}u_{i_1j_1k_1}u_{i_2i_1k_2}$ for $N = n$. As before, we have $E\{\Sigma_{(S)}u_{i_1j_1k_1}u_{i_2i_1k_2}\} = \Sigma_{(S)}E\{u_{i_1j_1k_1}u_{i_2i_1k_2}\}$. Using (7.9) and (7.15), we obtain

$$E\{u_{i_1j_1k_1}u_{i_2i_1k_2}\} = q^6[E\{A\}]^4 E\{A^2\} E\{f_{i_1j_1k_1}f_{i_2i_1k_2} \times E(W_{i_1 1}W_{j_1 2}W_{k_1 3}W_{i_2 1}W_{i_1 2}W_{k_2 3}\,|\,(z))\}, \quad (7.30)$$

where $E\{\cdot\,|(Z)\}$ is calculated for fixed $z_{i_1}, z_{j_1}, z_{k_1}, z_{i_2}$ and z_{k_2} such that $z_{j_1} < z_{i_1} < z_{k_1}, z_{i_1} < z_{i_2} < z_{k_2}$. By Lemma 3.1 we have

$$E\{W_{i_1 1}W_{j_1 2}W_{k_1 3}W_{i_2 1}W_{i_1 2}W_{k_2 3}\,|\,(Z)\} = \frac{1}{(a+2b)^5} I_5(a, Z), \quad (7.31)$$

where

$$I_5(a, Z) = \Phi(0, y^{(1)}_{j_1 2})\,\Phi(0, y^{(1)}_{k_1 3})\,\Phi(0, y^{(1)}_{i_2 1})\,\Phi(0, y^{(1)}_{k_2 3}) \times \Phi(|y^{(1)}_{i_1 2} - y^{(1)}_{i_1 2}|,\ \max\{y^{(1)}_{i_1 1}, y^{(1)}_{i_1 2}\}).$$

On substituting (7.31) into (7.30), we have

$$E\{u_{i_1j_1k_1}u_{i_2i_1k_2}\} = \frac{q^6E\{A^2\}[E\{A\}]^4}{(a+2b)^5}\int_{(Z)} f_{i_1j_1k_1}f_{i_2i_1k_2}I_5(a, Z)\,g(z_{j_1}, z_{i_1}, z_{k_1}, z_{i_2}, z_{k_2})\,dZ.$$

The domain of integration (Z) is obviously completely determined by the domain of summation S. In fact, the joint density $g(z_{j_1}, z_{i_1}, z_{k_1}, z_{i_2}, z_{k_2})$ is defined by (6.5) with $F(x) = x$. But (6.5) is defined for ordered values of x and we must therefore subdivide S into the subdomains

$$S_1:\{0<j_1<i_1<i_2<k_1<k_2\},$$
$$S_2:\{0<j_1<i_1<i_2<k_2<k_1\},$$
$$S_3:\{0<j_1<i_1<k_1<i_2<k_2\}.$$

Let us consider S_1. In this case the joint density has the form

$$g(z_{j_1}, z_{i_1}, z_{i_2}, z_{k_1}, z_{k_2})=C_{j_1i_1i_2k_1k_2}(n)\frac{1}{H^n}z_{j_1}^{j_1-1}(z_{i_1}-z_{j_1})^{i_1-j_1-1}$$
$$\times(z_{i_2}-z_{i_1})^{i_2-i_1-1}(z_{k_1}-z_{i_2})^{k_1-i_2-1}(z_{k_2}-z_{k_1})^{k_2-k_1-1}(H-z_{k_2})^{n-k_2}.$$

On putting $z = Hz$ and noting that

$$\sum_{(S)}C_{j_1i_1i_2k_1k_2}(n)x_1^{j_1-1}(x_2-x_1)^{i_1-j_1-1}(x_4-x_2)^{i_2-i_1-1}(x_3-x_4)^{k_1-i_2-1}$$
$$\times(x_5-x_3)^{k_2-k_1-1}(1-x_5)^{n-k_2}=n(n-1)\dots(n-4)=(n)_5,$$

we obtain

$$\sum_{(S_1)}E\{u_{i_1j_1k_1}u_{i_2i_1k_2}\}=\frac{q^6E\{A^2\}E^4\{A\}(n)_5}{(a+2b)^5}\int_{\mathfrak{D}_1}f\left(t-\frac{2}{v}\sqrt{b^2+H^2(x_2-x_1+x_3)^2}\right)$$
$$\times f\left(t-\frac{2}{v}\sqrt{b^2+H^2(x_4-x_2+x_5)^2}\right)I_5(a,\ X)\,dx_1\dots dx_5,$$

where $\mathfrak{D}_1=\{0<x_1<x_2<x_4<x_3<x_5<1;\ 0\le x_2-x_1+x_3\le 1;\ 0\le x_4-x_2+x_5\le 1\}$.

In exactly the same way we can obtain the expectation for summation over S_2 and S_3 leading to the corresponding subdomains of integration $\mathfrak{D}_2$ and $\mathfrak{D}_3$. A direct calculation shows that in all three cases the integrands are identical. Therefore, on putting $\mathfrak{D}_1\cup\mathfrak{D}_2\cup\mathfrak{D}_3=\mathfrak{D}$, we obtain

$$\sum_{(S)}E\{u_{i_1j_1k_1}u_{i_2i_1k_2}\}=(n)_5\frac{q^6E\{A^2\}E^4\{A\}}{(a+2b)^5}\int_{\mathfrak{D}}f\left(t-\frac{2}{v}\sqrt{b^2+H^2(x_2-x_1+x_3)^2}\right)$$
$$\times f\left(t-\frac{2}{v}\sqrt{b^2+H^2(x_4-x_2+x_5)^2}\right)I_5(a,\ X)\,dx_1\dots dx_5,\qquad(7.32)$$

where $\mathfrak{D}=\{0<x_1<x_2<x_3<1;\ 0<x_2<x_4<x_5<1;\ 0\le x_2-x_1+x_3\le 1;\ 0\le x_4-x_2+x_5\le 1\}$.

By such calculations we see that whenever one number from the triple of different numbers (i_1, j_1, k_1) is equal to one number from the triple of different numbers (i_2, j_2, k_2) the expectation of $\Sigma u_{i_1j_1k_1}u_{i_2j_2k_2}$ has the form (7.32).

Therefore

$$E\left\{\sum u_{i_1j_1k_1}u_{i_2j_2k_2}\,|\,n\right\}=\frac{(n)_5}{(a+2b)^5}\,C_5(a,\ t), \tag{7.33}$$

where $C_5(a, t)$ is a function of the time t and the parameter a, and is independent of n.

To obtain an exact formula for the expectation of the random variable (7.25) we need to calculate for fixed n all the conditional expectations of sums of the type (7.33), substitute them into (7.26) and sum over all n. This procedure applied to (7.33) leads to the formula for unconditional expectation:

$$E\left\{\sum u_{i_1j_1k_1}u_{i_2j_2k_2}\right\}=\lambda^5 C_5(a,\ t).$$

In general,

$$E\{[U_1+U_2]^2\}=\lambda C_1(a,\ t)+\ldots+\lambda^6 C_6(a,\ t),$$

where the $C_\nu(a, t)$ are functions of time and the parameter a.

From the formula for the variance (7.24) we see that we still need to know $(E\{U_1+U_2\})^2$. By (7.23) we see that the square of the expectation is a polynomial of the sixth degree in λ, and by (7.22) the coefficient in the leading term is $(qE\{A\})^6 I_3^2(a)$. On the other hand, it can be shown that

$$C_6(a,\ t)=(qE\{A\})^6\int_{\mathscr{D}} f\left(t-\frac{2}{v}\sqrt{b^2+H^2(x_2-x_1+x_3)^2}\right)$$
$$\times f\left(t-\frac{2}{v}\sqrt{b^2+H^2(x_5-x_4+x_6)^3}\right)I_6(a,\ X)\,dx_1\ldots dx_6,$$

where

$$I_6(a,\ X)=\Phi(0,\ y_{21}^{(1)})\,\Phi(0,\ y_{12}^{(1)})\ldots\Phi(0,\ y_{63}^{(1)}),$$

and the domain of integration is

$$\mathscr{D}\colon \{0<x_1<x_2,\ x_3<1;\ 0<x_4<x_5,\ x_6<1;\ 0\leqslant x_2-x_1+x_3\leqslant 1;$$
$$0\leqslant x_5-x_4+x_6\leqslant 1\}.$$

From this we obtain

$$C_6(a,\ t)=(qE\{A\})^6 I_3^2(a).$$

The variance of the sum wave U_1+U_2 is therefore a polynomial in λ of the fifth degree, with the coefficient in the leading term equal to

$$d_5(a,\ t)=q^6(E\ \{A\})^4 E\{A^2\}\sum_{\nu=1}^{9}\varphi_\nu, \tag{7.34}$$

where ϕ_ν is an integral of the signal function.

On passing to the limit as $a \to \infty$, we finally obtain

$$\lim_{a\to\infty} \operatorname{var}\{U_1 + U_2\} = \lambda d_1(t) + \ldots \lambda^5 d_5(t),$$

where

$$d_5(t) = \lim_{a\to\infty} d_5(a, t).$$

The other coefficients $d_\nu(t)$ $(\nu = 1, \cdots, 4)$ have approximately the same form as $d_5(t)$.

As a result, we conclude that the expectation and variance of the sum wave made up of single and double waves can be expressed in the form of polynomials in λ of the third and fifth degree with coefficients which are functions of time. A knowledge of the mean and variance of course gives only a very rough idea of the character of a stochastic process. In principle it is clearly possible to extend the calculations to higher-order moments, but the computational difficulties increase sharply. In the following section we find the univariate asymptotic distribution of $U_1 + U_2$ as $\lambda \to \infty$.

§8. The asymptotic distribution of the sum wave taking account of double reflections

The aim of this section is to investigate the asymptotic behavior of the process $U_1 + U_2$ as $\lambda \to \infty$.

Finding the asymptotic distribution is split into two stages:

1) We show that $U_1 + U_2$ is asymptotically normal as $n \to \infty$ for $N = n$ $(n = 1, 2, \cdots)$; the proof of this will be based on two theorems generalizing the results of Z. A. Lomnicki and S. K. Zaremba [5].

2) By using a modified version of a theorem of R. L. Dobrušin [6] we show that $U_1 + U_2$ is asymptotically normal as $\lambda \to \infty$.

For stage 1) we consider the random variables v^{ijk}, v^{ik} and v^i $(i, j, k = 1, 2, \cdots)$ and suppose that they satisfy the following conditions.

C1. Two finite groups of elements composed of the random variables v^{ijk}, v^{ik} and v^i are mutually independent if none of the values of the indices of any element of the first group is encountered among the values of the indices of the elements of the second group.

C2. $E\{v^{ijk}\} = E\{v^{ik}\} = E\{v^i\} = 0$ for all i, j, k.

C3. All the product moments of the second order are equal to

$$E\{v^{i_1 j_1 k_1} v^{i_2 j_2 k_2}\} = \sigma_{rs} \quad (r, s = 1, 2, 3).$$

Here σ_{rs} are constants and there are five different indices in the product

$v^{i_1 j_1 k_1} v^{i_1 j_2 k_2}$. The index standing in the rth place in $v^{i_1 j_1 k_1}$ is equal to the index in the sth place in $v^{i_2 j_2 k_2}$.

C4. The random variables v^{ijk}, v^{ik}, v^{i} have jointly bounded moments of up to the mth order inclusive, given by

$$E\left\{v^{i_1 j_1 k_1} \dots v^{i_p j_p k_p} v^{i_{p+1} k_{p+1}} \dots v^{i_{p+q} k_{p+q}} v^{i_{p+q+1}} \dots v^{i_{p+q+l}}\right\} < c,$$
$$p,\ q,\ l = 1,\ \dots,\ m_0,\ p+q+l = m_0.$$

Theorem 8.1. *If conditions* C1–C4 *are satisfied, the random variable*

$$V_n = \frac{1}{n^{5/2}}\left[\sum_{i,j,k=1}^{n} v^{ijk} + \sum_{i,k=1}^{n} v^{ik} + \sum_{i=1}^{n} v^{i}\right]$$

for any $m \leq m_0$ *has the limit moments*

$$\lim_{n\to\infty} E\{V_n^m\} = \begin{cases} \dfrac{m!}{2^{\frac{m}{2}}\left(\frac{m}{2}\right)!}\sigma^{\frac{m}{2}} & \text{for } m \text{ even}, \\ 0 & \text{for } m \text{ odd}, \end{cases} \tag{8.1}$$

where $\sigma = \sum_{r,s=1}^{3} \sigma_{rs} \neq 0$.

Proof. For V_n^m we have

$$E\{V_n^m\} = \frac{1}{n^{\frac{5m}{2}}} E\left\{\left[\sum_{i,j,k=1}^{n} v^{ijk} + \sum_{i,k=1}^{n} v^{ik} + \sum_{i=1}^{n} v^{i}\right]^m\right\}$$
$$= n^{-\frac{5m}{2}} \sum E\left\{v^{i_1 j_1 k_1} \dots v^{i_p j_q k_p} v^{i_{p+1} k_{p+1}} \dots v^{i_{p+q} k_{p+q}} v^{i_{p+q+1}} \dots v^{i_{p+q+l}}\right\},$$

where the summation is taken over all $i, j, k = 1, \cdots, n$ such that $p + l + q = m$.

Each product

$$v^{i_1 j_1 k_1} \dots v^{i_p j_p k_p} v^{i_{p+1} k_{p+1}} \dots v^{i_{p+q} k_{p+q}} v^{i_{p+p+1}} \dots v^{i_{p+q+l}} \tag{8.2}$$

is decomposed into independent groups of factors, which can be done by C1. If the product is split into α groups containing $p_1, \cdots, p_\alpha$ $(\sum_{\nu=1}^{\alpha} p_\nu = m)$ factors, the number of different indices in the product will not exceed $\sum_{\nu=1}^{\alpha}(2p_\nu + 1) = 2m + \alpha$. The number of decompositions of products (8.2) which can be decomposed into not more than α groups will be of order $O(n^{2m+\alpha})$. But by C4 the expectation of such products is jointly bounded, and therefore the contribution to the total expectation $E\{V_n^m\}$ from all the products which are decomposed into not more than α groups will be of order $O(n^{\alpha - m/2})$.

If $\alpha > m/2$ there will be one element in at least one of the groups of factors.

Then by C2 the expectation of the whole product is zero.

Let $\alpha \leq m/2$ and m be odd. Then $\alpha \leq (m-1)/2$ and therefore

$$E\{V_n^m\} = O\left(n^{\frac{m-1}{2} - \frac{m}{2}}\right) = O\left(\frac{1}{\sqrt{n}}\right)$$

and $E\{V_n^m\} \longrightarrow 0$ as $n \longrightarrow \infty$.

Let m be even. Then for $\alpha \leq m/2 - 1$ the contribution from all the products (8.2) to the total expectation $E\{V_n^m\}$ will be of order $O(1/n)$ and will tend to zero as $n \longrightarrow \infty$.

Let $\alpha = m/2$. Then (8.2) is split into paired factors. If in at least one of these pairs there are fewer than five different indices, there will be fewer than $5m/2$ different indices in the whole product, and therefore the contribution to $E\{V_n^m\}$ from all such products will be of order $O(n^{-\beta})$, where $\beta > 0$. Hence as $n \longrightarrow \infty$ this contribution will tend to zero.

We consider finally the products (8.2) decomposed into pairs so that in each pair there are five different indices. The pairs will obviously be comprised of the random variables v^{ijk}. Depending on the position of the common index in the pair $v^{i_1 j_1 k_1} v^{i_2 j_2 k_2}$, nine different variants are possible.

On the other hand, when m is even $m/2$ groups can be formed from m elements with two elements in each group in $R_m = (1/(m/2)!)\, C_m^2 C_{m-2}^2 \cdots C_2^2$ ways.

As a result, by C3 we obtain

$$E\{V_n^m\} = \frac{c^{\frac{m}{2}}}{n^{\frac{5m}{2}}} R_m \sum (1) + O\left(\frac{1}{n}\right),$$

where the summation is taken over all different $i(\nu), j(\nu), k(\nu) = 1, \cdots, n$ and $\nu = 1, 2, \cdots, m/2$. But then $\Sigma(1) = (n)_{5m/2}$. Therefore

$$E\{V_n^m\} = R_m \frac{c^{\frac{m}{2}} (n)_{\frac{5m}{2}}}{n^{\frac{5m}{2}}} + O\left(\frac{1}{n}\right).$$

On substituting the value of R_m and letting $n \longrightarrow \infty$, we obtain (8.1). The theorem is proved.

Theorem 8.2. *If conditions C1–C3 are satisfied, together with condition C5 that $E\{(v^{ijk})^2\}$ converges uniformly with respect to i, j, k to a constant $\mu > 0$,* *then the random variable*

$$V_n = \frac{1}{n^{5/2}} \left[\sum_{i,\, j,\, k=1}^{n} v^{ijk} + \sum_{i,\, k=1}^{n} v^{ik} + \sum_{i=1}^{n} v^{i} \right]$$

is asymptotically normal as $n \longrightarrow \infty$.

Proof. We introduce auxiliary random variables with the corresponding indices i, j and k:

$$\omega(n) = \begin{cases} v, & \text{if} \quad |v| < \ln n, \\ 0 & \text{otherwise}. \end{cases}$$

$$\eta(n) = \omega(n) - E\{\omega(n)\},$$

$$\zeta(n) = v - \eta(n).$$

It is easy to see that $E\{\eta(n)\} = E\{\zeta(n)\} = 0$ for all i, j, k and n.

We shall prove that

$$\lim_{n\to\infty} E\{\eta^{i_1 j_1 k_1}(n)\, \eta^{i_1 j_2 k_2}(n)\} = \sigma_{11} \tag{8.3}$$

uniformly with respect to all i, j, k. By C3 we have

$$E\{\eta^{i_1 j_1 k_1}(n)\, \eta^{i_1 j_2 k_2}(n)\} = \sigma_{11} - E\{v^{i_1 j_1 k_1} \zeta^{i_1 j_2 k_2}(n)\}$$
$$- E\{v^{i_1 j_2 k_2} \zeta^{i_1 j_1 k_1}(n)\} + E\{\zeta^{i_1 j_1 k_1}(n)\, \zeta^{i_1 j_2 k_2}(n)\}.$$

By Schwarz's inequality we have

$$|E\{v^{i_1 j_1 k_1} \zeta^{i_1 j_2 k_2}(n)\}| \leqslant [E\{(v^{i_1 j_1 k_1})^2\}\, E\{(\zeta^{i_1 j_2 k_2}(n))^2\}]^{1/2}.$$

We shall show that $\lim_{n\to\infty} E\{(\zeta^{i_1, j_2, k_2}(n))^2\} = 0$ uniformly with respect to all i, j and k. From this it will follow that

$$\lim_{n\to\infty} E\{v^{i_1 j_1 k_1} \zeta^{i_1 j_2 k_2}(n)\} = 0.$$

We have $E\{(\eta(n))^2\} = E\{(\omega(n))^2\} - (E\{\omega(n)\})^2$. By the definition of $\omega(n)$ it follows that

$$E\{(\omega(n))^2\} = \int_{-\ln n}^{\ln n} x^2 dF_v(x) \text{ and } E\{\omega(n)\} = \int_{-\ln n}^{\ln n} x\, dF_v(x),$$

where $F_v(x)$ is the distribution function of the random variable v. Therefore by condition C5

$$\lim_{n\to\infty} E\{(\eta(n))^2\}$$
$$=\lim_{n\to\infty}\int_{-\ln n}^{\ln n} x^2 dF_v(x) - \lim_{n\to\infty}\left(\int_{-\ln n}^{\ln n} x\,dF_v(x)\right)^2 = \mu \tag{8.4}$$

uniformly with respect to all i, j and k.

Also,

$$E\{(\zeta(n))^2\} = E\{(v-\eta(n))^2\} = E\{v^2\} - 2E\{v\eta(n)\} + E\{(\eta(n))^2\}.$$

From the condition of $\eta(n)$

$$E\{v\eta(n)\} = \int_{-\ln n}^{\ln n} x(x - E\{\omega(n)\})\,dF_v(x)$$

and passing to the limit as $n \to \infty$, we see that

$$\lim_{n\to\infty} E\{v\eta(n)\} = \mu \tag{8.5}$$

uniformly with respect to all i, j and k.

It follows from (8.4) and (8.5) that $\lim_{n\to\infty} E\{(\zeta(n))^2\} = 0$ uniformly with respect to all i, j and k. In exactly the same way it can be shown that

$$\lim_{n\to\infty} E\{v^{i_1 j_2 k_2}\zeta^{i_1 j_1 k_1}(n)\} = 0,$$
$$\lim_{n\to\infty} E\{\zeta^{i_1 j_1 k_1}(n)\,\zeta^{i_1 j_2 k_2}(n)\} = 0,$$

uniformly with respect to i, j and k, and (8.3) follows.

By similar calculations we can show that

$$\lim_{n\to\infty} E\{\eta^{i_1 j_1 k_1}(n)\,\eta^{i_2 j_2 k_2}(n)\} = \sigma_{rs} \tag{8.6}$$

uniformly with respect to all i, j and k.

As before, the index in the rth place of η^{i_1, j_1, k_1} is equal to the index in the sth place of η^{i_2, j_2, k_2}.

We also note that for any given n condition C1 is satisfied for each of the sets of random variables $\eta(n)$ and $\zeta(n)$.

We now introduce the random variables

$$\eta_n = n^{-5/2}\left[\sum_{i,\,j,\,k=1}^{n} \eta^{ijk}(n) + \sum_{i,\,k=1}^{n} \eta^{ik}(n) + \sum_{i=1}^{n} \eta^{i}(n)\right],$$

$$\zeta_n = n^{-5/2}\left[\sum_{i,\,j,\,k=1}^{n} \zeta^{ijk}(n) + \sum_{i,\,k=1}^{n} \zeta^{ik}(n) + \sum_{i=1}^{n} \zeta^{i}(n)\right].$$

Then $v_n = \eta_n + \zeta_n$. The random variable η_n has finite moments of any order which can be calculated in the same way as the moments of v_n in the proof of Theorem 8.1. But the moments

$$E\{\eta^{i_1j_1k_1}(n)\ldots\eta^{i_pj_pk_p}(n)\,\eta^{i_{p+1}k_{p+1}}(n)\ldots\eta^{i_{p+q}k_{p+q}}(n)\,\eta^{i_{p+q+1}}(n)\ldots\eta^{i_{p+q+l}}(n)\}$$

depend on n. Nevertheless, we have

$$|E\{\eta^{i_1j_1k_1}(n)\ldots\eta^{i_pj_pk_p}(n)\,\eta^{i_{p+1}k_{p+1}}(n)\ldots\eta^{i_{p+q}k_{p+q}}(n)\,\eta^{i_{p+q+1}}(n)\ldots$$
$$\ldots\eta^{i_{p+q+l}}(n)\}| \leqslant (2\ln n)^m, \tag{8.7}$$

where $l + p + q = m$; that is, for a given m and any finite n the moments are bounded. By (8.7) we see that, when m is odd,

$$E\{\eta_n^m\} = O\left(\frac{(\ln n)^m}{\sqrt{n}}\right)$$

and therefore

$$\lim_{n\to\infty} E\{\eta_n^m\} = 0. \tag{8.8}$$

As in the proof of Theorem 8.1, we find that for even m, using (8.6),

$$\lim_{n\to\infty} E\{\eta_n^m\} = \frac{m!}{2^{\frac{m}{2}}\left(\frac{m}{2}\right)!}\,\sigma^{\frac{m}{2}}, \tag{8.9}$$

where $\sigma = \sum_{r,\,s=1}^{3} \sigma_{rs}$.

It therefore follows from (8.8) and (8.9) that all the moments of η_n tend to moments of the normal distribution as $n \to \infty$. In accordance with [7] the distribution of η_n is then asymptotically normal.

We now consider the behavior of ζ_n. We can find $E\{\zeta_n^2\}$ in the same way as $E\{V_n^2\}$, except that whereas previously terms in $E\{V_n^2\}$ were equal to σ_{rs}, the corresponding terms in $E\{\zeta_n^2\}$ will tend to zero uniformly with respect to i, j and k as $n \to \infty$, and therefore $\lim_{n\to\infty} E\{\zeta_n^2\} = 0$. But since $E\{\zeta_n\} = 0$ it follows that $\operatorname{p\,lim}_{n\to\infty} \zeta_n = 0$. Therefore, by Cramér's theorem ([8], Russian p. 281), the random variable $V_n = \eta_n + \zeta_n$ is asymptotically normal as $n \to \infty$. The theorem is proved.

We use these two theorems to prove that the univariate distribution function

function of the process $U_1 + U_2$ is asymptotically normal as $n \to \infty$. We first make the following remark.

We introduce the random variables u^i, u^{i0k} and u^{ijk}, which for all $i, j, k = 1, \cdots, n$ are equal to the corresponding random variables u_i, u_{i0k} and u_{ijk} when $z_{(j)} < z_{(i)}, z_{(k)}$ and zero otherwise. Since $z_{(i)}, z_{(j)}$ and $z_{(k)}$ are not ordered, they are independent, and therefore u^i, u^{i0k} and u^{ijk} are also independent if they have no common indices. The sum wave can therefore be expressed in terms of u^i, u^{i0k} and u^{ijk}, so that

$$U_1 + U_2 = \sum_{i,j,k=1}^{n} u^{ijk} + \sum_{i,k=1}^{n} u^{i0k} + \sum_{i=1}^{n} u^i .$$

We use the notation

$$\left.\begin{aligned} v^{ijk} &= u^{ijk} - E\{u^{ijk}\}, \\ v^{ik} &= u^{i0k} - E\{u^{i0k}\}, \\ v^{i} &= u^{i} - E\{u^{i}\}. \end{aligned}\right\} \tag{8.10}$$

Then

$$V_n = (U_1 + U_2)\,|\,n - E\{(U_1 + U_2)\,|\,n\} = \sum_{i,j,k=1}^{n} v^{ijk} + \sum_{i,k=1}^{n} v^{ik} + \sum_{i=1}^{n} v^i .$$

It is easy to verify that C1 and C2 are satisfied for the random variables (8.10). If the signal function $f(t - \tau)$ is such that C3 and C5 are satisfied it follows from Theorem 8.2 that the random variable $\overline{V}_n = V_n/n^{5/2}$ is asymptotically normal as $n \to \infty$. But the random variable $\overline{U}_n = [(U_1 + U_2)|n]/n^{5/2}$ is then also asymptotically normal as $n \to \infty$. Therefore the univariate distribution of the process $(U_1 + U_2)|n$ is also asymptotically normal as $n \to \infty$. Its expectation is a polynomial in n of the third degree with the coefficient of n^3 by (7.21) equal to

$$c_3 = \left[\frac{qE\{A\}}{a + 2b}\right]^3 I_3(a),$$

and the variance can be shown to be a polynomial in n of the fifth degree. We assume that $c_3 \neq 0$ and that the coefficient of n^5 in the variance is also nonzero.

Before proceeding to the final step in the proof that $U_1 + U_2$ is asymptotically normal, we state the following theorem.

Theorem 8.3. *Let:* a) $U^*(n)$ *be a sequence of random variables such that as* $n \to \infty$

$$K_n(x) = P\left\{\frac{U^*(n) - c_3 n^3 - c_2 n^2 - c_1 n}{(d_5 n^5 + d_4 n^4 + \ldots + d_1 n)^{1/2}} < x\right\} \to \frac{1}{\sqrt{2\pi}} \int_{-\infty}^{x} e^{-\frac{y^2}{2}}\, dy = \mathcal{K}(x);$$

b) $N(p\lambda)$ *be a sequence of random variables not dependent on* $U^*(n)$ *and such that as* $\lambda \to \infty$

$$R_\lambda(x) = P\left\{\frac{N(\lambda p) - \lambda p}{\sqrt{\lambda p}} < x\right\} \to \frac{1}{\sqrt{2\pi}} \int_{-\infty}^{x} e^{-\frac{y^2}{2}} dy = R(x).$$

Then as $\lambda \to \infty$

$$P\left\{\frac{U^*[N(\lambda p)] - c_3\lambda^3 p^3}{(p\lambda)^{5/2}\sqrt{9c_3^2 + d_5}} < x\right\} \to \frac{1}{\sqrt{2\pi}} \int_{-\infty}^{x} e^{-\frac{y^2}{2}} dy.$$

The proof of the theorem is similar to that given by R. L. Dobrušin [6], and is therefore omitted.

We apply this result to $U_1 + U_2$. The process $(U_1 + U_2)|n$ has already been shown to be asymptotically normal as $n \to \infty$; that is, condition a) of Theorem 8.3 is satisfied. Furthermore, we know that the Poisson process $N(a + 2b)$ is asymptotically normal as $\lambda \to \infty$ with mean and variance equal to $(a + 2b)\lambda$, so that condition b) of Theorem 8.3 is also satisfied. Therefore by Theorem 8.3 the process $U_1 + U_2$ is asymptotically normal as $\lambda \to \infty$ with mean $(\lambda qEA)^3 I_3(a) + o(\lambda^3)$ (by (7.22)) and variance $\lambda^5(a + 2b)^5 d_5(a) + o(\lambda^5)$.

If we consider a steady-state process (that is, the process $\lim_{a\to\infty}(U_1 + U_2)$), it will also be asymptotically normal as $\lambda \to \infty$.

The results of this section show that the sum wave consisting of once- and twice-reflected waves for a sufficiently high intensity of the inhomogeneous medium λ behaves as a stochastic process whose univariate distribution function is normal.

§9. Conclusion

The theory developed in this paper enables us to solve a number of seismic problems. Basically they are of two types: problems of testing hypotheses and parametric problems. The first category contains, it seems to us, a fundamental problem of seismic prospecting: to test on the basis of observed data the hypothesis that there is an extended boundary in a given stratified medium. A problem of this kind is solved in [9], where it is assumed that the useful wave is that wave reflected from an extended determinate boundary and that everything else is noise. Noise is considered to be additive with the useful signal, independent of it and represented by a random stationary normal process $U(t, b)$ with zero mean. The process is correlated with time (Markov correlation) but not with distance between detectors (that is, directions).

The models of the medium we have proposed lead to the following conditions for a solution of problems of the first type. If the useful signal is a wave reflected from a determinate extended boundary and noise is a single sum wave (taking no account of multiple reflections), the seismogram is of the form

$$U(t, b) = A_0 qf\left(t - \frac{2}{v}\sqrt{b^2 + z_0^2}\right) + U_1(t, b),$$

where A_0 and z_0 are characteristics of the unknown extended boundary and $U_1(t, b)$ is the single sum wave (see (3.6)). In this case we are concerned with additive noise, independent of the useful signal, whose probabilistic characteristics are studied in §§3 and 4. But, in contrast to the assumptions of [9], the noise is correlated with distance (direction), there is asymptotic normality as $\lambda \to \infty$, and the correlations are not Markov.

It can be surmised that in seismic prospecting a study of the stratified media themselves is of interest. From the results we have obtained we can assert that the sum wave of single reflections under certain assumptions behaves as a Gaussian process. This enables us to investigate the second type of problem and estimate parameters of the medium such as the mean extent of a discontinuity s or the intensity λ of the medium.

If we take account of double reflections in our investigation of the sum wave we can suppose that in the medium there is an extended determinate boundary, the signal from which is mixed with signals from discontinuities in the medium. In this case the seismogram is of the form

$$U(t, b) = A_0 qf\left(t - \frac{2}{v}\sqrt{b^2 + z_0^2}\right) + U_1(t, b) + U_2(t, b) + U_0(t, b, A_0, z_0),$$

where $A_0 qf(t - 2\sqrt{b^2 + z_0^2}/v)$ is the useful sign $U_1(t, b) + U_2(t, b) + U_0(t, b, A_0, z_0)$ is the noise, $U_1(t, b)$ and $U_2(t, b)$ being the single and double sum waves generated by all the discontinuities except for the basic extended layer; and $U_0(t, b, A_0, z_0)$ is the double sum wave in whose formation the basic extended boundary takes part. It follows that the noise $U_1(t, b) + U_2(t, b) + U_0(t, b, A_0, z_0)$ depends on the useful signal. In this paper we have investigated the probabilistic behavior of the noise. It is shown, in particular, that the univariate distribution of noise is asymptotically normal as $\lambda \to \infty$.

Furthermore, it seems to us that it is possible to investigate the noise when the amplitude of the wave has more specific stochastic and asymptotic properties. A problem of this kind is considered in §4, where the amplitude is normally distributed with expectation zero and variance A_0^2 tending to zero as the intensity λ of the discontinuities increases in such a way that $A_0^2 \to \sigma^2$.

We also remark that, except in §6, we have considered the boundaries to be "transparent"; that is, no account has been taken of the absorption of energy of the wave by refraction. With this approach the principal part of the sum wave for large λ, if we take account of double reflections, is composed of double waves. In fact, the expectation of the sum wave is given in the form of a polynomial in λ of the third degree. The term (see §7) containing λ^3 characterizes the double waves and is the principal part of the sum wave as $\lambda \to \infty$. In practice, however, it is assumed in most cases that single reflections make the basic contribution in the formation of the sum wave. This is true in the case of the models we have used if we assume that $\lambda q < 1$. But if $\lambda \to \infty$ for constant q, the principal part of the sum wave will be composed not of single but of multiple waves.

We have stated the assumption that this result follows from the fact that we took no account of the absorption of energy of the wave due to refractions on passing through discontinuities. We therefore investigated in §6 the behavior of the sum wave with account taken of refractive indices.

More detailed problems can be posed only in accordance with the requirements of seismic prospecting, whether it be obtaining information on noise or the signal or estimating unknown parameters. In conclusion we note that §2 was written jointly by L. A. Zolotuhina and K. P. Latyšev, §§3–5 by K. P. Latyšev, §6 by L. A. Zolotuhina, and the other sections jointly by L. A. Zolotuhina, K. P. Latyšev and V. N. Čugueva.

BIBLIOGRAPHY

[1] J. L. Doob, *Stochastic processes*, Wiley, New York and Chapman & Hall, London, 1953; Russian transl., IL, Moscow, 1956. MR 15, 445; 19, 71.

[2] L. A. Zolotuhina, *Probabilistic characteristics and asymptotic distributions of once-reflected sum waves in a stochastic model of a stratified medium applied to seismic prospecting*, Trudy Mat. Inst. Steklov. 95 (1968), 21–41 = Proc. Steklov Inst. Math. 95 (1968), 23–46.

[3] András Prékopa, *On secondary processes generated by a random point distribution of Poisson type*, Ann. Univ. Sci. Budapest. Eötvös. Sect. Math. 1 (1958), 153–170. MR 22 #10009.

[4] S. S. Wilks, *Mathematical statistics*, Wiley, New York, 1962. MR 26#1949.

[5] Z. A. Lomnicki and S. K. Zaremba, *A further instance of the Central Limit Theorem for dependent random variables*, Math. Z. 66 (1957), 490–494. MR 20 #6152.

[6] R. L. Dobrušin, *Lemma on the limit of compound random functions*, Uspehi Mat. Nauk 10(1955), no. 2 (64), 157–159. (Russian) MR 17, 48.

[7] M. Fréchet and J. Shohat, *A proof of the generalized second-limit theorem in the theory of probability*, Trans. Amer. Math. Soc. 33 (1931), 533–543.

[8] H. Cramér, *Mathematical methods of statistics,* Princeton Math. Series, vol. 9, Princeton Univ. Press, Princeton, N. J., 1946; Russian transl., IL, Moscow, 1948. MR 8, 39.

[9] F. M. Gol'cman, L. A. Zolotuhina, K. P. Latyšev, L. A. Halfin, N. M. Halfina and V. N. Čugueva, *Statistical problems in the interpretation of seismic data,* Trudy Mat. Inst. Steklov. 79 (1965), 160–181 = Proc. Steklov Inst. Math. 79 (1965), 178–201. MR 34 #7194.

Trudy Mat. Inst. Steklov.
95 (1968)

Proc. Steklov Inst. Math.
95 (1968)

HEAD WAVES IN A MEDIUM WITH A HIGH-SPEED LAYER

UDC 534.131

P. V. KRAUKLIS

Abstract. The first part of the paper is concerned with the formation of a head wave in a thin elastic layer having a higher propagation volocity than the surrounding medium. The required solution is given by the sum of damped waves. The complex roots of the dispersion equation are investigated and curves found for the phase and group velocities and the damping. Amplitude functions describing the intensity of the head waves also found. It is deduced that head waves cannot exist if the ratio of the layer thickness to the wave length is less than 0.5. Some numerical results are given in the form of graphs.

In the second part of the paper the effect of screening head waves by a thin layer of increased velocity is studied. A simple method is suggested for calculating the screening and numerical results are given for one model of the medium.

Bibliography 6 items. 4 figures.

1. Despite the considerable number of experimental and theoretical papers on the dynamic characteristics of refracted (head) waves in media containing thin layers, there has so far been no specific presentation on the frequency composition, the nature of the damping with distance and the dispersion of these waves. It was found in some experiments that the apparent velocity of the head wave depends on the ratio of the wavelength to the thickness of the layer and the amplitude decreases with distance according to a law different from the r^{-2} law. The velocity as a function of the frequency has a minimum in the low-frequency part of the dispersion curve. Unfortunately, the validity of the results of the modeling for ratios h/λ in the range 0.1–0.7 is open to question because with these values of h/λ the thickness of the layer is commensurable with the thickness of the two-dimensional model [1]. Attempts have been made in theoretical papers to calculate the head waves by the superposition of separate waves calculated by the ray method, but here also the results have greatest validity when h/λ is large [2]. In studying the wave pattern in the low-frequency region it is more convenient to obtain solutions in the form of a decomposition with respect to the damped waves. This method is used in the literature, for example in the study of PL waves in seismic recordings [3]. For damped waves (modes) the group velocity is not exactly equal to the rate of energy transfer as in the case of normal waves; nevertheless, if the damping on the wavelength is small and varies slowly with frequency, the group velocity u defined by $u = \partial\omega_n / \partial k$ (k is the wave number and ω_n the real part of the frequency) is close to the pulse rate.

This assumption is supported by the results of some experimental work on propagation in an absorbent medium [4].

2. Let us consider a medium consisting of two half-spaces $0(z<0)$ and $2(z>h)$ separated by an elastic layer $1(0<z<h)$. We denote by $a_0^{-1}, a_1^{-1}, a_2^{-1}$ and $b_0^{-1}, b_1^{-1}, b_2^{-1}$ respectively the velocities of the longitudinal and transverse waves in the media 0, 1 and 2, and by ρ_0, ρ_1, and ρ_2 the densities of the media. The source of the perturbations is situated at the point with coordinates $r=0$, $z=-H$, in the half-space 0, and the perturbations are defined by the potential

$$\varphi_0^0=\int_0^\infty \frac{J_0(kr)\,dk}{2\pi i}\int_{\sigma-i\infty}^{\sigma+i\infty} X_0^0(k,\ \eta)\, e^{k\left[t\eta b_0^{-1}\mp\sqrt{1+(a_0\eta b_0^{-1})^2}(z+H)\right]}\,d\eta, \tag{1}$$

where J_0 is the zero-order Bessel function and $X_0^0(k,\eta)$ depends on the type of source and the nature of its variation with time. If the source is a center of dilatation and $\epsilon(t)$ is the function of time we shall have, for example,

$$X_0^0(k,\ \eta)=[4\pi(\lambda+2\mu)\,\eta\alpha_0]^{-1}, \tag{2}$$

where λ and μ are elastic constants and $\alpha_0=\sqrt{1+(a_0\eta b_0^{-1})^2}$. In each medium the longitudinal and transverse potentials satisfy the equations

$$\Delta\varphi=a^2\frac{\partial^2\varphi}{\partial t^2},\quad \Delta\psi=b^2\frac{\partial^2\psi}{\partial t^2}, \tag{3}$$

with the zero initial data $\phi_i=\dot{\phi}_i=\dot{\psi}_i=\psi_i=0$ (for $t=0$ and $i=0,1,2$) and continuity conditions on the interfaces of the displacement and stress vectors. We shall assume at first that the elastic properties of the half-spaces are identical. From now on we consider only a head wave travelling in the upper medium as a longitudinal wave, and therefore in writing out the solution of the problem we confine ourselves to formulas for the longitudinal potential at the point of observation with coordinates $(r,-H)$ [5]:

$$\varphi_0=\int_0^\infty \frac{J_0(kr)\,dk}{2\pi i}\int_{\sigma-i\infty}^{\sigma+i\infty} X_0 e^{k[t\eta b_0^{-1}-2H\alpha_0]}\,d\eta. \tag{4}$$

Here

$$X_0=\frac{\Delta_1}{\Delta_0}X_0^0,\quad \frac{\Delta_1}{\Delta_0}=\frac{1}{2}(M_1+N_1),$$

where

$$M_1=-\frac{(A+C)\,\mathrm{th}\,\frac{kh\alpha_1}{2}-(B+D)\,\mathrm{th}\,\frac{kh\beta_1}{2}-E+F\,\mathrm{th}\,\frac{kh\alpha_1}{2}\,\mathrm{th}\,\frac{kh\beta_1}{2}}{(A-C)\,\mathrm{th}\,\frac{kh\alpha_1}{2}-(B-D)\,\mathrm{th}\,\frac{kh\beta_1}{2}-E-F\,\mathrm{th}\,\frac{kh\alpha_1}{2}\,\mathrm{th}\,\frac{kh\beta_1}{2}},$$

$$N_1=\frac{(A+C)\,\mathrm{cth}\,\frac{kh\alpha_1}{2}-(B+D)\,\mathrm{cth}\,\frac{kh\beta_1}{2}-E+F\,\mathrm{cth}\,\frac{kh\alpha_1}{2}\,\mathrm{cth}\,\frac{kh\beta_1}{2}}{(A-C)\,\mathrm{cth}\,\frac{kh\alpha_1}{2}-(B-D)\,\mathrm{cth}\,\frac{kh\beta_1}{2}-E-F\,\mathrm{cth}\,\frac{kh\alpha_1}{2}\,\mathrm{cth}\,\frac{kh\beta_1}{2}},$$

$$A=(g_1-\sigma_{01}g_0)^2,\ D=4\alpha_0\beta_0\alpha_1\beta_1(1-\sigma_{01})^2,\ g_i=2+\delta_i^2\eta^2,\ \delta_i=\frac{b_i}{b_0},$$
$$B=\alpha_1\beta_1(\sigma_{01}g_0-2)^2,\ E=\alpha_1\beta_0\sigma_{01}\delta_1^2\eta^4,\ \alpha_i=\sqrt{1+\gamma_i^2\eta^2},\ \gamma_i=\frac{a_i}{b_0}, \tag{5}$$
$$C=\alpha_0\beta_0(g_1-2\sigma_{01})^2,\ F=\alpha_0\beta_1\sigma_{01}\delta_1^2\eta^4,\ \beta_i=\sqrt{1+\delta_i^2\eta^2},\ \sigma_{01}=\frac{\rho_0}{\rho_1}\delta_1^2.$$

The branches of the radicals are specified by the condition $\arg\alpha_i=\arg\beta_i=0$ for $\eta>0$, and the branch cuts are drawn from the branching points into the left half-plane parallel to the real axis.

The integrand in (4) is expressed in the form of parts of the field which are symmetric and skewsymmetric with respect to the plane $z=h/2$.

3. In the investigation of the solution (4) and the isolation of that part of the field corresponding to the head wave we use the saddle-point method to calculate the inner integral. We do not discuss this method here (its characteristics can be found from [6]), but merely note that the location of the saddle-points is determined by the coordinates of the source and detector and by the time t. If the observation points are moved away from the source parallel to the plane $z=h/2$, then the saddle points move along the imaginary axis from infinity to the points $\pm i\gamma_0^{-1}$. Since the contours and the roots of the dispersion equations are symmetrically situated about the real axis, we shall consider the upper half-plane only. A stationary contour passing through a saddle point can intersect the poles of the integrand in the left half-plane η. As kh increases these poles monotonically descend, and as $kh\to\infty$ they tend to the point $\delta_1^{-1}i$. For values of time satisfying

$$t>\frac{2a_0H}{\sqrt{1-(\gamma_1\gamma_0^{-1})^2}},$$

the saddle point is below the point $i\gamma_1^{-1}$. If, however,

$$t>\frac{2a_0H}{\sqrt{1-(\delta_1\gamma_0^{-1})^2}}$$

and $\delta_1^{-1}>\gamma_0^{-1}$, then a stationary contour lies below $i\delta_1^{-1}$ and the inner integral is the sum of two terms, one of which is obtained by integration along the stationary contour and the other is the sum of the residues in the complex roots of the equations

$$\Delta_3\equiv(A-C)\operatorname{th}\frac{kh\alpha_1}{2}-(B-D)\operatorname{th}\frac{kh\beta_1}{2}-E-F\operatorname{th}\frac{kh\alpha_1}{2}\operatorname{th}\frac{kh\beta_1}{2}=0, \tag{6}$$
$$\Delta_4\equiv(A-C)\operatorname{cth}\frac{kh\alpha_1}{2}-(B-D)\operatorname{cth}\frac{kh\beta_1}{2}-E-F\operatorname{cth}\frac{kh\alpha_1}{2}\operatorname{cth}\frac{kh\beta_1}{2}=0. \tag{7}$$

The dispersive part of the solution is given by the second term. After substituting this into (4) and interchanging the order of integration and summation we obtain the interference field in the form of an expansion with respect to the damped waves:

$$U=\sum_{n=1}^{\infty} 2\,\mathrm{Re}\int_{0}^{\infty} J_0(kr)\,G(k,\ \eta_n)\exp k\,[t\eta_n b_0^{-1}-2H\alpha_{0n}]\,X_0^0 dk. \tag{8}$$

The function $G(k,\ \eta_n)$, in the integrand of (8) is called the excitation function. We can transform (8) to a more convenient form by using the asymptotic form of Bessel's function, valid for $kr \gg 1$, the notation

$$\frac{k\,\mathrm{Re}\,\eta_n}{b_0}=-\varkappa_n,\qquad \frac{k\,\mathrm{Im}\,\eta_n}{b_0}=\omega_n$$

and changing to the variable ω_n. Limiting ourselves to those waves proceeding from the source, we obtain

$$U=\sqrt{\frac{2}{\pi kr}}\sum_{n=1}^{\infty}\mathrm{Re}\int_{\omega_n^*}^{\infty} G(\omega_n)\exp(-\varkappa_n t-2Hk_n\,\mathrm{Re}\,\alpha_{0n})$$
$$\times\exp i\,[\omega_n t-k_n\,\mathrm{Im}\,\alpha_{0n}2H-k_n r]\,X_0^0 d\omega_n, \tag{9}$$

where ω_n^* denotes the boundary values of ω_n for each mode.

The use of the stationary phase method in investigating (9) is more complicated than in the case of normal waves, because the integrand contains a factor describing the exponential damping of the field with time. However, numerical calculations show that the real parts of the roots of the dispersion equations are, first, small compared with the imaginary parts and, second, vary with the frequency in many cases much more slowly than the imaginary parts. From the physical point of view this must lead to a small change in the form of the pulse on the wave length (having in mind here the wave length corresponding to the dominant frequency), which enables us to use the concept of the group velocity and boundary frequency of the vibrations. From the equation determining the saddle point, given by

$$\frac{\partial}{\partial\omega_n}\{\omega_n t-k_n\,\mathrm{Im}\,\alpha_{0n}2H-k_n r\}=0 \tag{10}$$

we can obtain, after simplification, the following relationship between the time t and the coordinates of the source and observation point:

$$ut-r-2H\frac{uc-a_0^{-2}}{a_0^{-1}\sqrt{c^2-a_0^{-2}}}=0. \tag{11}$$

Equation (11) has a simple meaning: it defines the time of passage of vibrations falling on the layer with phase velocity c, propagated through the layer with velocity $u=\partial\omega_n/\partial k$ and arriving at the source with the same phase velocity.

To determine the quantitative characteristics of the field a model of the medium was taken with layer parameters $a_1^{-1} = 4200$ meters/second, $b_1^{-1} = 2120$ meters/second and $\rho_1 = 8.4$ grams/cm^3, and parameters of the surrounding medium $a_0^{-1} = 2540$ meters/second, $b_0^{-1} = 1320$ meters/second and $\rho_0 = 1.2$ grams/cm^3. These parameters belong to a medium frequently used in experiments consisting of a sheet of brass placed in an epoxy resin. To find the complex roots of the dispersion equations (6) and (7) corresponding to this problem, a computer program was compiled enabling the real and imaginary parts of the roots to be found to any desired degree of accuracy by an iteration process. Figure 1 shows

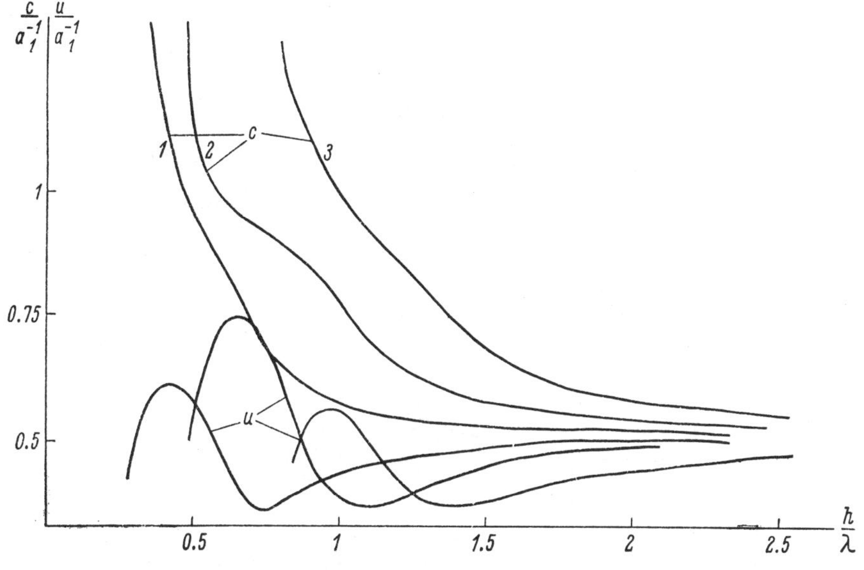

Figure 1

the graphs of c and u as functions of h/λ (where λ is the wave length in the medium) for the first three damped modes. We note that the curves are frequency bounded below and deduce that the conditions of constructive interference for wave lengths $h/\lambda \leq 0.5$ cannot be satisfied. The maximum values of the group velocity are located in the domain of incidence at a limiting angle arc sin $a_1 a_0^{-1}$, and the values of u are greatest for symmetric modes (20–25% below the velocity of the longitudinal waves in the material of the layer). It is easy to show that vibrations corresponding to the segments of maximum values of u are characterized by a slower damping distance $(r^{-5/6})$ than for other parts of the dispersion curve (r^{-1}). As we know, this situation also arises for minimum values of u. The qualitative difference from the last case is that the vibrations have a strong beginning (the Airy phase) and then, on dispersing, split into low-frequency and high-frequency parts. The numerical results support the assumption that the dispersion properties are primarily determined by the velocities in the media and depend little on the density differentials. Figure 2, for example, shows how c and u depend on the real part of the frequency for two density differentials:

$\rho_0 \rho_1^{-1} = 0.2$ (solid line) and $\rho \rho_1^{-1} = 0.7$ (dotted line). The absolute value of the excitation function $A = |G|$ and the absorption coefficient κ caused by the radiation of energy from the layer depend to a large degree on the ratio of the the densities, which, for example, is illustrated in Figure 3, where the indices

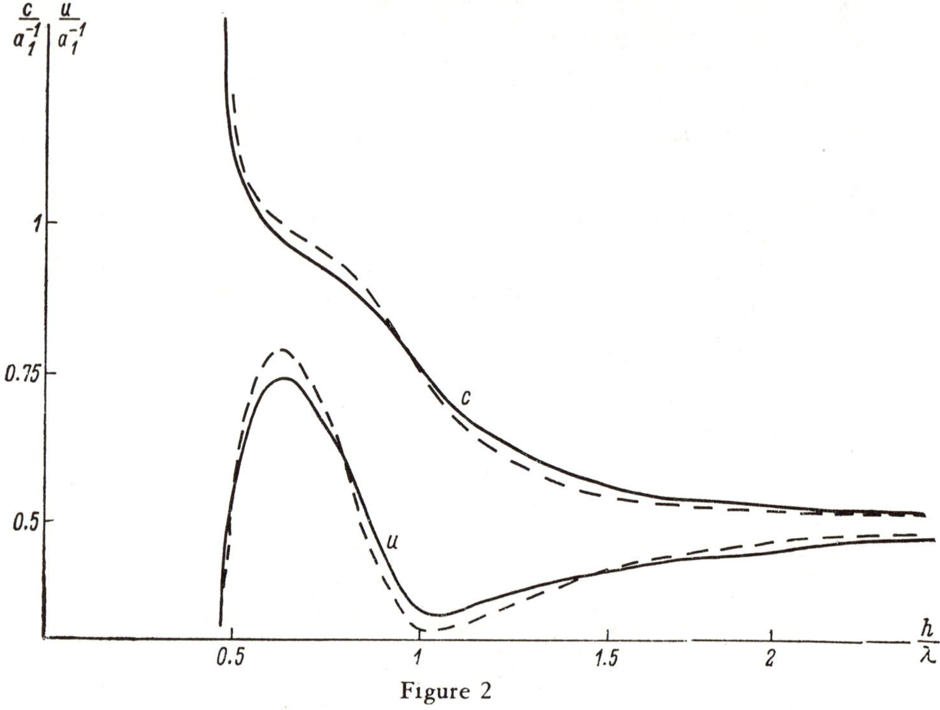

Figure 2

1 and 2 correspond to the cases $\rho_0 \rho_1^{-1} = 0.7$ and $\rho_0 \rho_1^{-1} = 0.2$ respectively. The graphs are drawn for the case of the first symmetric mode.

From the graph of κ we can determine the damping on the wavelength in the layer. For a wave which corresponds to the maximum of the first symmetric wave (for $h/\lambda \approx 0.7$) the amplitude is diminished by a factor 1.08. This comparatively slight damping indicates the possibility of retaining the form of the wave for small distances from the initial point. At the same time, the considerable difference in magnitude between the group and phase velocities must lead to a transfer of the pulse energy from the later phases to the earlier. This effect was predicted earlier in [2]. Finally, the considerable variation in the first arrival velocities depending on the transmission band of the recording channel may result from the narrow width of the extrema on the group velocity curve.

4. We have considered the case when the parameters of the surrounding half-spaces coincide. The method of investigating the head wave in a layer is no different in principle when there are different elastic constants. In such a system, however, other head waves can occur and the study of these is of independent interest. The most curious case occurs when a head wave resulting from refraction in the lower half-space is screened by a thin layer. We show how the amplitude

in the frequency function can easily be found numerically. Let us consider the

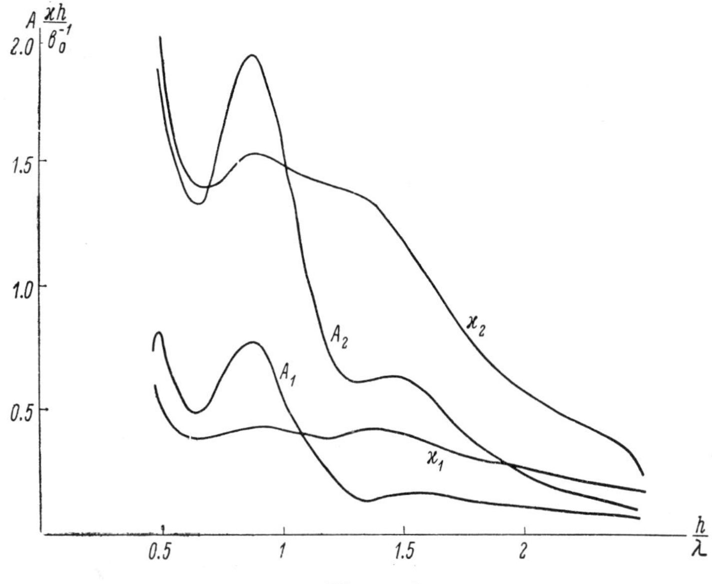

Figure 3

case of a strong screen, where we have

$$a_0^{-1} < a_2^{-1} < b_1^{-1}. \tag{12}$$

The solution of the problem is given by (4), where

$$\left.\begin{aligned} X_0 &= \frac{\Delta_1}{\Delta_0} X_0^0, \\ \Delta_0 &= -RRU_1 - Q_4Q_4U_2 + Q_3^+Q_3U_3 + Q_1^+Q_1U_4 - 2\alpha_1^{-1}\beta_1^{-1}Q_2^+Q_2, \\ \Delta_1 &= Q_1R_1U_1 + Q_3Q_4U_2 - Q_5Q_3U_3 - Q_6Q_1U_4 + 2\alpha_1^{-1}\beta_1^{-1}Q_7^+Q_2. \end{aligned}\right\} \tag{13}$$

In the products RR, Q_4Q_4 and so on, the first functions depend on the arguments $(0, 1, \eta)$ and the second on $(1, 2, \eta)$. We have

$$R(i, k, \eta) = A - B - C + D - E - F,$$
$$Q_4(i, k, \eta) = A - B - C + D + E + F,$$
$$Q_3^+(i, k, \eta) = A + B - C - D - E + F,$$
$$Q_3(i, k, \eta) = A - B + C - D + E - F,$$
$$Q_1^+ = A + B - C - D + E - F, \quad Q_1 = A - B + C - D - E + F,$$
$$Q_5 = A + B + C + D - E - F, \quad Q_6 = A + B + C + D + E + F,$$
$$Q_2 = G - H, \quad Q_2^+ = -K + L, \quad Q_7^+ = K - L,$$
$$G = 2\alpha_i\beta_i(g_k - 2\sigma_{ik})(g_k - \sigma_{ik}g_i), \quad K = -2\alpha_k\beta_k(g_k - \sigma_{ik}g_i)(\sigma_{ik}g_i - 2),$$
$$H = -4\alpha_i\beta_i\alpha_k\beta_k(1 - \sigma_{ik})(\sigma_{ik}g_i - 2),$$
$$L = 4\alpha_i\beta_i\alpha_k\beta_k(1 - \sigma_{ik})(g_k - 2\sigma_{ik}),$$

$$U_1 = \exp[kh(\alpha_1 + \beta_1)], \quad U_2 = \exp[-kh(\alpha_1 + \beta_1)],$$
$$U_3 = \exp[kh(\alpha_1 - \beta_1)], \quad U_4 = \exp[kh(\beta_1 - \alpha_1)]. \tag{14}$$

The other functions in (14) can easily be found explicitly with the help of (5).

The head wave propagated with velocity a_2^{-1} is found by integrating along a cut leading from the point i/γ_2. If (12) holds it can be shown that the integrand has no poles close to the cut, and therefore the usual scheme for calculating an integral along a cut can be used. To do this we put the slowly varying part of the integrand in the form

$$\frac{M + N\sqrt{1 + \gamma_2^2\eta^2}}{\mathcal{E} + \mathcal{F}\sqrt{1 + \gamma_2^2\eta^2}}$$

and then expand it as a power series in α_2. The coefficient of α_2 is

$$\Gamma(\omega) = \frac{\mathcal{E}N - M\mathcal{F}}{\mathcal{E}^2}, \tag{15}$$

where

$$\begin{gathered}
\mathcal{F} = -R_1(-\tilde{B} + \tilde{D} - \tilde{E})U_1 - Q_4(-\tilde{B} + \tilde{D} + \tilde{E})U_2 + \\
+ Q_3^+(-\tilde{B} - \tilde{D} + \tilde{E})U_3 + Q_1^+(-\tilde{B} - \tilde{D} - \tilde{E})U_4 + 2Q_2^+H\,\alpha_1^{-1}\beta_1^{-1}, \\
\mathcal{E} = -R(A - C - F)U_1 - Q_4(A - C + F)U_2 + \\
+ Q_3^+(A + C - F)U_3 + Q_1^+(A + C + F) - 2Q_2^+G,\ \alpha_1^{-1}\beta_1^{-1}, \\
M = Q_1(A - C - F)U_1 + Q_3(A - C + F)U_2 - \\
- Q_5(A + C - F)U_3 - Q_6(A + C + F)U_4 + 2Q_7^+G\,\alpha_1^{-1}\beta_1^{-1}, \\
N = Q_1(-\tilde{B} + \tilde{D} - \tilde{E})U_1 + Q_3(-\tilde{B} + \tilde{D} + \tilde{E})U_2 - \\
- Q_5(-\tilde{B} - \tilde{D} + \tilde{E})U_3 - Q_6(-\tilde{B} - \tilde{D} - \tilde{E})U_4 - 2Q_7^+H\,\alpha_1^{-1}\beta_1^{-1}.
\end{gathered} \tag{16}$$

The sign $\sim$ in (16) indicates a function with no term in α_2.

Expression (15) is the so-called coefficient of formation of a screened wave. If we put $h = 0$ we obtain the coefficient of formation of a head wave occurring at the interface between two half-spaces. It is easy to see that the coefficients of the functions U_i^2, in the numerator of (15) are zero. It follows that the coefficient of formation of a head wave decreases exponentially as $h/\lambda \to \infty$. As an example we took the model of a medium described by the parameters $a_0^{-1} =$ 3000 meters/second, $b_0^{-1} = 1800$ meters/second, $a_1^{-1} = 6000$ meters/second, $b_1^{-1} = 4200$ meters/second, $a_2^{-1} = 4000$ meters/second, $b_2^{-1} = 2400$ meters/second. In one case the densities of the media were assumed to be identical, and in the other the density of the layer was higher. The graph of the modulus of the coefficient of formation as a function of h/λ_2 is shown in Figure 4.

The curves fall sharply for small h/λ_2, and an increase in the density differential gives rise to a negligible widening of the transmission band for low frequencies. The formulas therefore enable us without difficulty to take account of the effect of screening head waves. To obtain the field of the head wave we need to multiply the coefficient of formation by the spectral function of the source, take the Fourier transform and consider the geometric divergence which will be the same as for a wave without the screen.

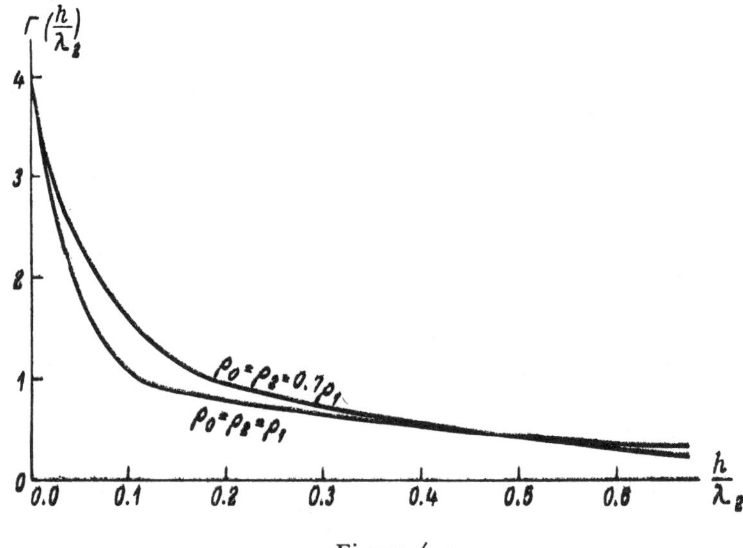

Figure 4

In conclusion I wish to thank L.A. Molotkov for discussion of the results, and L.A. Krauklis for assistance in compiling the computer program for calculating the roots.

BIBLIOGRAPHY

[1] F. K. Levin and J. D. Ingram, *Head waves from a bed of finite thickness*, Geophysics **27** (1962), 753–765.

[2] T. W. Spencer, *Refraction along a layer*, Geophysics **30** (1965), 369–388.

[3] Robert A. Phinney, *Leaking modes in the crustal wave guide*, J. Geophys. Res. 66 (1961), no. 5, 1445–1470.

[4] H. P. Bucker and Halcyon E. Morris, *Normal-mode intensity calculation for a constant depth shallow-water channel*, J. Acoust. Soc. Amer. **36** (1965), no. 6, 1010–1017.

[5] L. A. Molotkov, *On the propagation of elastic waves in media containing thin plane-parallel layers*, Problems of the Dynamic Theory of the Propagation of Seismic Waves, vol. 5, Izdat. Leningrad. Univ., Leningrad, 1961, pp. 240–280. (Russian)

[6] G. O. Petrašen, *Propagation of elastic waves in stratified isotropic media separated by parallel planes*, Leningrad. Gos. Univ. Uč. Zap. **162** Ser. Mat. Nauk 26 (1952). (Russian) MR 17, 1157.

Trudy Mat. Inst. Steklov.
95 (1968)

Proc. Steklov Inst. Math.
95 (1968)

CONSTRUCTION OF AN ASYMPTOTIC SERIES OF EIGENFUNCTIONS OF THE "BOUNCING BALL" TYPE

UDC 534.21

V. F. LAZUTKIN

Abstract. Asymptotic series are found for certain subsequences of eigenfunctions of the problem

$$(\Delta + k^2)\, u(x, y) = 0, \quad u\,|_S = 0$$

for a two-dimensional domain with smooth boundary S. The eigenfunctions from these subsequences are concentrated near a diameter AB of the domain which is stable with respect to the first approximation (for a system of rays of geometric optics). With each diameter can be associated a number

$$\alpha = \arccos \sqrt{\left(1 - \frac{d}{R_A}\right)\left(1 - \frac{d}{R_B}\right)},$$

where d is the diameter length and R_A and R_B are the radii of curvature of the boundary at the points A and B. For a stable diameter $0 < \alpha < \pi$.

When α/π is irrational, infinite asymptotic series are found both for the eigenfunctions and for the eigenvalues. The series for the eigenvalues has the form

$$k_{np} = \frac{1}{d}\left\{ n\pi + \left(p + \frac{1}{2}\right)\alpha + \sum_{l=1}^{\infty} \frac{\alpha_{lp}}{n^l} \right\}.$$

Here $n \gg 1$ and p is bounded.

When $\alpha/\pi = m_1/m_2$ is an irreducible fraction, only a finite segment of an asymptotic series can be found, increasing in size as m_2 increases.

Bibliography 5 items. 1 figure.

Introduction

Let Ω be a domain on a plane with a sufficiently smooth boundary S. We consider the eigenfunction problem for the Laplace operator with zero conditions

$$\left.\begin{aligned} (\Delta + k^2)\, u(x, y) &= 0, \\ u\,|_S &= 0. \end{aligned}\right\} \tag{0.1}$$

A number of papers have been devoted to finding the asymptotic behavior of the eigenfunctions for the problem (0.1) (see [1]–[5]). In this paper we find an asymptotic series for "bouncing ball" eigenfunctions; that is, for a subsequence of eigenfunctions close to a segment AB which has ends at the points A and B on S. All the points of this segment, except for A and B, must, of course, belong to Ω. We assume that the equations of the boundary in the neighborhoods of A and B are, respectively, of the form

$$\left.\begin{aligned} x &= \sum_{s=2}^{\infty} A_s y^s, \\ x &= d - \sum_{s=2}^{\infty} B_s y^s. \end{aligned}\right\} \tag{0.2}$$

Here (x, y) are Cartesian coordinates with origin at A and x-axis along AB; α is the leingth of AB; and $A_2 = 1/2R_A$ and $B_2 = 1/2R_B$, where R_A and R_B are the radii of curvature of the boundary at A and B respectively.

The first term of the asymptotic form for eigenfunctions concentrated about a segment such as AB was found in a paper by V. S. Buldyrev [2]. Necessary conditions on R_A and R_B were found for a subsequence of such functions to exist. The asymptotic form obtained in [2] for the eigenvalues is

$$k_{n\varkappa} = \frac{1}{d}\left\{n\pi + \left(\varkappa + \frac{1}{2}\right)\alpha(R_A, R_B) + o(1)\right\}. \tag{0.3}$$

Here n and κ are integers; $n \to +\infty$ but $\kappa \geq 0$ is bounded. A formula for $\alpha(R_A, R_B)$ is given in §4 of this paper together with a geometric interpretation of this quantity. We always have $0 < \alpha(R_A, R_B) < \pi$.

We see that when $\alpha(R_A, R_B)/\pi$ is rational there is a large spectral "degeneration" as $n \to \infty$. More precisely, some of the eigenvalues for large n are grouped into families of numbers which are very close to one another. One of these families has eigenvalues $k_{n\kappa}$, with indices n and κ satisfying

$$nm_2 + \varkappa m_1 = \text{const},$$

if $\alpha/\pi = m_1/m_2$ is an irreducible fraction. The proximity of the numbers to each other in the family increases as n increases, so that this is not a true but in some sense an "asymptotic" degeneration. It can be postulated that this asymptotic degeneration must somehow affect the asymptotic behavior; and this in fact is the case.

The greatest degeneration occurs when $\alpha = \pi/2$. This is the so-called confocal case when even the first term of the asymptotic form is not unique. If $\alpha/\pi = m_1 m_2$, where m_1/m_2 is an irreducible fraction, we can find, by the method used in this paper, sequences of functions $u_{n\kappa}(x, y)$ and numbers $k_{n\kappa}(n \to \infty, \kappa \leq C)$ such that

$$(\Delta + k_{n\varkappa}^2)\, u_{n\varkappa} = O\left(\frac{1}{n^{\frac{1}{2}(m_2-4)}}\right), \qquad u_{n\varkappa}|_S = 0. \tag{0.4}$$

For the eigenvalues we have a segment of the asymptotic series

$$k_{n\varkappa} = \frac{1}{d}\left\{n\pi + \left(\varkappa + \frac{1}{2}\right)\alpha\,(R_A,\ R_B) + \sum_{l=1}^{\left[\frac{1}{2}(m_2-5)\right]} \frac{a_{l\varkappa}}{n^l} + O\left(\frac{1}{n^{\frac{1}{2}(m_2-4)}}\right)\right\} \tag{0.5}$$

Further approximations in the form in which they are required here cannot be found in this case.[1] The asymptotic form, beginning with a fully determinate term, must therefore change abruptly.

If $\alpha(R_A, R_B)/\pi$ is irrational the standard method enables us to find infinite asymptotic series for the eigenfunctions and eigenvalues.

All the constructions in this paper are formal. We do not concern ourselves with the convergence of the series and do not prove estimates of the form (0.4). However, the asymptotic form can, if desired, be proved by methods roughly corresponding to those of [4]. If $m_2 > 8$ the closeness of the constructed functions to the eigenfunctions of $C(\overline{\Omega})$ can be proved ($C(\overline{\Omega})$ is the Banach space of continuous functions on $\overline{\Omega}$ with the usual norm).

§1. The coordinates (ξ, η)

It is convenient to introduce a large parameter $N = k^{1/2}$ and consider $\mu - Ny$ in place of y. The coordinates ξ and η are defined by

$$\xi = x + \sum_{s=2}^{\infty} \frac{1}{N^s}\, \hat{\varkappa}_s(x,\ \mu), \qquad \eta = \sum_{p=1}^{\infty} \frac{1}{N^{p-1}}\, \hat{\varphi}_p(x,\ \mu). \tag{1.1}$$

1) *Added in proof.* The author has obtained new results in this direction. It can be shown that, even in the case of rational α/π, infinite asymptotic series can be found (by another method) for eigenfunctions and eigenvalues, provided m_2 is so large that (0.5) contains a term removing the degeneracy. It is sufficient that α should not assume the following exceptional values:

$$\frac{\pi}{2},\ \frac{\pi}{3},\ \frac{2\pi}{3},\ \frac{\pi}{4},\ \frac{3\pi}{4},\ \frac{\pi}{6},\ \frac{5\pi}{6}.$$

Here $\hat{\kappa}_s(x, \mu)$ and $\hat{\psi}_p(x, \mu)$ are polynomials of the form $\hat{\kappa}_s(x, \mu) = \sum_{l=0}^{s} \kappa_{sl}(x)\mu^l$. If the index s is even, only the coefficients of even powers of μ do not vanish identically, and if s is odd only the coeffi- of odd powers of μ do not vanish identically. The coefficients $\kappa_{sl}(x)$ are functions which are analytic in a domain $\mathfrak{D}$ of the complex plane containing the segment $[0, d]$.

We shall always denote polynomials of this form by capped letters with a lower index.

We suppose that $\psi_{11}(x)$ has no zeros in $\mathfrak{D}$. Then the inverse formulas expressing the variables x and μ in terms of ξ and η are given by:

$$x = \xi + \sum_{s=2}^{\infty} \frac{1}{N^s} \dot{\lambda}_s(\xi, \eta),$$
$$\mu = \sum_{p=1}^{\infty} \frac{1}{N^{p-1}} \dot{\theta}_p(\xi, \eta). \tag{1.2}$$

Here $\dot{\lambda}_s(\xi, \eta)$ and $\dot{\theta}_p(\xi, \eta)$ are polynomials of the same form as $\hat{\kappa}_s$ and $\hat{\psi}_p$ with ξ playing the part of x and η the part of μ. These polynomials are denoted by dotted letters.

Suppose that we have an analytic function $f(x, \mu)$. Consider the superposition

$$f_1\left(\xi, \eta, \frac{1}{N}\right) = f\left(x\left(\xi, \eta, \frac{1}{N}\right), \mu\left(\xi, \eta, \frac{1}{N}\right)\right).$$

The formal Taylor series in inverse powers of N for the function f_1 is of the form

$$f_1\left(\xi, \eta, \frac{1}{N}\right) = \sum_{q=0}^{\infty} \frac{1}{N^q} \gamma_q(f), \tag{1.3}$$

where

$$\gamma_q(f) = \sum_{l=0}^{q} \sum_{n=0}^{q-l} \sum_{m=0}^{\left[\frac{l}{2}\right]} \dot{\alpha}_{q-l+n}^{(n)} \dot{\beta}_l^{(m)} \frac{1}{n!\, m!} \frac{\partial^{n+m}}{\partial \mu^n \partial x^m} f \Big|_{\substack{x=\xi \\ \mu=\dot{\theta}_1}}, \tag{1.4}$$

$$\dot{\alpha}_0^{(0)} = 1, \quad \dot{\alpha}_s^{(0)} = 0 \quad \text{for} \quad s \geqslant 1,$$

$$\dot{\alpha}_s^{(n)} = \sum_{\substack{\sum_{i=1}^{n} k_i = s \\ k_i \geqslant 2}} \prod_{i=1}^{n} \dot{\theta}_{k_i}, \quad 2 \leqslant 2n \leqslant s,$$

$$\dot{\beta}_0^{(0)}=1,\ \dot{\beta}_s^{(0)}=0 \quad \text{for} \quad s\geqslant 1,$$

$$\dot{\beta}_s^{(n)}=\sum_{\substack{\sum_{i=1}^{n} k_i=s \\ k_i\geqslant 2}} \prod_{i=1}^{n} \dot{\lambda}_{k_i}, \quad 2\leqslant 2n\leqslant s. \tag{1.5}$$

For example,

$$\gamma_0(f)=f(x,\ \mu)\Big|_{\substack{x=\xi\\ \mu=\dot{\theta}_1}},\ \gamma_1(f)=\dot{\theta}_2\frac{\partial f}{\partial \mu}\Big|_{\substack{x=\xi\\ \mu=\dot{\theta}_1}},$$

$$\gamma_2(f)=\dot{\theta}_3\frac{\partial f}{\partial \mu}\Big|_{\substack{x=\xi\\ \mu=\dot{\theta}_1}}+\frac{1}{2}\dot{\theta}_2^2\frac{\partial^2 f}{\partial \mu^2}\Big|_{\substack{x=\xi\\ \mu=\dot{\theta}_1}}+\dot{\lambda}_2\frac{\partial f}{\partial x}\Big|_{\substack{x=\xi\\ \mu=\dot{\theta}_1}}.$$

Using (1.4), we obtain the equations relating $\hat{\kappa}_s$ and $\hat{\psi}_p$ with $\dot{\lambda}_q$ and $\dot{\theta}_q$:

$$\begin{gathered}\gamma_0(\hat{\psi}_1)=\eta,\\ \sum_{p=1}^{q+1}\gamma_{q-p+1}(\hat{\psi}_p)=0,\quad q\geqslant 1,\\ \dot{\lambda}_q+\sum_{s=2}^{q}\gamma_{q-s}(\hat{\kappa}_s)=0,\quad q\geqslant 2\end{gathered} \tag{1.6}$$

If all the $\hat{\kappa}_s$ and $\hat{\psi}_p$ are known it is easy to show all the $\dot{\theta}_q$ and $\dot{\lambda}_q$ are uniquely determined by (1.6). Conversely $\hat{\kappa}_s$ and $\hat{\psi}_p$ are uniquely defined when $\dot{\theta}_q$ and $\dot{\lambda}_q$ are given. The specific form of these functions is determined later.

We now require in addition that the equations $\xi=0$ and $\xi=d$ should coincide with the equations of the boundary in the neighborhood of A and B respectively. It follows from (0.2) and (1.3) that the following relations are then necessary:

$$x\,|_{\xi=0}=\sum_{q=2}^{\infty}\frac{1}{N^q}\left(\sum_{s=2}^{q}\gamma_{q-s}(A_s\mu^s)\,|_{\xi=0}\right),$$

$$x\,|_{\xi=d}=d-\sum_{q=2}^{\infty}\frac{1}{N^q}\left(\sum_{s=2}^{q}\gamma_{q-s}(B_s\mu^s)\,|_{\xi=d}\right).$$

Using (1.2), we obtain

$$\left.\begin{aligned}\dot{\lambda}_q(0,\ \eta)&=\sum_{s=2}^{q}\gamma_{q-s}(A_s\mu^s)\,|_{\xi=0},\\ \dot{\lambda}_q(d,\ \eta)&=-\sum_{s=2}^{q}\gamma_{q-s}(B_s\mu^s)\,|_{\xi=d}.\end{aligned}\right\} \tag{1.7}$$

This is the first boundary condition imposed on our functions. The second boundary condition follows from the requirement that $\operatorname{Im}\eta(x, y) = 0$ if (x, y) is a boundary point of Ω. Since x and μ are then real, it follows from (1.2) that the coefficients of the polynomials $\dot{\theta}_p(0, \eta)$, $\dot{\theta}_p(d, \eta)$, $\dot{\lambda}_s(0, \eta)$ and $\dot{\lambda}_s(d, \eta)$ must be real. It is sufficient to require that the coefficients of $\dot{\theta}_p(0, \eta)$ and $\dot{\theta}_p(d, \eta)$ be real, for then (see (1.7)) $\dot{\lambda}_s$ will also satisfy this condition.

The condition

$$\operatorname{Im}\theta_{pl}(0) = \operatorname{Im}\theta_{pl}(d) = 0 \tag{1.8}$$

is a sufficient condition for $\operatorname{Im}\eta(x, y) = 0$ on the boundary if we require that η should be uniquely determined from

$$x = \sum_{s=2}^{\infty} \frac{1}{N^s} \dot{\lambda}_s(0, \eta), \qquad \mu = \sum_{p=1}^{\infty} \frac{1}{N^{p-1}} \dot{\theta}_p(0, \eta) \tag{1.9}$$

and from

$$x = d + \sum_{s=2}^{\infty} \frac{1}{N^s} \dot{\lambda}_s(d, \eta), \qquad \mu = \sum_{p=1}^{\infty} \frac{1}{N^{p-1}} \dot{\theta}_p(d, \eta) \tag{1.10}$$

and, in addition, it is a sufficient condition for the right sides of these equations to be analytic functions of η.

§2. The Laplace operator in (ξ, η) coordinates

The Laplace operator in (ξ, η) coordinates has the form

$$\Delta = (\nabla\xi)^2 \frac{\partial^2}{\partial\xi^2} + (\nabla\eta)^2 \frac{\partial^2}{\partial\eta^2} + \Delta\xi \frac{\partial}{\partial\xi} + \Delta\eta \frac{\partial}{\partial\eta} + 2(\nabla\xi, \nabla\eta)\frac{\partial^2}{\partial\xi\partial\eta}. \tag{2.1}$$

The functions $(\nabla\xi)^2$, $(\nabla\eta)^2$, $\Delta\xi$, $\Delta\eta$ and $(\nabla\xi, \nabla\eta)$ depend on N, and their inverse powers of N are

$$(\nabla\xi)^2 = 1 + \sum_{s=2}^{\infty} \frac{1}{N^s} \hat{a}_s, \qquad (\nabla\eta)^2 = \sum_{s=0}^{\infty} \frac{1}{N^{s-2}} \hat{b}_s, \tag{2.2}$$

$$\Delta\xi = \sum_{s=0}^{\infty} \frac{1}{N^s} \hat{c}_s,$$

$$\Delta\eta = \sum_{s=0}^{\infty} \frac{1}{N^{s-1}} \hat{d}_s,$$

$$(\nabla\xi,\ \nabla\eta) = \sum_{s=1}^{\infty} \frac{1}{N^{s-1}} \hat{e}_s.$$

Here

$$\begin{aligned}
\hat{a}_s &= 2\frac{\partial \hat{\chi}_s}{\partial x} + \sum_{\substack{l+m=s+2\\ l,\, m\geqslant 2}} \frac{\partial \hat{\chi}_l}{\partial \mu}\cdot\frac{\partial \hat{\chi}_m}{\partial \mu} + \sum_{\substack{l+m=s\\ l,\, m\geqslant 2}} \frac{\partial \hat{\chi}_l}{\partial x}\cdot\frac{\partial \hat{\chi}_m}{\partial x},\\
\hat{b}_s &= \sum_{\substack{l+m=s+2\\ l,\, m\geqslant 1}} \frac{\partial \hat{\psi}_l}{\partial \mu}\cdot\frac{\partial \hat{\psi}_m}{\partial \mu} + \sum_{\substack{l+m=s\\ l,\, m\geqslant 1}} \frac{\partial \hat{\psi}_l}{\partial x}\cdot\frac{\partial \hat{\psi}_m}{\partial x},\\
\hat{c}_s &= \frac{\partial^2 \hat{\chi}_{s+2}}{\partial \mu^2} + \frac{\partial^2 \hat{\chi}_s}{\partial x^2},\\
\hat{d}_s &= \frac{\partial^2 \hat{\psi}_{s+2}}{\partial \mu^2} + \frac{\partial^2 \hat{\psi}_s}{\partial x^2},\\
\hat{e}_s &= \frac{\partial \hat{\psi}_s}{\partial x} + \sum_{\substack{l+m=s+2\\ l\geqslant 2,\, m\geqslant 1}} \frac{\partial \hat{\chi}_l}{\partial \mu}\cdot\frac{\partial \hat{\psi}_m}{\partial \mu} + \sum_{\substack{l+m=s\\ l\geqslant 2,\, m\geqslant 1}} \frac{\partial \hat{\chi}_l}{\partial x}\cdot\frac{\partial \hat{\psi}_m}{\partial x}.
\end{aligned} \tag{2.3}$$

We now pass to the variables (ξ, η) by formula (1.3), thus obtaining

$$\begin{aligned}
(\nabla\xi)^2 &= 1 + \sum_{q=2}^{\infty} \frac{1}{N^q} \dot{a}_q,\\
(\nabla\eta)^2 &= \sum_{q=0}^{\infty} \frac{1}{N^{q-2}} \dot{b}_q,\\
\Delta\xi &= \sum_{q=0}^{\infty} \frac{1}{N^q} \dot{c}_q,\\
\Delta\eta &= \sum_{q=0}^{\infty} \frac{1}{N^{q-1}} \dot{d}_q,\\
(\nabla\xi,\ \nabla\eta) &= \sum_{q=1}^{\infty} \frac{1}{N^{q-1}} \dot{e}_q,
\end{aligned} \tag{2.4}$$

where

$$\dot{a}_q = \sum_{s=2}^{q} \gamma_{q-s}(\hat{a}_s),$$

$$\dot{b}_q = \sum_{s=0}^{q} \gamma_{q-s}(\hat{b}_s), \tag{2.5}$$

$$\dot{c}_q = \sum_{s=0}^{q} \gamma_{q-s}(\hat{c}_s),$$

$$\left.\begin{aligned} \dot{d}_q &= \sum_{s=0}^{q} \gamma_{q-s}(\hat{d}_s), \\ \dot{e}_q &= \sum_{s=1}^{q} \gamma_{q-s}(\hat{e}_s). \end{aligned}\right\}$$

§3. Substitution into the equation

We seek a solution in the form

$$u_\kappa(\xi, \eta, N) = e^{iN^2\xi + \sum_{l=0}^{\infty} \frac{1}{N^l}\varphi_l(\xi, \eta)} D_\kappa(\eta), \tag{3.1}$$

where κ is a fixed positive integer and

$$D_\kappa(\eta) = e^{-\frac{\eta^2}{2}} H_\kappa(\eta),$$

$H_\kappa(\eta)$ being the Hermite polynomial of degree κ. The function $D_\kappa(\eta)$ satisfies

$$D_\kappa'' = (\eta^2 - t_\kappa) D_\kappa \tag{3.2}$$

where $t_\kappa = 2\kappa + 1$.

On using (2.1) and (2.4), we obtain, after some calculations,

$$(\Delta + k^2) u_\kappa = \left[\left(\sum_{r=-2}^{\infty} \frac{1}{N^r} \dot{\delta}_{r+4}\right) D_\kappa(\eta) + \right.$$

$$\left. + \left(\sum_{r=-2}^{\infty} \frac{1}{N^r} \dot{\varepsilon}_{r+3}\right) D_\kappa'(\eta)\right] e^{iN^2\xi + \sum_{l=0}^{\infty} \frac{1}{N^l}\varphi_l}. \tag{3.3}$$

Here

$$\dot{\delta}_{r+4} = 2i\frac{\partial \varphi_{r+2}}{\partial \xi} + \sum_{\substack{l+m=r \\ l,\, m \geqslant 0}} \frac{\partial \varphi_l}{\partial \xi} \cdot \frac{\partial \varphi_m}{\partial \xi} + \frac{\partial^2 \varphi_r}{\partial \xi^2} - \dot{a}_{r+4} + \sum_{\substack{l+s=r+2 \\ s \geqslant 2,\, l \geqslant 0}} 2i\dot{a}_s \frac{\partial \varphi_l}{\partial \xi} +$$

$$+\sum_{\substack{l+m+s=r\\ l,\,m\geqslant 0,\,s\geqslant 2}} \dot a_s \frac{\partial\varphi_l}{\partial\xi}\frac{\partial\varphi_m}{\partial\xi}+\sum_{\substack{l+s=r\\ l\geqslant 0,\,s\geqslant 2}} \dot a_s\frac{\partial^2\varphi_l}{\partial\xi^2}+\sum_{\substack{s+l+m=r+2\\ s\geqslant 0,\,l,\,m\geqslant 1}} \dot b_s\frac{\partial\varphi_l}{\partial\eta}\frac{\partial\varphi_m}{\partial\eta}$$

$$+\sum_{\substack{s+l=r+2\\ s\geqslant 0,\,l\geqslant 2}} \dot b_s\frac{\partial^2\varphi_l}{\partial\eta^2}+(\eta^2-t_\varkappa)\dot b_{r+2}+i\dot c_{r+2}+\sum_{\substack{s+l=r\\ s,\,l\geqslant 0}}\dot c_s\frac{\partial\varphi_l}{\partial\xi}+\sum_{\substack{l+s=r+1\\ s\geqslant 0,\,l\geqslant 1}}\dot d_s\frac{\partial\varphi_l}{\partial\eta}$$

$$+\sum_{\substack{s+m=r+3\\ s,\,m\geqslant 1}} 2i\dot e_s\frac{\partial\varphi_m}{\partial\eta}+\sum_{\substack{s+l+m=r+1\\ s,\,m\geqslant 1,\,l\geqslant 0}} 2\dot e_s\frac{\partial\varphi_l}{\partial\xi}\frac{\partial\varphi_m}{\partial\xi}+\sum_{\substack{s+l=r+1\\ s,\,l\geqslant 1}} 2\dot e_s\frac{\partial^2\varphi_l}{\partial\xi\partial\eta}, \tag{3.4}$$

$$\dot\varepsilon_{r+3}=\sum_{\substack{s+l=r+2\\ s\geqslant 0,\,l\geqslant 0}} 2\dot b_s\frac{\partial\varphi_l}{\partial\eta}+2i\dot e_{r+3}+\sum_{\substack{s+l=r+1\\ s\geqslant 1,\,l\geqslant 0}} 2\dot e_s\frac{\partial\varphi_l}{\partial\xi}+\dot d_{r+1}. \tag{3.5}$$

The function u_K will formally satisfy the Helmholtz equation if $\dot\delta_{r+4}=\dot\epsilon_{r+3}=0$ for all r. We take $\operatorname{Im} u_K$ as the asymptotic expression for the eigenfunctions. Let us determine when $\operatorname{Im} u_K$ satisfies the boundary conditions. Since $\operatorname{Im}\eta(x, y)=0$ on the boundary, we have

$$\operatorname{Im} u_\varkappa|_{\xi=0}=D_\varkappa(\eta)e^{\operatorname{Re}\left(\sum_{l=0}^{\infty}\frac{1}{N^l}\varphi_l(0,\,\eta)\right)}\sin\left(\sum_{l=0}^{\infty}\frac{1}{N^l}\operatorname{Im}\varphi_l(0,\,\eta)\right).$$

Since η is arbitrary the sine will vanish if for all l

$$\operatorname{Im}\varphi_{lm}(0)=0 \quad \text{for} \quad 0<m\leqslant l. \tag{3.6}$$

Without loss of generality we put

$$\varphi_{2s,\,0}(0)=0. \tag{3.7}$$

Then $\operatorname{Im} u_K|_{\xi=0}=0$. In order that $\operatorname{Im} u_K$ should vanish for $\xi=d$ also, we must have

$$\operatorname{Im}\varphi_{lm}(d)=0 \quad \text{for} \quad 0<m\leqslant l, \tag{3.8}$$

$$N^2d+\sum_{s=0}^{\infty}\frac{1}{N^{2s}}\operatorname{Im}\varphi_{2s,\,0}(d)=n\pi. \tag{3.9}$$

The asymptotic expansion of the eigenvalues $k_{nK}=N^2_{nK}$ in inverse powers of n is determined from (3.9).

It only remains to find functions $\dot\psi_p$, $\hat\kappa_s$ and $\dot\phi_l$ such that $\dot\delta_{r+4}=0$, $\dot\epsilon_{r+3}=0$ and the boundary conditions (1.7), (1.8) and (3.6)–(3.8) are satisfied.

§4. First approximation

From (3.4) and (3.5) we obtain

$$\dot{\delta}_2 = 2i\frac{\partial\dot{\varphi}_0}{\partial\xi} - \dot{a}_2 + (\eta^2 - t_\varkappa)\dot{b}_0 + i\dot{c}_0, \tag{4.1}$$

$$\dot{\varepsilon}_1 = 2i\dot{e}_1. \tag{4.2}$$

We put $\dot{\delta}_2 = \dot{\epsilon}_1 = 0$. Then from (4.2) it follows that

$$\dot{e}_1 = \gamma_0(\hat{e}_1) = 0. \tag{4.3}$$

From (1.6) we obtain

$$\theta_{11}(\xi) = \frac{1}{\psi_{11}(\xi)}. \tag{4.4}$$

Hence θ_{11} also has no zeros, and therefore

$$\hat{e}_1 = \frac{\partial\hat{\psi}_1}{\partial x} + \frac{\partial\hat{\varkappa}_2}{\partial\mu}\cdot\frac{\partial\hat{\psi}_1}{\partial\mu} = 0. \tag{4.5}$$

Rewriting (4.5) for the coefficients of the polynomials, we obtain

$$\varkappa_{22} = -\frac{1}{2}\cdot\frac{\psi_{11}'}{\psi_{11}} = \frac{1}{2}\cdot\frac{\theta_{11}'}{\theta_{11}}. \tag{4.6}$$

We now write $\dot{\delta}_2 = 0$ for the coefficients of the polynomials; then

$$-a_{22} + b_{00} = 0, \tag{4.7}$$

$$2i\varphi_{00}' - a_{20} - t_\varkappa b_{00} + ic_{00} = 0. \tag{4.8}$$

By using (2.3) and (2.5) we can obtain

$$\begin{aligned} a_{22} &= (2\varkappa_{22}' + 4\varkappa_{22}^2)\,\theta_{11}^2, \\ a_{20} &= 2\varkappa_{20}', \\ b_{00} &= \psi_{11}^2, \\ c_{00} &= 2\varkappa_{22}. \end{aligned} \tag{4.9}$$

We express a_{22} and b_{00} in terms of θ_{11} and substitute into (4.7), to obtain

$$\theta_{11}'' = \frac{1}{\theta_{11}^3}. \tag{4.10}$$

The solution of equation (4.10) has the form

$$\theta_{12}(\xi) = \sqrt{l + \frac{(\xi - x_0)^2}{l}}, \tag{4.11}$$

where $l > 0$ and x_0 are constants. To determine these constants we use the first boundary condition (1.7), which for the first approximation is

$$\begin{gathered} \varkappa_{22}\big|_{x=0} = -\frac{1}{2R_A}, \\ \varkappa_{22}\big|_{x=d} = \frac{1}{2R_B}, \\ \varkappa_{20}\big|_{x=0} = \varkappa_{20}\big|_{x=d} = 0. \end{gathered} \tag{4.12}$$

We obtain the boundary conditions

$$\frac{\theta_{11}'}{\theta_{11}}\bigg|_{\xi=0} = -\frac{1}{R_A}, \quad \frac{\theta_{11}'}{\theta_{11}}\bigg|_{\xi=d} = \frac{1}{R_B}. \tag{4.13}$$

From (4.13) we obtain the following expressions for l:

$$l = \sqrt{x_0(R_A - x_0)}, \quad l = \sqrt{(d - x_0)(R_B - d + x_0)}. \tag{4.14}$$

The graphs of these functions are semicircles of radii $R_A/2$ and $R_B/2$, and the two cases are shown in the diagram. We see that they do not always have a point of intersection C in the upper half-plane. The conditions for such a point to exist are also the necessary conditions for the formation of eigenfunctions of the "bouncing ball" type. It is found that they coincide with the stability conditions for the system of rays with respect to the first approximation (see [2]). These conditions are conveniently formulated if we use the quantities

$$g_A = 1 - \frac{d}{R_A}, \quad g_B = 1 - \frac{d}{R_B},$$

expressing the deviation from the so-called confocal case in which $g_A = g_B = 0$.

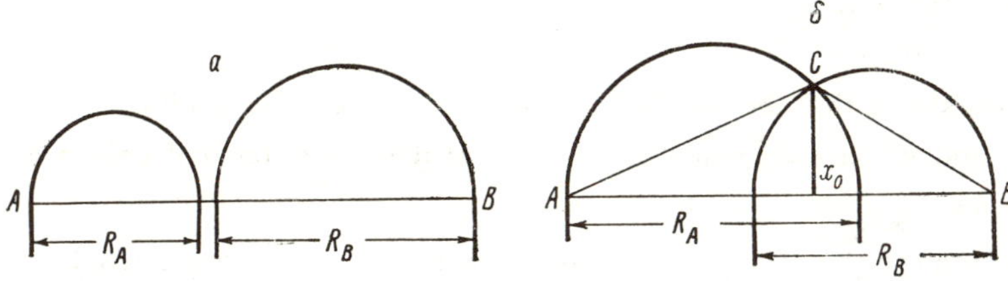

The quantities x_0 and l are determined in two cases:

1) $g_A = g_B = 0$ (the confocal case); in this case x_0 and l are uniquely determined because the semicircles representing the graphs of the functions (4.14) coincide.

2) $0 < g_A g_B < 1$; in this case x_0 and l are uniquely determined.

The set of stable states of the resonator (on the $(g_A g_B)$ plane) therefore consists of two connected open sets and one point. We note that R_A and R_B can take both positive and negative values. Both cases shown in the diagram correspond to positive radii.

From (4.8) and (3.7) we obtain

$$\varphi_{00}(\xi) = -i\varkappa_{20}(\xi) - i\frac{t_x}{2}\int_0^{\xi}\frac{d\xi}{\theta_{11}^2(\xi)} - \frac{1}{2}\ln\theta_{11}. \tag{4.15}$$

The formula for the eigenvalues coincides with (0.3), where for α we must take

$$\alpha(R_A, R_B) = \int_0^d \frac{d\xi}{\theta_{11}^2} = \arccos\sqrt{g_A g_B}. \tag{4.16}$$

A simple geometric meaning can be given to $\alpha(R_A, R_B)$: it is equal to the angle ACB in diagram (b). We see that $0 < \alpha(R_A, R_B) < \pi$ always holds and that $\alpha = \pi/2$ in the confocal case.

We keep the function $\kappa_{20}(x)$ arbitrary and satisfying only the boundary conditions (4.12).

§5. Subsequent approximations

Suppose that we have found all the functions $\hat{\psi}_p$, $\dot{\theta}_p$, $\dot{\lambda}_s$, $\hat{\kappa}_s$ and $\dot{\phi}_l$ for $s \leq r+3$, $p \leq r+2$ and $l \leq r+1$, and that δ_{q+4} and ϵ_{q+3} vanish for $q \leq r-1$ and the following boundary conditions hold:

(1.7) for $q \leq r+3$,

(1.8) for $b \leq r+2$,

(3.6)–(3.8) for $l \leq r+1$.

We also assume that the leading coefficients in the polynomials $\dot{\theta}_p$, $\hat{\psi}_p$, $\dot{\lambda}_s$ and $\hat{\kappa}_s$ already constructed are real functions (for real values of the arguments).

We now require that $\dot{\delta}_{r+4} = \dot{\epsilon}_{r+3} = 0$ and first find the leading coefficients of the polynomials $\dot{\theta}_{r+3}$ and $\hat{\kappa}_{r+4}$. From (3.4) and (3.5) we obtain

$$a_{r+4,\, r+4} = b_{r+2,\, r+2}, \tag{5.1}$$

$$e_{r+3,\, r+3} = 0. \tag{5.2}$$

It can be shown that

$$\gamma_{r+2}(f) = \dot{\theta}_{r+3} \frac{\partial}{\partial \mu} f \Big|_{\substack{x=\xi \\ \mu=\theta_1}} + F(f), \tag{5.3}$$

where $F(f)$ depends only on the functions already found, and on f. In this section we denote by F a function depending only on the quantities already found. We can then write

$$\dot{a}_{r+4} = \left(2 \frac{\partial \hat{\varkappa}_{r+4}}{\partial x} + 2 \frac{\partial \hat{\varkappa}_{r+4}}{\partial \mu} \cdot \frac{\partial \hat{\varkappa}_2}{\partial \mu}\right)\Big|_{\substack{x=\xi \\ \mu=\theta_1}} + \dot{\theta}_{r+3} \frac{\partial}{\partial \mu} \hat{a}_2 \Big|_{\substack{x=\xi \\ \mu=\theta_1}} + F, \tag{5.4}$$

$$\dot{b}_{r+2} = 2 \frac{\partial \hat{\psi}_{r+3}}{\partial \mu} \cdot \frac{\partial \hat{\psi}_1}{\partial \mu} \Big|_{\substack{x=\xi \\ \mu=\theta_1}} + F, \tag{5.5}$$

$$\dot{e}_{r+3} = \left(\frac{\partial \hat{\psi}_{r+3}}{\partial x} + \frac{\partial \hat{\varkappa}_{r+4}}{\partial \mu} \cdot \frac{\partial \hat{\psi}_1}{\partial \mu} + \frac{\partial \psi_{r+3}}{\partial \mu} \cdot \frac{\partial \hat{\varkappa}_2}{\partial \mu}\right)\Big|_{\substack{x=\xi \\ \mu=\dot{\theta}_1}} + F. \tag{5.6}$$

Using (1.6), we obtain from (5.2) and (5.6), after some calculations,

$$\varkappa_{r+4,\ r+4} = \frac{1}{r+4} \cdot \frac{1}{\theta^{r+3}} \left(\theta'_{r+3,\ r+3} - \frac{\theta'}{\theta} \theta_{r+3,\ r+3}\right) + F \tag{5.7}$$

For brevity we put $\theta = \theta_{11}$ here.

On expressing $a_{r+4,\ r+4}$ and $b_{r+2,\ r+2}$ in terms of $\theta_{r+3,\ r+3}$ (here we must use (5.4), (5.5), (1.6) and (5.7)) and equating the expressions obtained, we have

$$\theta''_{r+3,\ r+3} + \frac{(r+4)^2 - 1}{\theta^4} \theta_{r+3,\ r+3} = F. \tag{5.8}$$

The function F on the right side of (4.8) depends on the functions already obtained, and in fact on their leading coefficients only, and is therefore real because all the operations leading to (5.8) are real also.

We easily obtain from (1.7) the boundary conditions

$$\varkappa_{r+4,\ r+4} \Big|_{\substack{x=0 \\ x=d}} = F, \tag{5.9}$$

The function F on the right side is also real. In combination with (5.7) this gives the boundary conditions for the Sturm-Liouville problem, from which which (in the case when we do not have an eigenvalue) we can determine $\theta_{r+3,\ r+3}$ and the other leading coefficients of the approximation in question apart from $\phi_{r+2,\ r+2}$.

We write the equations $\dot{\delta}_{r+4} = 0$ and $\dot{\epsilon}_{r+3} = 0$ in the following form:

$$\dot{\delta}_{r+4} = 2i\frac{\partial\dot{\varphi}_{r+2}}{\partial\xi} + \dot{b}_0\frac{\partial^2\dot{\varphi}_{r+2}}{\partial\eta^2} - \dot{a}_{r+4} + \dot{b}_{r+2}(\eta^2 - t_x) + i\dot{c}_{r+2} + F = 0, \quad (5.10)$$

$$\dot{\varepsilon}_{r+3} = 2\dot{b}_0\frac{\partial\dot{\varphi}_{r+2}}{\partial\eta} + 2ie_{r+3} + \dot{d}_{r+1} + F = 0. \quad (5.11)$$

It is clear that $\phi_{r+2,\,r+2}$ is determined from (5.11). This is the only leading coefficient of the approximation in question which turns out to be complex.

We suppose that the following coefficients are already determined:

$$\theta_{r+3,\,l},\quad \psi_{r+3,\,l} \text{ with } l \geqslant m,$$

$$\varkappa_{r+4},_{\,l},\quad \lambda_{r+4,\,l} \text{ with } l \geqslant m+1.$$

We recall that it was agreed that F should denote functions depending on quantities already obtained, and therefore we may include the above coefficients.

On using (5.4)–(5.6) and analogous formulas, we can obtain from (5.11) (for $m>0$)

$$\varphi_{r+2,\,m} = \frac{i\theta}{m}\left(\theta'_{r+3,\,m-1} - \frac{\theta'}{\theta}\theta_{r+3,\,m-1}\right) - i\theta^m\varkappa_{r+4,\,m} + F. \quad (5.12)$$

On equating the mth coefficient in (5.10) to zero and using (5.12), we can, exactly as in (5.8), obtain an equation for $\theta_{r+3,\,m-1}$ when $m>0$:

$$\theta''_{r+3,\,m-1} + \frac{m^2-1}{\theta^4}\theta_{r+3,\,m-1} = F. \quad (5.13)$$

We attach to (5.13) the boundary conditions (1.8), (3.6) and (3.8), which together with (5.12) give two Sturm-Liouville problems: for $\operatorname{Im}\theta_{r+3,\,m-1}$ and for $\operatorname{Re}\theta_{r+3,\,m-1}$ seperately. The right side of (5.13) is in general complex.

All the coefficients of the polynomial $\dot{\theta}_{r+3}$ can thus be found. In the polynomial $\hat{\kappa}_{r+4}$ only the leading coefficient is uniquely determined; the others must satisfy only certain boundary conditions which are obtained from (1.7).

When r is odd the construction of the approximation comes to an end, since all $\phi_{r+2,\,m}$ are determined from (5.12). If, however, r is even, it is still necessary to find $\phi_{r+2,\,0}$, which can be determined from (5.10). It is necessary to equate to zero the coefficient of the term in (5.10) of zero degree in η and take account of the boundary condition (3.7)

§6. Up to what stage is the construction of an asymptotic series possible?

The construction of an asymptotic series is possible until Sturm-Liouville problems of the type (5.8) with the corresponding boundary conditions are solvable; that is, until for some r we encounter an eigenvalue in this problem.

Let us determine when zero is an eigenvalue of the operator L_m given by

$$L_m f = \frac{d^2}{dx^2} f + \frac{m^2 - 1}{\theta^4(x)} f \tag{6.1}$$

on the interval $[0, d]$ with boundary conditions

$$\left(f' - \frac{\theta'}{\theta} f\right)\Big|_{\substack{x=0\\x=d}} = 0 \quad \left(\text{or} \quad f\big|_{\substack{x=0\\x=d}} = 0\right). \tag{6.2}$$

The change of variables

$$z = \operatorname{arc\,tg} \frac{x - x_0}{l} \tag{6.3}$$

and the substitution $f = \theta g$ reduces the equation $L_m f = 0$ to

$$g'' + m^2 g = 0 \tag{6.4}$$

on the interval $[z_1, z_2]$ with boundary conditions

$$g'\big|_{\substack{z=z_1\\z=z_2}} = 0 \quad \left(\text{or} \quad g\big|_{\substack{z=z_1\\z=z_2}} = 0\right). \tag{6.5}$$

Here

$$z_1 = -\operatorname{arc\,tg} \frac{x_0}{l}, \quad z_2 = \operatorname{arc\,tg} \frac{d - x_0}{l}. \tag{6.6}$$

We see that zero is an eigenvalue of L_m if and only if

$$\sin m(z_2 - z_1) = 0.$$

It is easy to verify that

$$z_2 - z_1 = \alpha(R_A, R_B). \tag{6.7}$$

We deduce from this that zero is an eigenvalue of L_m if and only if $\alpha(R_A, R_B)/\pi = m_1/m_2$, where m_1 is an integer.

BIBLIOGRAPHY

[1] J. B. Keller and S. I. Rubinow, *Asymptotic solution of eigenvalue problems*, Ann. Physics 9 (1960), 24–75.

[2] V. S. Buldyrev, *Asymptotics of eigenfunctions of the Helmholtz equation for plane convex regions*, Vestnik Leningrad. Univ. 20 (1965), no. 22, 38–51. (Russian) MR 32 #8692.

[3] L. A. Vaĭnšteĭn, *Open resonators with cylindrical mirrors*, Ž. Tehn. Fiz. 34 (1964), 205–217 = Soviet Physics Tech. Phys. 9 (1964), 166–174. MR 31 #5463.

[4] V. F. Lazutkin, *The asymptotic behavior of the eigenfunctions of Laplace's operator which are concentrated near the boundary of the domain*, Ž. Vyčisl. Mat. i Mat. Fiz. 7 (1967), 1237–1249. (Russian) MR 36 #5493.

[5] V. F. Lazutkin, *Parabolic equations and asymptotic eigenfunctions of the Helmholtz equations for three-dimensional regions*, Vestnik Leningrad. Univ. 20 (1965), no. 22, 52–57. (Russian) MR 34 #1732.

Trudy Mat. Inst. Steklov.
95 (1968)

Proc. Steklov Inst. Math.
95 (1968)

GREEN'S FUNCTION FOR THE DIFFRACTION PROBLEM ON A CONVEX CYLINDER WITH VARIABLE IMPEDANCE

UDC 534.26

I. A. MOLOTKOV

Abstract. The paper discusses the plane problem of finding the Green's function $G(M_0, M, k)$ of the Helmholtz operator $\Delta + k^2$ outside an arbitrary smooth convex contour Γ with the condition

$$\frac{\partial G}{\partial n} + ikg(s)\, G\,|_\Gamma = 0, \quad g(s) > 0$$

and with limiting absorption at infinity; S is the arc length on Γ. The asymptotic behavior of $G(M_0, M, k)$ $(k \to \infty)$ is investigated for various positions of the points M_0 and M in the case of an arbitrary smooth function $g(s)$.

Several terms of a ray series are found for the function $G(M_0, M, k)$ in the lit region. The asymptotic form of G in the shadow and half-shadow regions in the boundary layer adjoining Γ is investigated by finding eigenfunctions—solutions of the homogeneous problem—in the form of the product of an exponential and an Airy function. The arguments of the exponential and the Airy function are asymptotic series in powers of $k^{-1/3}$. The eigenfunctions found in the boundary layer may be continued to any distance from Γ, where their representations are transformed into ray series. A new asymptotic representation for Green's function is found in the form of a superposition of eigenfunctions. The application of Watson summation and the saddle-point method shows that in the lit region this new representation is transformed into a ray series.

Bibliography 9 items. 3 figures.

§1. Statement of the problem

We consider the following plane problem:

$$\begin{gathered}(\Delta + k^2)\, G(M_0, M, k) = -\delta(M - M_0),\\ \frac{\partial G}{\partial n} + ikg(s)\, G\,|_\Gamma = 0, \\ \lim_{r\to\infty} \sqrt{r}\left(\frac{\partial G}{\partial r} - ikG\right) = 0, \quad r = \sqrt{x_M^2 + y_M^2}.\end{gathered} \tag{1.1}$$

Here Γ is a sufficiently smooth convex curve given by the natural equation $\rho = a(s)$; the points M_0 and M are situated outside Γ; δ is the Dirac delta function; s is the arc length of Γ measured from a certain point; n is the length of the external normal to Γ; and k is the wave number. The function $g(s)$ is assumed to be sufficiently smooth and positive. In acoustics and electrodynamics its reciprocal $g^{-1}(s)$ is known as the normal impedance. If the contour Γ departs to infinity the principle of limiting absorption replaces the radiation condition in problem (1.1).

We require to find the high-frequency asymptotic behavior (as $k \to \infty$) of Green's function $G(M_0, M, k)$ for various positions of M_0 and M.

The principal term in the asymptotic form for arbitrary smooth functions $a(s)$ and $g(s)$ was found in [1]. Prior to this the particular case $a = \text{const}$, $g = \text{const}$ was investigated in [2] and the case $a^{1/3}(s)g(s) = \text{const}$ in [3]. A generalization of the known principle of locality was suggested in [4] and used as a basis for a hypothesis concerning the principal term of the asymptotic form of the solution of (1.1). The asymptotic formulas obtained by different methods in [4] and [1] turned out to be coincident. But neither of these papers enabled questions to be answered about the subsequent terms of the asymptotic series.

In this paper the asymptotic form of the solution of (1.1) is derived by a more systematic method (in comparison with [1]) which is independent of exact solutions of particular problems. This enables us to obtain the subsequent terms in the asymptotic expansion as $k \to \infty$ for the solution of (1.1).

Many problems in the theory of elasticity, electrodynamics and acoustics (see, for example, [5] and [6] reduce to problem (1.1) in the high-frequency approximation. The possibility of determining correction terms in the asymptotic form for these problems enables additional physical effects to be obtained.

The principal aim of this study is to obtain asymptotic formulas for the Green's function in the shadow and half-shadow zones. The derivation of these formulas requires more precise representations for Green's function in the lit region also; §2 is devoted to this.

§2. The lit region

Suppose that the observer M is situated in that part of the plane which is lit by the source M_0. Then the Green's function of problem (1.1) consists of two components

$$G = G_i + G_r. \tag{2.1}$$

The component G_i corresponds to the primary wave formed by the action of a point source in unbounded space and has the form

$$G_i = \frac{i}{4} H_0^{(1)}(kr) \sim \frac{e^{ikr+\frac{i\pi}{4}}}{2\sqrt{2\pi kr}} \sum_{m=0}^{\infty} \frac{a_m}{(ikr)^m}, \tag{2.2}$$

where

$$r = |\overline{M_0 M}|, \qquad a_m = \frac{\Gamma^2\left(m+\frac{1}{2}\right)}{\pi \cdot 2^m m!}.$$

The component G_r corresponds to the reflected wave, which, in accordance with the ray method, is sought in the form of an asymptotic series

$$G_r \sim \frac{e^{ik\tau+i\frac{\pi}{4}}}{2\sqrt{2\pi k}} \sum_{m=0}^{\infty} \frac{b_m(\tau, \alpha)}{(ik)^m}. \tag{2.3}$$

Here τ and α are orthogonal radial coordinates. The coordinate $\tau = l_0 + l$ is the length of the ray proceeding from the source to the observer and reflected from the boundary Γ (Figure 1). The coordinate α is the angle of departure of the ray from M_0 measured from a certain direction. The coefficients $b_m(t, \alpha)$ must be such that (2.1) satisfies the conditions of problem (1.1).

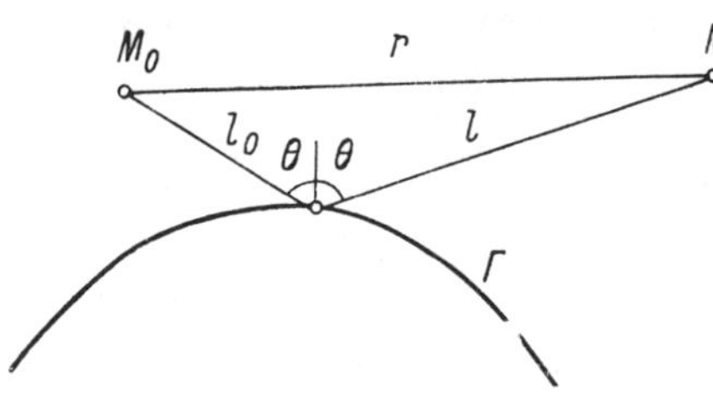

Figure 1

Recurrence formulas (see [7]) are obtained from the Helmholtz equation for the functions $b_m(\tau, \alpha)$ and are given by

$$b_m(\tau, \alpha) = \frac{\psi_m(\alpha)}{\sqrt{J(\tau, \alpha)}} - \frac{1}{\sqrt{J(\tau, \alpha)}} \int_0^{\tau} \sqrt{J(\tau', \alpha)}\, \Delta b_{m-1}(\tau', d)\, d\tau$$

$$(m = 0, 1, 2, \ldots),$$

$$b_{-1} \equiv 0, \tag{2.4}$$

where $J(\tau, \alpha)$ is the Lamé coefficient corresponding to α. The function $\psi_m(\alpha)$ must be determined from the boundary condition.

Let us fix the position of the source. From geometric considerations we can easily find the changes in the coordinates of the observer, the segment l_0 and the angle of reflection θ corresponding to small changes in α and τ:

$$\frac{d\theta}{d\alpha}=1+\frac{l_0}{a(s)\cos\theta},\quad \frac{dl_0}{d\alpha}=l_0\,\mathrm{tg}\,\theta,$$
$$\frac{da(s)}{d\alpha}=a'(s)\frac{l_0}{\cos\theta},\quad \frac{dg(s)}{d\alpha}=g'(s)\frac{l_0}{\cos\theta}, \tag{2.5}$$

$$J(\tau,\alpha)=l_0+l+\frac{2l_0l}{a(s)\cos\theta}=A(\alpha)\tau+B(\alpha), \tag{2.6}$$

$$A(\alpha)=1+\frac{2l_0(\alpha)}{a[s(\alpha)]\cos\theta(\alpha)},\quad B(\alpha)=-\frac{2l_0^2(\alpha)}{a[s(\alpha)]\cos\theta(\alpha)}.$$

If we use the expression in orthogonal coordinates for the Laplace operator and take into account that the Lamé coefficient with respect of τ is unity, then the integrand in (2.4) is given by

$$\sqrt{J}\,\Delta b_{m-1}=\frac{1}{\sqrt{J}}\left[\frac{\partial}{\partial\alpha}\left(\frac{1}{J}\cdot\frac{\partial b_{m-1}}{\partial\alpha}\right)+\frac{\partial}{\partial\tau}\left(J\frac{\partial b_{m-1}}{\partial\tau}\right)\right]. \tag{2.7}$$

Since $J(\tau,\alpha)$ is a linear function of τ, the expression (2.7) is a rational function of τ. The integrals in (2.4) are therefore easily obtained, even though they are very complicated functions of α. We obtain

$$b_0(\tau,\alpha)=\frac{\psi_0(\alpha)}{\sqrt{J}},$$
$$b_1(\tau,\alpha)=\frac{\psi_1(\alpha)}{\sqrt{J}}+\frac{1}{8A^3J^{7/2}}[C(\alpha)J^2+D(\alpha)J+E(\alpha)], \tag{2.8}$$
$$\cdots\cdots\cdots\cdots\cdots,$$

where

$$C(\alpha)=A^2(4\psi_0''+A^2)-2A(\psi_0A''+4\psi_0'A')+5\psi_0A'^2,$$
$$D(\alpha)=A\psi_0(A'B-AB')'+(4A\psi_0'-5A'\psi_0)(A'B-AB'),$$
$$E(\alpha)=\frac{5}{3}\psi_0(A'B-AB')^2.$$

The derived functions $A(\alpha)$ and $B(\alpha)$ here are easily calculated with the help of (2.5).

Let us also find the functions $\psi_m(\alpha)$, $m=0,1,2,\cdots$. It is necessary to express differentiation along the normal in the boundary conditions in terms of differentiation with respect to τ (in the case G_i) and with respect to τ and α (in the case G_r). The calculations lead to the following expansions, which are valid on Γ:

$$\frac{\partial G_i}{\partial n}=-\cos\theta\frac{\partial G_i}{\partial r}-\frac{\sin^2\theta}{2r\cos\theta}G_i+\ldots,$$
$$\frac{\partial G_r}{\partial n}=\cos\theta\frac{\partial G_r}{\partial\tau}+\mathrm{tg}^2\,\theta\left(\frac{1}{a}+\frac{\cos\theta}{2r}\right)G_r-\frac{\sin\theta}{r}\frac{\partial G_r}{\partial\alpha}+\ldots. \tag{2.9}$$

As $k \to \infty$

$$\frac{\partial G_i}{\partial r} = ikG_i\,[1 + O(k^{-1})],$$

$$\frac{\partial G_r}{\partial \tau} = ikG_r\,[1 + O(k^{-1})], \quad \frac{\partial G_r}{\partial \alpha} = O(G_r).$$

Equations (2.9) therefore have the character of expansions in inverse powers of k. We substitute (2.2), (2.3) and (2.9) into the boundary condi tion (1.1) and take into account that τ, r and J on Γ become l_0. On successively equating to zero the coefficients of different powers of k, we obtain

$$\psi_0(\alpha) = \frac{K-1}{K+1}, \quad K \equiv \frac{\cos\theta}{g},$$

$$\psi_1(\alpha) = -\frac{Cl_0^2 + Dl_0 + E}{8A^3 l_0^3} + (\mathrm{tg}^2\,\theta - 1)\frac{Kl_0^{-1} + (1-K)(ag)^{-1}}{(1+K)^2}$$
$$+ \frac{\sin\theta}{l_0 g(1+K)}\psi_0' + \frac{\psi_0}{8l_0},$$

. . . .

As a result,

$$G_r = \frac{e^{ik\tau + \frac{i\pi}{4}}}{2\sqrt{2\pi kJ}}\Big\{\psi_0 + \frac{1}{ik}\Big[\frac{\psi_0}{8l_0} + \frac{Kl_0^{-1} + (1-K)(ag)^{-1}}{(1+K)^2} + \frac{\psi_0'\sin\theta}{gl_0(1+K)}$$
$$+ \frac{CJ^2 + DJ + E}{8A^3J^3} - \frac{Cl_0^2 + Dl_0 + E}{8A^3l_0^3}\Big] + O\Big(\frac{1}{k^2}\Big)\Big\}. \tag{2.10}$$

By a and g in (2.10) we understand the values of $a(s)$ and $g(s)$ at the point of reflection. The correction term of order k^{-1} on the right side of (2.10) depends not only on a and g but also on the values of a', a'', g' and g'' at the point of reflection.

§3. The asymptotic behavior of "eigenfunctions" in the boundary layer

Nontrivial solutions (as $k \to \infty$) of (1.1) with zero on the right side in place of the δ-function will be called eigenfunctions of the problem (1.1).

The derivation of asymptotic formulas for Green's function in the shadow and half-shadow zones is composed of three stages:

1) determination of the asymptotic behavior of eigenfunctions in the boundary layer, that is, close to the contour Γ (the thickness of the boundary layer tends to zero as $k \to \infty$);

2) continuation of the eigenfunctions to any distances from the boundary Γ;

3) construction of Green's functions as the superposition of eigenfunctions.

We now proceed to search for eigenfunctions in the surface layer. We can use orthogonal coordinates s, n (along and across rays moving along Γ and trailing from it along the tangent to the observer). The amplitude of the field varies rapidly with n and comparatively smoothly with s. Instead of n it is therefore convenient to consider the coordinate

$$\nu = nk^q, \quad 0 < q < 1. \tag{3.1}$$

In the boundary layer ν can then be regarded as finite. In the new variables the Helmholtz equation and boundary condition are:

$$\left(k^{2q} + \frac{k^q \nu}{a}\right)\frac{\partial^2 u}{\partial \nu^2} + \frac{k^q}{a}\cdot\frac{\partial u}{\partial \nu} + \frac{1}{1 + \dfrac{\nu}{k^q a}}\frac{\partial^2 u}{\partial s^2}$$

$$+ \frac{\nu a' k^q}{(ak^q + \nu)^2}\cdot\frac{\partial u}{\partial s} + k^2\left(1 + \frac{\nu}{k^q a}\right)u = 0, \tag{3.2}$$

$$u - \frac{i}{k^{1-q} g(s)}\cdot\frac{\partial u}{\partial \nu}\bigg|_{\nu=0} = 0. \tag{3.3}$$

Solutions of (3.2) are sought in the form[1)]

$$\begin{gathered} u = \exp\left[ik\tau(s) + k^p\varphi(s) + \Lambda(s, \nu, k)\right]\Phi\left[L(s, \nu, k)\right], \quad 0 \leqslant p < q, \\ \Lambda \sim \alpha(s, \nu) + \frac{\beta(s, \nu)}{k^p} + \frac{\gamma(s, \nu)}{k^{2p}} + \dots, \\ L \sim a(s, \nu) + \frac{b(s, \nu)}{k^p} + \frac{c(s, \nu)}{k^{2p}} + \dots . \end{gathered} \tag{3.4}$$

It is assumed here that the function Φ satisfies the linear equation of the second order

$$\Phi''(L) = Q(L)\,\Phi(L), \tag{3.5}$$

and that the functions $\alpha(s, \nu)$, $a(s, \nu)$, $\beta(s, \nu)$, $b(s, \nu)$, $\cdots$ are polynomials in ν with coefficients depending on s. These assumptions enable us to determine uniquely the unknown functions $\tau(s)$, $\phi(s)$, $\alpha(s, \nu)$, $a(s, \nu)$, $\beta(s, \nu)$, $b(s, \nu)$, $\cdots$ and the numbers p and q from the conditions (3.2), (3.3) and the condition as $\nu \to \infty$.

On substituting (3.4) into (3.2) and equating to zero the coefficients

---

1)Expansions of the form (3.4) appeared first in [8] and generalize formula (11) of [1].

of different powers of k (the multipliers of Φ and Φ' must be separately equated to zero), we obtain a system of differential equations for determining the unknown functions.

The first equation of this system (formed by terms containing $k^2\Phi$) is the eikonal equation which gives

$$\tau(s) = s.$$

Terms containing $k^{1+p}\Phi$, $k^{2-q}\Phi$, $k^{2q}\Phi$ also lay claim to seniority. A study of the possible variants shows that the solution of the system is nontrivial only if $1 + p = 2 - q = 2q$; that is, if $p = 1/3$ and $q = 2/3$. Then, isolating the terms containing $k^{4/3}\Phi$ and $k^{4/3}\Phi'$ in the Helmholtz equation, we obtain the system of equations

$$\begin{gathered} 2\frac{\partial \alpha}{\partial \nu}\cdot\frac{\partial a}{\partial \nu} + \frac{\partial^2 a}{\partial \nu^2} = 0, \\ \left(\frac{\partial \alpha}{\partial \nu}\right)^2 + \frac{\partial^2 \alpha}{\partial \nu^2} + Q\,[a(s,\ \nu)]\left(\frac{\partial a}{\partial \nu}\right)^2 + \frac{2\nu}{a(s)} + 2i\varphi' = 0. \end{gathered} \tag{3.6}$$

The left sides in (3.6) contain polynomials in ν with coefficients depending on s. The degrees of these polynomials are uniquely determined. It turns out that $Q(a)$ and $a(s,\nu)$ are polynomials of the first degree, and α does not depends on ν. The first equation in (3.6) is then satisfied identically, and the second splits into an equation for ν^1 and an equation for ν^0. Equating the free term in Q to zero, we obtain

$$\varphi(s) = \frac{i}{2^{1/2}}\int_0^s \frac{\zeta(s)}{a^{2/3}(s)}\,ds,$$

$$a(s,\ \nu) = -\eta(s)\,\nu + \zeta(s), \quad \eta(s) = \left(\frac{2}{a(s)}\right)^{1/3}.$$

The function Φ must be one of the Airy functions. The free term $\zeta(s)$ of the polynomial $a(s,\nu)$ is for the present arbitrary (it is determined from the boundary condition).

We now consider the system of two equations which arises when we single out the terms containing $k\Phi$ and $k\Phi'$ from the Helmholtz equation. This system contains new unknown functions $b(s,\nu)$ and $\beta(s,\nu)$. It turns out that $b(s,\nu)$ does not depend on ν and $\beta(s,\nu)$ is a polynomial in ν of the second degree. The system of equations gives explicit formulas for $\alpha(s,\nu)$, $b(s,\nu)$ and $\beta(s,\nu)$ in terms of another arbitrary function $\kappa(s)$, which is found later from the boundary condition. This process of determining the coefficients of the asymptotic series (3.4) can be continued. At each step we obtain a system of two equations of the type (3.6) containing a new unknown function of s, which is later uniqly determined from the boundary condition (3.3). As a reeult, we obtain

$$\alpha(s,\nu)=\frac{1}{2^{1/3}}\int_0^s\frac{\varkappa(s)}{a^{2/3}(s)}\,ds-\frac{1}{6}\ln a(s),$$

$$b(s,\nu)=-i\varkappa(s),$$

$$\beta(s,\nu)=i\left[\frac{\nu^2}{6}\cdot\frac{a'(s)}{a(s)}+\nu\frac{\zeta'(s)}{\eta(s)}+\chi(s)\right],$$

$$c(s,\nu)=\nu^2\mu(s)+\nu\psi(s)+\xi(s),$$

$$\gamma(s,\nu)=\nu\rho(s)+\vartheta(s),$$

$$\cdots\cdots\cdots$$

Explicit formulas, which are not given here on account of their complexity, are obtained for the functions $\chi(s)$, $\mu(s)$, $\psi(s)$, $\vartheta(s)$ and $\rho(s)$. The functions $\zeta(s)$, $\kappa(s)$ and $\xi(s)$ are for the present arbitrary.

The radiation condition (limiting absorption condition) is satisfied, since among the possible solutions of (3.5), which in this problem is the Airy equation $w''(L)=Lw(L)$, a function $w_1(L)$ having zeros for $\arg L=\pi/3$ is chosen.

Finally, substituting (3.4) into the boundary condition (3.3) gives

$$w_1[\zeta(s)]=0,$$

$$\varkappa(s)=\frac{\eta(s)}{g(s)},$$

$$\xi(s)=0,$$

$$\cdots\cdots$$

The function $\zeta(s)$ is therefore constant and equal to one of the complex numbers $\zeta_1, \zeta_2, \zeta_3, \cdots$ which are the zeros of the Airy function $w_1(\zeta)$. As a result, we obtain the following denumerable set of eigenfunctions of problem (1.1) in the boundary layer (for finite ν):

$$u_j=\frac{C_j}{a^{1/6}(s)}\exp\left[ik(s-s_0)+i\zeta_j m\int_{s_0}^{s}\frac{ds}{a^{2/3}(s)}+\int_{s_0}^{s}\frac{ds}{a(s)\,g(s)}+\frac{ia'(s)\,\nu^2}{6\cdot 2^{1/3}ma(s)}\right.$$
$$\left.+\frac{i\zeta_j^2\lambda(s)}{m}+\frac{c(s,\nu)}{2^{2/3}m^2}+O\left(\frac{1}{k}\right)\right]w_1[L(\zeta_j,M)],\quad j=1,2,3,\ldots. \tag{3.7}$$

Here

$$m=\left(\frac{k}{2}\right)^{1/3}, \quad \lambda(s)=\int_{s_0}^{s} a^{-4/3}(s)\left[\frac{1}{60}+\frac{4}{135}a'^2(s)-\frac{2}{45}a(s)a''(s)\right]ds,$$
$$\nu=nk^{2/3}, \quad L(\zeta, M)=\zeta-\nu\eta(s)-\frac{i}{ma^{1/3}(s)g(s)}+\ldots, \quad \eta=\left(\frac{2}{a(s)}\right)^{1/3}, \tag{3.8}$$

where C_j are arbitrary constants independent of the coordinates but possibly dependent on k, and s_0 is an arbitrary point on Γ.

The factor $a^{-1/6}(s)$ in (3.7) is obtained by the parabolic equation method in [9] where the case of the boundary condition $u|_\Gamma = 0$ is considered. In the case of an impedance boundary condition we see that we have an additional factor

$$\exp\left(\int_{s_0}^{s}\frac{ds}{a(s)g(s)}\right)$$

of the same order as $k \to \infty$.

§4. Continuation of eigenfunctions outside the boundary layer

We shall assume that formula (3.7) is valid not only for finite ν but also in the wider region

$$\nu \leqslant O(k^{\varepsilon}), \quad 0 \leqslant \varepsilon < \frac{1}{6}. \tag{4.1}$$

On the basis of this assumption, which is justified in the particular cases of problem (1.1) already investigated, we can obtain one more asymptotic formula for eigenfunctions valid on the boundary of the layer (4.1) and find the continuation of these eigenfunctions outside the boundary layer.

Let $\nu \sim k^{\epsilon}$ and $n \sim k^{2/3+\epsilon}$. Then to the Airy function in (3.7) we can apply the asymptotic formula

$$w_1(t)=it^{-1/4}\exp\left(-\frac{2}{3}t^{3/2}\right)\left[1-\frac{5}{48t^{3/2}}+O\left(\frac{1}{t^3}\right)\right] \tag{4.2}$$
$$\left(\frac{\pi}{3}+\delta\leqslant \arg t\leqslant\frac{7\pi}{3}-\delta, \quad \delta>0\right)$$

or

$$w_1[L(\zeta_j,\ M)] \sim \frac{e^{-i\frac{\pi}{4}}a^{1/4}(s)}{2^{1/12}\nu^{1/4}} \exp\left(\frac{2i}{3}\nu^{3/2}\eta^{3/2} - i\zeta_j\sqrt{\nu\eta}\right)$$

$$\times\left[1 + \frac{i}{mga^{1/3}}\left(i\sqrt{\nu\eta} - \frac{1}{4\nu\eta}\right) - \frac{5i}{48(\nu\eta)^{3/2}} + \frac{\zeta_j}{4\nu\eta} + \frac{i\zeta_j^2}{4\sqrt{\nu\eta}} + \dots\right]. \quad (4.3)$$

Also, from the observation point $M(s, n)$ we draw the tangent and normal to Γ (Figure 2). Let $s = s_1$ be the coordinate of the point of contact and r_1 the length of the tangent. The position of M can also be described by the ray coordinates $\tau = (s_1 - s_0) + r_1$ and s_1 which are related to the coordinates s, n of the boundary layer in the following way:

$$\begin{gathered}\tau = (s - s_0) + \frac{(s - s_1)^3}{3a^2(s_1)} - \frac{7}{24}\cdot\frac{a'(s_1)(s - s_1)^4}{a^3(s_1)} \\ + \frac{16 + 29a'^2(s_1) - 11a(s_1)a''(s_1)}{120a^4(s_1)}(s - s_1)^5 + \dots, \\ n = \frac{(s - s_1)^2}{2a(s_1)} - \frac{a'(s_1)(s - s_1)^3}{6a^2(s_1)} \\ + \frac{5 + 2a'^2(s_1) - a(s_1)a''(s_1)}{24a^3(s_1)}(s - s_1)^4 + \dots.\end{gathered} \quad (4.4)$$

Substituting (4.3) and (4.4) into (3.7) leads to very complicated double series, which, however, can be considerable simplified so that only one small parameter m^{-1} remains. We finally obtain

$$u_j(M) = -\frac{C_j e^{-i\frac{\pi}{4}}a^{1/6}(s_1)}{\sqrt{m}}\cdot\frac{q_j(s_1)e^{ik\tau}}{\sqrt{r_1}}\left[1 + O\left(\frac{1}{k}\right)\right], \quad (4.5)$$

$$j = 1,\ 2,\ \dots,$$

where

$$q_j(s) = \exp\left[i\zeta_j m\int_{s_0}^{s}\frac{ds}{a^{2/3}(s)} + \int_{s_0}^{s}\frac{ds}{a(s)g(s)} + \frac{i\zeta_j^2\lambda(s)}{m}\right]. \quad (4.6)$$

The tangent length r_1 has the meaning of a Lamé coefficient for the ray coordinate s_1. Therefore (4.5) and (4.6) determine the leading term of the ray series for the function $u_j(M)$. Since Γ is convex, (4.5) is valid provided

$$kn \gg 1 \quad (4.7)$$

(which prohibits approach to the unique caustic of Γ); that is, for arbitrarily large distances from Γ. Formula (4.5) realizes the continuation of eigenfunctions outside the boundary layer. In the annular region defined by (4.1) and (4.7) formulas (3.7) and (4.5) hold simultaneously. The fact that (4.5) and (4.6) are formulas of the ray method not containing the parameter ϵ from (4.1) demonstrates the advantage of the assumption made at the beginning of this section.

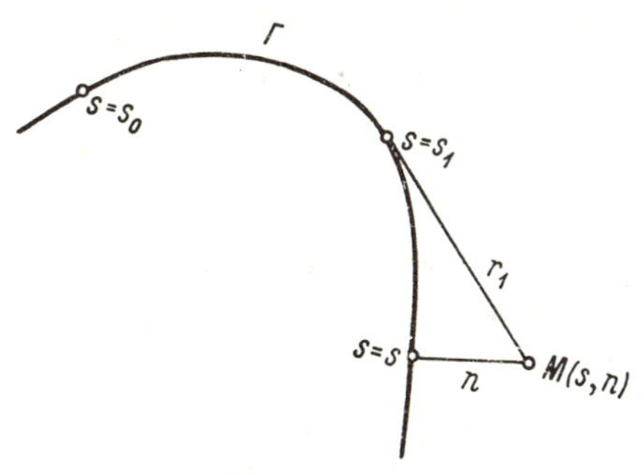

Figure 2

§5. Construction of the Green's function

In §4 it was shown that the eigenfunctions (3.7) become ray expressions on the boundary of the surface boundary layer, and that we can thereby obtain the continuation of the eigenfunctions to any distance from Γ. Similar arguments for the boundary of another boundary layer (surrounding the limit ray) enable us to construct Green's function for the problem (1.1) and trace the coupling of the wave field in the lit and shadow zones.

The Green's function $G(M_0, M, k)$ for problem (1.1) will be sought in the form of a superposition of functions u_j:

$$G(M_0, M, k) = \sum_{j=1}^{\infty} u_j. \tag{5.1}$$

It is now necessary to determine the coefficients $C_j(k)$ in u_j. Let the points $M(s, \nu)$ and $M_0(s_0, \nu_0)$ belong to the boundary layer (4.1). We assume that the expressions u_j are residues of a regular function of ζ with simple poles at $\zeta = \zeta_1, \zeta_2, \cdots$. We replace (5.1) by the integral

$$G(M_0, M, k) = [a(s)a(s_0)]^{-1/6} \exp\left[ik(s - s_0) + \int_{s_0}^{s} \frac{ds}{a(s)g(s)} + \frac{i}{6 \cdot 2^{1/3} m} \right.$$
$$\left. \times \left(\frac{a'(s)\nu^2}{a(s)} + \frac{a'(s_0)\nu_0^2}{a(s_0)}\right)\right] \int_C F(\zeta, k) w_1[L(\zeta, M_>)] \Big\{ w_2\big[L(\zeta, M_<)\big]$$
$$- \frac{w_2(\zeta)}{w_1(\zeta)} w_1[L(\zeta, M_<)]\Big\} \exp\left[im\zeta \int_{s_0}^{s} \frac{ds}{a^{2/3}(s)} + \frac{i\zeta^2\lambda(s)}{m}\right] d\zeta \tag{5.2}$$

along the contour C enclosing the zeros ζ_j, $j = 1, 2, 3, \cdots$. Here $F(\zeta, k)$ is a regular function on ζ inside C and related to C_j by

$$C_j(k) = -\frac{4\pi w_1 [L(\zeta_j, M_0)]}{[w_1'(\zeta_j)]^2 a^{1/6}(s_0)} F(\zeta_j, k) \exp\left(\frac{ia'(s_0)\nu_0^2}{6 \cdot 2^{1/3} ma(s_0)}\right); \tag{5.3}$$

the subscripts $>$ and $<$ denote the point (M_0 or M) at which $\nu\eta(s)$ has the greater (lesser) value; and the function $w_2(\zeta)$ is another solution of the Airy equation and is the complex conjugate of $w_1(\zeta)$ for real ζ. The rest of the notation is explained in (3.8).

Formula (5.2) satisfies all the conditions of (1.1) as $k \to \infty$ except that the inhomogeneous Helmholtz equation is not satisfied close to the source. In order to take account of the effect of the source we choose the function $F(\zeta, k)$ from the condition matching (5.2)[1)] with (2.1), (2.2) and (2.10) in the lit region close to the limit ray.

We transform (5.2) in the region

$$n \sim k^{-\frac{2}{3}+\varepsilon}, \quad s \sim k^{-\frac{1}{3}+\frac{\varepsilon}{2}} \quad \left(0 < \varepsilon \leqslant \frac{1}{6}\right), \tag{5.4}$$

so that the arc length s is measured from the point of reflection. These values of n and s correspond to angles of reflection θ close to the limit

$$\frac{\pi}{2} - \theta \sim k^{-\frac{1}{3}+\frac{\varepsilon}{2}}. \tag{5.5}$$

With condition (5.4) the Airy function can be replaced by the asymptotic expansions (4.2), (4.3). Similar substitutions apply in the case of second Airy functions. It is then convenient in (5.2) to single out explicitly a new large parameter $p = k^{\epsilon}$ and introduce new coordinates

$$N = n\frac{m^2}{p} \cdot \frac{2}{a^{1/3}}, \quad S = s\frac{m}{\sqrt{p}} \cdot \frac{1}{a^{2/3}} \quad (a \equiv a(0)), \tag{5.6}$$

which are finite in the region (5.4), and a new variable of integration $t = p^{-1}\zeta$. After these transformations the integrand in (5.2) takes the form

$$e^{p^{3/2} f_i(t)} \Psi_i(t, m, p) + e^{p^{3/2} f_r(t)} \Psi_r(t, m, p), \tag{5.7}$$

1) If Γ is finite this matching requires that we add to (5.2) sliding waves running several times round Γ and also waves round Γ in the opposite direction. These waves are described by formulas analogous to (5.2). However, in comparison with the fundamental sliding wave they are exponentially small as $k \to \infty$ and can be neglected in the future.

where

$$
\begin{aligned}
f_i(t) &= -\frac{2}{3}(t-N)^{3/2} + \frac{2}{3}(t-N_0)^{3/2} + it(S-S_0),\\
f_r(t) &= -\frac{2}{3}(t-N)^{3/2} - \frac{2}{3}(t-N_0)^{3/2} + \frac{4}{3}t^{3/2} + it(S-S_0),\\
\Psi_i(t, m, p) &= \frac{\sqrt{p}\,F_0(pt)}{\sqrt[4]{(N-t)(N_0-t)}} \exp\left\{-\frac{p^2}{m}\cdot\frac{ia'}{3a^{1/3}}\right.\\
&\quad\times\left.[NS\sqrt{N-t} - N_0S_0\sqrt{N_0-t} + t(S-S_0)^2]\right\}[1+\ldots],\\
\Psi_r(t, m, p) &= -\frac{\sqrt{p}\,F_0(pt)}{\sqrt[4]{(N-t)(N_0-t)}} \exp\left\{-\frac{p^2}{m}\cdot\frac{ia'}{3a^{1/3}}\right.\\
&\quad\times\left.[NS\sqrt{N-t} + N_0S_0\sqrt{N_0-t} + t(S-S_0)^2]\right\}[1+\ldots].
\end{aligned}
\tag{5.8}
$$

For brevity the subsequent terms in the series in the square brackets are not given. N_0 and S_0 are the values of the coordinates N and S at M_0. For definiteness the formulas are give in the case $N > N_0$.

The phase functions $f_i(t)$ and $f_r(t)$ have saddle points at $t = t_i$ and $t = t_r$ respectively, where

$$
t_i = \frac{N+N_0}{2} - \frac{(S-S_0)^4 + (N-N_0)^2}{4(S-S_0)^2},
$$

$$
t_r = -\left(\frac{N-S^2}{2S}\right)^2 = -\left(\frac{N_0-S_0^2}{2S_0}\right)^2.
$$

Equating the two expressions for t_r gives the law of reflection.

We retain for both terms (5.7) the common ends of the contour of integration at infinity and for a finite distance superpose the path of integration for each term onto the corresponding stationary contours. Since $f_k''(t_k) \neq 0$ and $\Psi_k(t_k) \neq 0$ for $k = i, r$ and the integrands are regular in the neighborhoods of the saddle points (which is true when (5.5) holds), the asymptotic calculation (5.2) reduces to a standard variant of the saddle-point method. Therefore

$$
\int_{L_0}\left[\Psi_0(t) + \frac{\Psi_1(t)}{q} + \ldots\right]e^{qf(t)}dt \sim e^{qf(t_0)+i\varphi}\Psi_0(t_0)\sqrt{\frac{2\pi}{q\,|f''(t_0)|}}\sum_{n=0}^{\infty}\frac{c_n}{q^n},
$$

where

$$
c_0 = 1,
$$

$$
c_1 = \frac{f^{(4)}(t_0)}{8f''^2(t_0)} + \frac{f'''(t_0)}{2f''^2(t_0)}\cdot\frac{\Psi_0'(t_0)}{\Psi_0(t_0} - \frac{\Psi_0''(t_0)}{2\Psi_0(t_0)\,f''(t_0)} - \frac{5f'''^2(t_0)}{24f''^3(t_0)} + \frac{\Psi_1(t_0)}{\Psi_0(t_0)},
$$

$$
\ldots\ldots\ldots\ldots\ldots\ldots\ldots\ldots\ldots\ldots\ldots;
$$

q is a large parameter; L_0 is part of the stationary contour belonging to a neighborhood of the saddle point t_0; and $\phi = \arg(t - t_0)$ on the path of integration from the saddle point.

After the asymptotic calculation of the integral (5.2) by the saddle-point method it is convenient to transfer to polar coordinates r, θ for the first term and to l_0, l, θ for the second term (Figure 1). This can be done with the help of the expansions

$$r=(s-s_0)+\frac{s-s_0}{2}\left(\frac{1}{a}+\frac{\cos\theta}{s-s_0}\right)^2(s^2+s_0^2)-\frac{(s-s_0)^3}{3a^2}+\dots,$$

$$l=s+\frac{1}{2s}\left(n-\frac{s^2}{2a}\right)^2+\frac{s}{a}\left(n-\frac{s^2}{2a}\right)+\frac{s^3}{3a^2}+\dots,$$

$$l_0=|s_0|+\frac{1}{2|s_0|}\left(n_0-\frac{s_0^2}{2a}\right)^2+\frac{|s_0|}{a}\left(n_0-\frac{s_0^2}{2a}\right)+\frac{|s_0|^3}{3a^2}+\dots,$$

$$\cos\theta=\frac{n-\frac{s^2}{2a}}{s}+\frac{a's^2}{6a^2}+\dots=\frac{n_0-\frac{s_0^2}{2a}}{|s_0|}+\frac{a's_0^2}{6a^2}+\dots,$$

relating the polar coordinates with coordinates of the boundary layer.

As a result of the above transformations of (5.2) we obtain Green's function in the lit region in the form (2.1). Each of the terms in this sum has the form of the ray series (2.2), (2.10), and does not depend on the parameter ϵ from (5.4). The expansions obtained will coincide completely with (2.2) and (2.10) if we put

$$F(\zeta,\ k)=\frac{i}{8\pi m}. \tag{5.9}$$

§6. Concluding formulas

Substituting (5.9) into (5.3) gives

$$C_j(k)=\frac{w_1[L(\zeta_j,\ M_0)]}{2ia^{1/6}(s_0)\,[w_1'(\zeta_j)]^2}\exp\left(\frac{ia'(s_0)\,\nu_0^2}{6\cdot 2^{1/3}ma(s_0)}\right) \tag{6.0}$$

for the coefficients of excitation of the various eigenfunctions by the point source. We can now write down the asymptotic formulas for Green's function for different positions of the source and observer in the shadow zone to complement formulas (2.2) and (2.10) in the lit zone.

If the points $M_0(s_0, \nu_0)$ and $M(s, \nu)$ belong to the boundary layer (ν_0

and ν are subject to (4.1)), we have

$$G(M_0, M, k)=\frac{i}{8\pi}\cdot\frac{\exp\left(ik(s-s_0)+\int\limits_{s_0}^{s}\frac{ds}{a(s)g(s)}\right)}{m\sqrt[6]{a(s)a(s_0)}}\int\limits_C w_1[L(\zeta, M_>)]$$

$$\times\left\{w_2[L(\zeta, M_<)]-\frac{w_2(\zeta)}{w_1(\zeta)}w_1[L(\zeta, M_<)]\right\}\exp\left(im\zeta\int\limits_{s_0}^{s}\frac{ds}{a^{2/3}(s)}+\frac{i\zeta^2\lambda(s)}{m}\right)d\zeta \tag{6.1}$$

or

$$G(M_0, M, k)=\frac{1}{2i}\cdot\frac{\exp\left(ik(s-s_0)+\int\limits_{s_0}^{s}\frac{ds}{a(s)g(s)}\right)}{m\sqrt[6]{a(s)a(s_0)}}$$

$$\times\sum_{j=1}^{\infty}\frac{\exp\left(im\zeta_j\int\limits_{s_0}^{s}\frac{ds}{a^{2/3}(s)}+\frac{i\zeta_j^2\lambda(s)}{m}\right)}{[w_1'(\zeta_j)]^2}w_1[L(\zeta_j, M)]w_1[L(\zeta_j, M_0)]. \tag{6.2}$$

Formulas (6.1) and (6.2) satisfy the principle of duality. Formula (6.1) is also valid at least in that part of the lit region which adjoins the limit ray. Formula (6.2) is most convenient for calculating the field in deep shadow, where the sum over j can be replaced by the first term (for $j = 1$).

If the superposition (5.1) is composed of functions (4.5), then, since the coefficients C_j are known, we obtain formulas for the Green's function in the case of points M_0 and M remote from the contour Γ. In this case the position of M_0 and M is described by the lengths of the tangents r_0 and r_1 drawn from these points to Γ, and by the points of contact s_0 and s_1 (Figure 3). In this notation we obtain

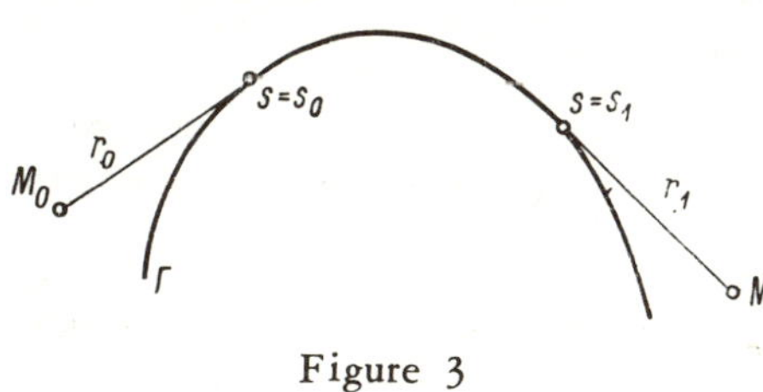

Figure 3

$$G(M_0, M, k) = \frac{\sqrt[6]{a(s_0)\,a(s_1)}}{2m^2\sqrt{r_0 r_1}} \exp\left[ik(r_0 + s_1 - s_0 + r_1) + \int_{s_0}^{s_1} \frac{ds}{a(s)\,g(s)}\right]$$

$$\times \sum_{j=1}^{\infty} [w_1'(\zeta_j)]^{-2} \exp\left(im\zeta_j \int_{s_0}^{s_1} \frac{ds}{a^{2/3}(s)} + \frac{i\zeta_j^2 \lambda(s)}{m}\right). \tag{6.3}$$

Formulas (6.2) and (6.3) clearly show the effect of different parts of the ray on the amplitude and phase of the wave field. A formula can similarly be found when one of the points M_0 and M lies the boundary layer and the other is outside it.

BIBLIOGRAPHY

[1] I. A. Molotkov, *Diffraction on a convex contour with smoothly varying radius of curvature and impedance*, Problems of Mathematical Physics, vol. 2, Izdat. Leningrad. Univ., Leningrad, 1967, pp. 124–132. (Russian)

[2] G. D. Maljužinec and L. A. Vaĭnšteĭn, *Transverse diffusion during diffraction on an impedance cylinder of large radius*, Radiotehn. i Èlektron. 6 (1961), no. 8, 1247–1258; no. 9, 1489–1495 = Radio Engrg. Electron. Phys.

[3] V. A. Fok and L. A. Vaĭnšteĭn, *Transverse diffusion during diffraction on a convex cylinder with smoothly varying curvature*, Radiotehn. i Èlektron. 8 (1963), no. 3, 363–388 = Radio Engrg. Electron. Phys.

[4] V. M. Babič, *On localizability considerations in short-wave diffraction problems*, Proc. Third All-Union Sympos. Diffraction of Waves, "Nauka", Moscow, 1964, pp. 78–89. (Russian)

[5] I. A. Molotkov, *Diffraction on a smooth convex cylinder with two propagation velocities*, Vestnik Leningrad. Univ. 22 (1967), no. 10, 69–79. (Russian)

[6] I. V. Muhina and I. A. Molotkov, *On the propagation of Rayleigh waves in an elastic half-space which is inhomogeneous with respect to two coordinates*, Izv. Akad. Nauk SSSR Ser. Fiz. Zemli 1967, no. 4, 3–8 = Izv. Akad. Sci. USSR Phys. Solid Earth

[7] V. M. Babič and A. S. Alekseev, *On the ray method of calculation of the intensity of wave fronts*, Izv. Akad. Nauk SSSR Ser. Geofiz. 1958, 17–31. (Russian) MR 20 #516.

[8] V. S. Buldyrev, *Wave propagation near the curved surface of an inhomogeneous body*, Problems of Mathematical Physics, no. 2, Spectral Theory, Diffraction Problems, Izdat. Leningrad. Univ., Leningrad, 1967, pp. 61–84. (Russian) MR 38 #6798.

[9] V. S. Buslaev, *Formulas for the short-wave asymptotic behavior in the diffraction problem on convex bodies*, Vestnik Leningrad. Univ. 17 (1962), no. 13, 5–21. (Russian) MR 25 #5693.

Trudy Mat. Inst. Steklov.
95 (1968)

Proc. Steklov Inst. Math.
95 (1968)

ON THE PROPAGATION OF ELASTIC WAVES FROM AN IRREGULARLY MOVING SOURCE

UDC 550.341

L. A. MOLOTKOV

Abstract. An investigation is made of propagation processes in an elastic half-space under the action of a source applied to an irregularly expanding circle. The case is considered in which the velocity of the source is retarded and the terminal velocity is less than the propagation velocities. The displacement field is found by solving Lamé's equation with zero initial data and boundary conditions defining the source. The solutions obtained are investigated in the front zones by asymptotic methods. The principal part of the displacement field is given by the sum of expressions representing groups of primary and secondary waves, each group consisting of longitudinal, transverse, head, Rayleigh and other waves. The study of individual waves determines the nature of a singularity on the front, and the damping of this wave along the rays is investigated. Special attention is given to the formation of waves at times when the velocity of the source coincides with the propagation velocity. It is shown how the character of the wave singularities is intensified if some first derivatives of the velocity vanish at the time of formation.

Bibliography 4 items. 9 figures.

This paper is devoted to a study of the processes of wave propagation in an elastic half-space under the action of a source applied to a circle which is expanding according to an arbitrary law. The displacement field which is formed is represented by three iterated integrals which are calculated close to the fronts by asymptotic methods [1, 2]. In the expressions obtained the principal part of the displacement field is divided into groups of primary and secondary waves, each group consisting of longitudinal, transverse, head, Rayleigh and other waves.

In contrast to [3], where the source velocity is taken to be strictly greater or strictly less than the propagation velocity, we impose no limitations on the source velocity. Special attention is given to the formation of waves at times when the source velocity coincides with the propagation velocity. It is shown how the character of a singularity of the waves changes if at the time of formation a number of first derivatives of the velocity vanish.

§1. Statement of the problem and solution

Suppose that in the cylindrical coordinate system r, θ, z a homogeneous elastic half-space $z \geq 0$ is given with density ρ and Lamé constants λ and μ.

The medium is in a state of rest for $t < 0$ and at $t = 0$ an axially symmetric source applied to the circle $r = s(t)$, $z = 0$ begins to act. The source is given by

$$t_{zz}|_{z=0} = \frac{f(t)\,\delta[r - s(t)]}{r}, \quad t_{zr}|_{z=0} = t_{z\theta}|_{z=0} = 0, \tag{1.1}$$

where t_{zz}, t_{zr} and $t_{z\theta}$ are components of the stress tensor and $\delta(x)$ is Dirac's function. The functions $f(t)$ and $s(t)$ define the intensity and motion of the source for positive t, have continuous derivatives and satisfy the conditions: $s(0) = 0$, $s'(t) > 0$, $s''(t) \leq 0$ and $|f(\infty)| < \infty$.

The displacement field $\mathbf{u} = q\mathbf{r}_1 + w\mathbf{k}_1$ which is formed satisfies the Lamé equations

$$\rho\frac{\partial^2\mathbf{u}}{\partial t^2} = (\lambda + 2\mu)\operatorname{grad}\operatorname{div}\mathbf{u} - \mu\operatorname{rot}\operatorname{rot}\mathbf{u} \tag{1.2}$$

with initial conditions

$$\mathbf{u}|_{t=0} = \dot{\mathbf{u}}|_{t=0} = 0 \tag{1.3}$$

and boundary conditions

$$\left(\frac{\partial q}{\partial z} + \frac{\partial w}{\partial r}\right)\Big|_{z=0} = 0,$$

$$\left[\lambda\left(\frac{\partial q}{\partial r} + \frac{q}{r}\right) + (\lambda + 2\mu)\frac{\partial w}{\partial z}\right]\Big|_{z=0} = -\frac{\delta[r - s(t)]}{r}f(t). \tag{1.4}$$

The problem (1.2)–(1.4) can be solved as in [1, 4] by using the Fourier-Bessel and Laplace integral transforms. As a result, we obtain the final expressions

$$w = \frac{1}{b\mu}\int_0^\infty \frac{kJ_0(kr)\,dk}{2\pi i}\int_{\sigma - i\infty}^{\sigma + i\infty} W(k, \eta)\,F(k, \eta)\,e^{ktb^{-1}\eta}d\eta,$$

$$q = \frac{1}{b\mu}\int_0^\infty \frac{kJ_1(kr)\,dk}{2\pi i}\int_{\sigma - i\infty}^{\sigma + i\infty} Q(k, \eta)\,F(k, \eta)\,e^{ktb^{-1}\eta}d\eta, \tag{1.5}$$

$$W_P = \frac{\alpha g}{R}, \quad W_S = -\frac{2\alpha}{R}, \quad Q_P = \frac{g}{R}, \quad Q_S = -\frac{2\alpha\beta}{R},$$

$$W = W_Pe^{-kz\alpha} + W_Se^{-kz\beta}, \quad Q = Q_Pe^{-kz\alpha} + Q_Se^{-kz\beta}, \tag{1.6}$$

$$F(k, \eta) = \int_0^\infty f(\tau)\,J_0[ks(\tau)]\exp(-ik\tau b^{-1}\eta)\,d\tau, \tag{1.7}$$

$$\left.\begin{aligned} &\alpha = \sqrt{1 + \gamma^2\eta^2}, \quad \beta = \sqrt{1 + \eta^2}, \quad g = 2 + \eta^2, \\ &D = g^2 - 4\alpha\beta, \quad a = \sqrt{\rho(\lambda + 2\mu)^{-1}}, \quad b = \sqrt{\rho\mu^{-1}}, \quad \gamma = ab^{-1}, \end{aligned}\right\} \tag{1.8}$$

determining the displacement vector $\mathbf{u}$. For the radicals α and β to be unique we draw branch cuts from the points $\pm i\gamma^{-1}$, and $\pm i$ into the left half-plane and specify the fundamental branches of the radicals by the conditions $\alpha > 0$ and $\beta > 0$ for $\eta > 0$.

We investigate the solution obtained by substituting (1.5) into (1.8) and applying the convolution theorem. Calculating one of the inner integrals, we obtain

$$
\begin{aligned}
w &= \frac{1}{b\mu}\int_0^\infty \frac{kJ_0(kr)\,dk}{2\pi i}\int_0^t f(\tau)\,J_0[ks(\tau)]\,d\tau\int_{\sigma-i\infty}^{\sigma+i\infty} We^{k(t-\tau)b^{-1}\eta}d\eta,\\
q &= \frac{1}{b\mu}\int_0^\infty \frac{kJ_1(kr)\,dk}{2\pi i}\int_0^t f(\tau)\,J_0[ks(\tau)]\,d\tau\int_{\sigma-i\infty}^{\sigma+i\infty} Qe^{k(t-\tau)b^{-1}\eta}d\eta.
\end{aligned}
\tag{1.9}
$$

We also put the displacement field (1.9) in the form

$$\mathbf{u} = \mathbf{u}_P + \mathbf{u}_S, \tag{1.10}$$

where the components of the vector $\mathbf{u}_l$ $(l = P, S)$ are given by

$$
\begin{aligned}
w_l &= \frac{1}{b\mu}\int_0^\infty \frac{kJ_0(kr)\,dk}{2\pi i}\int_0^{t_l} f(\tau)\,J_0[ks(\tau)]\,d\tau\int_{\sigma-i\infty}^{\sigma+i\infty} W_l e^{kf_l(\eta)}d\eta,\\
q_l &= \frac{1}{\beta\mu}\int_0^\infty \frac{kJ_1(kr)\,dk}{2\pi i}\int_0^{t_l} f(\tau)\,J_0[ks(\tau)]\,d\tau\int_{\sigma-i\infty}^{\sigma+i\infty} Q_l e^{kf_l(\eta)}d\eta,
\end{aligned}
\tag{1.11}
$$

$$f_P(\eta) = (t-\tau)\,b^{-1}\eta - z\alpha, \quad f_S(\eta) = (t-\tau)\,b^{-1}\eta - z\beta, \tag{1.12}$$

$$t_P = t - az, \quad t_S = t - bz. \tag{1.13}$$

§2. Calculation of the integrals with respect to η

In investigating the displacement field (1.10)–(1.13) we determine the location and displacement of the fronts, and study the singularities on the fronts and the variation of wave intensities with time. Special attention is given to the formation of waves at times when the source velocity $s'(t)$ passes through a value of the propagation velocity. We use asymptotic methods and calculate the inner integrals in (1.11) by the methods of the saddle point, steepest descent and stationary phase.

In studying the singularities of the displacement field we divide the path of integration into two intervals with respect to k: $[0, k_0]$ and $[k_0, \infty]$. Integrals over $[0, k_0]$ are, as in [4], regular functions, and integration over the finite interval $[0, k_0]$ is therefore of no interest. Since all the singularities of the displacement field are contained in the integrals over $[k_0, \infty)$, we concentrate our attention on this interval. It is helpful to choose k_0 so that the Bessel functions in the solution can be replaced by their asymptotic expressions and the inner integrals calculated by asymptotic methods.

We first calculate the Mellin integrals which we put in the single form

$$I_l = \frac{1}{2\pi i} \int_{\sigma - i\infty}^{\sigma + i\infty} X_l(\eta)\, e^{kf_l(\eta)} d\eta. \tag{2.1}$$

We draw branch cuts from the points $\pm i$, $\pm i\gamma^{-1}$ into the left half-plane along the lines

$$\operatorname{Im} f_l(\eta) = \operatorname{Im} f_l(\pm i), \quad \operatorname{Im} f_l(\eta) = \operatorname{Im} f_l(\pm i\gamma^{-1}) \tag{2.2}$$

and trace out the stationary contours λ_l of the functions $f_l(\eta)$ which will pass through the saddle points

$$\pm\eta_{P0} = \frac{\pm i(t-\tau)}{\gamma\sqrt{(t-\tau)^2 - a^2z^2}}, \quad \pm\eta_{S0} = \frac{\pm i(t-\tau)}{\sqrt{(t-\tau)^2 - b^2z^2}} \tag{2.3}$$

and satisfy the conditions

$$\operatorname{Im} f_l(\eta) = \operatorname{Im} f_l(\pm\eta_{l0}), \quad \operatorname{Re} f_l(\eta) \leqslant \operatorname{Re} f_l(\pm\eta_{l0}). \tag{2.4}$$

The position of the stationary contours and branch cuts is shown in Figures 1 and 2.

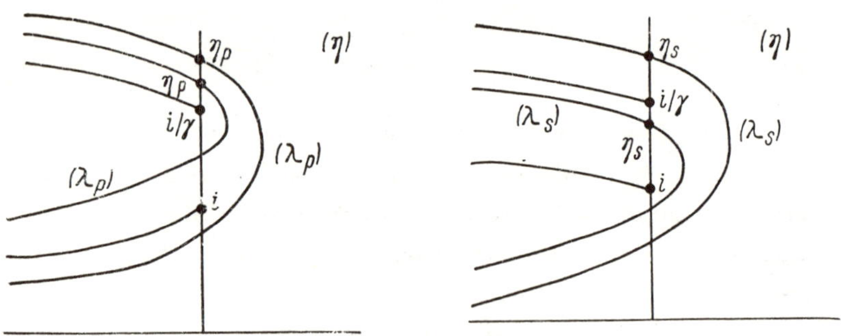

Figure 1 Figure 2

In going in formula (2.1) from the Mellin contour to the stationary contour (2.4) it is necessary to take into account the singularities of the integrands in the region lying between the contours. In this region we can have branch points $\pm i$ $(l = P)$ and $\pm i\gamma^{-1}$ $(l = S)$ and poles $\pm i\tau_R$ satisfying the equation $R = 0$. Investigations show that the poles $\pm i\tau_R$ and branch points $\pm i$ and $\pm i\gamma^{-1}$ must be taken into account if the following conditions are satisfied:

$$\left.\begin{array}{ll} az<(t-\tau)\sqrt{1-\gamma^2\tau_R^2} & (\eta=\pm i\tau_R,\ l=P), \\ bz<(t-\tau)\sqrt{1-\tau_R^2} & (\eta=\pm i\tau_R,\ l=S), \end{array}\right\} \tag{2.5}$$

$$\left.\begin{array}{ll} az<(t-\tau)\sqrt{1-\gamma^2} & (\eta=\pm i,\ l=P), \\ bz>(t-\tau)\sqrt{1-\gamma^2} & (\eta=\pm i\gamma^{-1},\ l=S). \end{array}\right\} \tag{2.6}$$

The Mellin integral (2.1) is therefore given by

$$I_l=I_{l0}+I_{lR}+\begin{cases} I_{P1}, & l=P, \\ I_{S\gamma}, & l=S, \end{cases} \tag{2.7}$$

where I_{l0} is the integral along the stationary contour. In the region (2.5) I_{lR} is equal to the sum of the residues at the poles $\pm i\tau_R$, and if conditions (2.6) are satisfied I_{P1} and $I_{S\gamma}$ are given by integrals along contours enclosing the corresponding branch points $\pm i$ and $\pm i\gamma^{-1}$. If conditions (2.5) or (2.6) are not satisfied these quantities will vanish.

To calculate I_{l0}, I_{P1} and $I_{S\gamma}$ we use the methods of the saddle point and steepest descent, the conditions of applicability being given by

$$k(t-\tau)b^{-1}\gg 1. \tag{2.8}$$

If conditions (2.8) are satisfied we obtain the following asymptotic relationships in the regions (2.5) and (2.6):

$$\begin{aligned} I_{l0}&=\sqrt{\frac{2}{\pi k A_l}}\operatorname{Re}\left\{X_l(\eta_l)\exp\left[i\left(k\omega_l+\frac{\pi}{4}\right)\right]\right\}, \\ I_{lR}&=-2\sin\left[kv_R(t-\tau)\right]\operatorname{Im}\operatorname{Res}\big|_{i\tau_R}X_l(\eta)\begin{cases}\exp(-kz\alpha_R), & l=P, \\ \exp(-kz\beta_R), & l=S,\end{cases} \\ I_{S\gamma}&=-\frac{\sqrt{2\gamma}}{kA_\gamma\sqrt{\pi kA_\gamma}}\operatorname{Re}\left\{Y_S(i\gamma^{-1})\exp\left[i\left(k\omega_\gamma+\frac{\pi}{4}\right)\right]\right\}, \\ I_{P1}&=-\frac{\sqrt{2}\exp(-kz\sqrt{1-\gamma^2})}{kA_1\sqrt{\pi kA_1}}\operatorname{Re}\left[Z_P(i)\exp(i\varphi_1)\right], \end{aligned} \tag{2.9}$$

where

$$\begin{aligned} &\omega_P=a^{-1}\sqrt{(t-\tau)^2-a^2z^2}, \quad \omega_S=b^{-1}\sqrt{(t-\tau)^2-b^2z^2}, \\ &A_l=|f''(\eta_l)|, \quad \alpha_R=\sqrt{1-\gamma^2\tau_R^2}, \quad \beta_R=\sqrt{1-\tau_R^2}, \\ &\omega_\gamma=a^{-1}\left[(t-\tau)-bz\sqrt{1-\gamma^2}\right], \quad A_\gamma=|f_S'(i\gamma^{-1})|, \\ &A_1=|f_P'(i)|, \quad \varphi_1=kb^{-1}(t-\tau)+\frac{\pi}{4}-\frac{3}{2}\arg f_P'(i). \end{aligned} \tag{2.10}$$

Expressions for the functions $Y_S(\eta)$ and $Z_P(\eta)$ can be obtained if $X_P(\eta)$ and $X_S(\eta)$ are given by

$$X_P(\eta)=\overline{Z_P(\eta)}+\beta Z_P(\eta), \quad X_S(\eta)=\overline{Y_S(\eta)}+\alpha Y_S(\eta), \tag{2.11}$$

where $\overline{Y_S(\eta)}$ and $Y_S(\eta)$ are regular in the neighborhoods of $\pm i\gamma^{-1}$, and $\overline{Z_P(\eta)}$ and $Z_P(\eta)$ are regular near the points $\pm i$. These expressions are given by

$$Y_S(\eta) = -2g^{-2}, \quad Z_P(\eta) = 4\alpha^2 g^{-3} \tag{2.12}$$

in the case of the components w, and by

$$Y_S(\eta) = -2\beta g^{-2}, \quad Z_P(\eta) = 4\alpha g^{-3} \tag{2.13}$$

for the components q.

If conditions (2.8) are satisfied, the Mellin integrals in (1.11) are therefore calculated and expressed by (2.7), (2.9) and (2.10). In accordance with (2.7) the field of longitudinal and transverse displacements $\mathbf{u}_P$ and $\mathbf{u}_S$ is given by

$$\mathbf{u}_P = \mathbf{u}_{P0} + \mathbf{u}_{PR} + \mathbf{u}_{P1}, \quad \mathbf{u}_S = \mathbf{u}_{S0} + \mathbf{u}_{SR} + \mathbf{u}_{S\gamma}, \tag{2.14}$$

where the components are expressed in terms of I_{l0}, I_{lR}, I_{P1} and $I_{S\gamma}$.

§3. Calculation of the integrals with respect to τ

To calculate the integrals with respect to τ we need to consider the integrals

$$\begin{aligned} K_1 &= \int_0^{t_l} f(\tau)\, J_0[ks(\tau)] \exp[ik\omega_l(\tau)]\, d\tau, \\ K_2 &= \int_0^{t_l} f(\tau)\, J_0[ks(\tau)] \exp(-ikv\tau)\, d\tau \end{aligned} \tag{3.1}$$

where k is large and v is equal to one of the propagation velocities a^{-1}, b^{-1} and v_R. For this purpose we divide the path of integration into two intervals: $[0, \epsilon]$, $[\epsilon, t_l]$ and choose ϵ sufficiently small so that in the first interval the functions $f(\tau)$, $s(\tau)$ and $\omega_l(\tau)$ can be approximated by the first two terms of Maclaurin's series.

To calculate the integrals over $[\epsilon, t_l]$ we assume that $k_0 s(\epsilon) \gg 1$ and replace the Bessel function $J_0[ks(\tau)]$ by its asymptotic expression. The integrals obtained, namely

$$\begin{aligned} \bar{K}_1 &= \frac{1}{\sqrt{2\pi k}} \int_\varepsilon^{t_l} \frac{f(\tau)}{\sqrt{s(\tau)}} \left[e^{ikf_1(\tau) - i\frac{\pi}{4}} + e^{ikf_2(\tau) + i\frac{\pi}{4}} \right] d\tau, \\ \bar{K}_2 &= \frac{1}{\sqrt{2\pi k}} \int_\varepsilon^{t_l} \frac{f(\tau)}{\sqrt{s(\tau)}} \left[e^{ikf_3(\tau) - i\frac{\pi}{4}} + e^{ikf_4(\tau) + i\frac{\pi}{4}} \right] d\tau, \end{aligned} \tag{3.2}$$

where

$$f_{1,2}(\tau)=\omega_l(\tau)\pm s(\tau),\quad f_{3,4}(\tau)=-v\tau+s(\tau), \tag{3.3}$$

are calculated by the stationary phase method. Only the functions $f_1(\tau)$ and $f_3(\tau)$ can have saddle points in the interval of integration. To determine these points we need to solve the equations

$$s'(\tau)\sqrt{v_l^2(t-\tau)^2-z^2}=v_l^2(t-\tau), \tag{3.4}$$

$$s'(\tau)=v_l. \tag{3.5}$$

Since the motion of the source is retarded and $s''(t)\le 0$, unique roots of (3.4) and (3.5) exist in $[0,t_l]$ if and only if

$$\left.\begin{array}{c} s'(0)>v_l,\quad s'(0)z<v_l t\sqrt{s'(0)^2-v_l^2},\\ s'(0)>v>s'(t_l). \end{array}\right\} \tag{3.6}$$

If (3.6) holds, the integrals (3.2) calculated by the stationary phase method have the final expressions

$$\begin{aligned} \overline{K}_2&=\frac{f(\tau_0)\exp[ikf_1(\tau_0)]}{ik\sqrt{-s(\tau_0)f_1''(\tau_0)}}+O\left(\frac{1}{k^{3/2}}\right),\\ \overline{K}_2&=\frac{f(\tau_0)\exp[ikf_3(\tau_0)]}{ik\sqrt{-s(\tau_0)s''(\tau_0)}}+O\left(\frac{1}{k^{3/2}}\right), \end{aligned} \tag{3.7}$$

where τ_0 denotes the saddle points. If (3.6) does not hold, then $K_1, K_2\sim O(k^{-3/2})$.

We now consider the particular case when $s''(\tau_0)=0,\ \cdots,\ s^{(n-1)}(\tau_0)=0$, $s^{(n)}(\tau_0)\ne 0$ $(n=2,3,\cdots)$. In this case $\overline{K}_2$ is given not by the second formula of (3.7) but by

$$\overline{K}_2=\frac{f(\tau_0)\,e^{i\left[kf_3(\tau_0)-\frac{\pi}{4}\right]}}{\sqrt{2\pi ks(\tau_0)}}\int_{\tau_0-\varepsilon}^{\tau_0+\varepsilon}\exp\left[ik\frac{s^{(n)}(\tau_0)(\tau-\tau_0)^n}{\Gamma(n+1)}\right]d\tau. \tag{3.8}$$

On introducing the new variable of integration

$$x=(\tau-\tau_0)\sqrt[n]{\frac{-ks^{(n)}(\tau_0)}{\Gamma(n+1)}},$$

replacing the limits of integration $\pm\epsilon^n\sqrt{-ks^{(n)}(\tau_0)\Gamma(n+1)^{-1}}$ by $\pm\infty$ and using the formulas

$$\int_{-\infty}^{\infty}e^{-ix^n}dx=\frac{2}{n}\Gamma\left(\frac{1}{n}\right)\left\{\begin{array}{ll}\exp\left(-\frac{i\pi}{2n}\right), & n \text{ even},\\ \cos\left(\frac{\pi}{2n}\right), & n \text{ odd},\end{array}\right. \tag{3.9}$$

we finally obtain

$$\bar{K}_2 = \frac{f(\tau_0)\sqrt{2}\,\Gamma(n^{-1})\,e^{i\left[kf_3(\tau_0)-\frac{\pi}{4}\right]}}{n\sqrt{\pi ks(\tau_0)}}\sqrt[n]{\frac{\Gamma(n+1)}{k\,|s^{(n)}(\tau_0)|}}$$

$$\times\begin{cases}\exp\left(-\dfrac{i\pi}{2n}\right), & n=2m,\\ \cos\dfrac{\pi}{2n}, & n=2m+1\end{cases}\qquad (m=1,\ 2,\ \ldots). \tag{3.10}$$

Let us now discuss the calculation of $\bar{K}_1$ in the case $s'(0) < v_l$ when (3.4) has no real roots. If we assume $s(\tau)$ and $f(\tau)$ to be analytic and consider the equation (3.4) on the complex plane τ, then, provided $z \ll v_l t$ and $z \ll s(t)$, we obtain the approximate expression

$$\tau_0 = t \pm \frac{ia^2 z}{\sqrt{1-a^2 s'(t)^2}} \tag{3.11}$$

for the complex saddle point. On replacing the contour of integration with respect to τ by the stationary contour passing through one of the points (3.11) and calculating the integral by the stationary phase method, we obtain the asymptotic equation

$$\bar{K}_1 = \frac{f(\tau_0)\exp\left[iks(t) - kz\sqrt{1-v_l^{-2}s'(t)^2} - i\frac{\pi}{4}\right]}{k\sqrt{s(t)\,f_1''(\tau_0)}}, \tag{3.12}$$

which is valid for $z > 0$.

Calculation of the integrals $\overset{*}{K}_1$ and $\overset{*}{K}_2$ over the interval $[0, \epsilon]$ for $s'(0) \neq v_l$ is reduced to substituting the approximate relations

$$s(\tau) = s'(0)\tau, \quad f(\tau) = f(0), \quad \omega_l(\tau) = \omega_l(0) + \omega_l'(0)\tau, \tag{3.13}$$

replacing the limits of integration $\pm k\epsilon b^{-1}$ by $\pm\infty$ and calculating

$$\int_0^\infty J_0(\alpha x)\,e^{-i\beta x}dx = \begin{cases}(\alpha^2-\beta^2)^{-1/2}, & \alpha>\beta>0,\\ -i(\beta^2-\alpha^2)^{-1/2}, & \beta>\alpha>0.\end{cases} \tag{3.14}$$

As a result, we obtain the final expressions

$$\begin{gathered}K_1^* = \frac{f(0)\,e^{ik\omega_l(0)}}{k}\begin{cases}[s'(0)^2-\omega_l'(0)^2]^{-1/2}, & s'(0)+\omega_l'(0)>0,\\ -i[\omega_l'(0)^2-s'(0)^2]^{-1/2}, & s'(0)+\omega_l'(0)<0,\end{cases}\\ K_2^* = \frac{f(0)}{k}\begin{cases}[s'(0)^2-v^2]^{-1/2}, & s'(0)>v,\\ -i[v^2-s'(0)^2]^{-1/2}, & s'(0)<v.\end{cases}\end{gathered} \tag{3.15}$$

We now consider the case when $s'(0) = v$ and (3.15) is not applicable to K_2^*. For generality we put $s''(0) = s'''(0) = \cdots = s^{(n-1)}(0) = 0$, $s^{(n)}(0) \neq 0 \, (n \geq 2)$. The function $x = s(\tau)$ and its inverse $\tau = f(x)$ should in this case be approximated by

$$s(\tau) = v\tau + \frac{s^{(n)}(0)\,\tau^n}{n!}, \quad f(x) = \frac{x}{v} - \frac{s^{(n)}(0)\,x^n}{n!\,v^{n+1}}. \tag{3.16}$$

On introducing a new variable of integration $y = ks(\tau)$, we obtain

$$K_2^* = \frac{f(0)}{kv} \int_0^{ks(\varepsilon)} j_0(y) \exp\left\{-i\left[y - \frac{s^{(n)}(0)\,y^n}{n!\,v^n k^{n-1}}\right]\right\} dy. \tag{3.17}$$

This can be simplified if the Bessel function $j_0(y)$ is replaced by its asymptotic expression and only the term $\tfrac{1}{2}\sqrt{2/\pi y}\,\exp[i(y-\pi/4)]$ is considered. The error in these substitutions and approximations (3.16) is of order $O(k^{-1})$. As a result, we obtain

$$K_2^* = \frac{e^{-i\frac{\pi}{4}} f(0)}{\sqrt{2\pi}\,kv} \int_0^{ks(\varepsilon)} \exp\left[\frac{is^{(n)}(0)\,y^n}{n!\,k^{n-1}v^n}\right] \frac{dy}{\sqrt{y}}. \tag{3.18}$$

On replacing the variable of integration y by $u = \sqrt{yv^{-1}\sqrt[n]{-s^{(n)}(0)\,k^{1-n}(n!)^{-1}}}$ we obtain

$$K_2^* = \sqrt{\frac{\pi}{2}}\, \frac{(n!)^{\frac{1}{2n}} k^{-\frac{n+1}{2n}} e^{-i\frac{\pi}{4}} f(0)}{\sqrt{v}\,\sqrt[2n]{-s^{(n)}(0)}} \int_0^{R(k)} \exp[-iu^{2n}]\, du, \tag{3.19}$$

where $R(k) = \sqrt{s(\epsilon)v^{-1}\sqrt[2n]{-k(n!)^{-1}s^{(n)}(0)}}$. It is easy to see that for large k and finite n the integral from (3.19) is calculated approximately in accordance with the second equation of (3.9). The final expression for K_2^* has the form

$$K_2^* = \frac{e^{-i\frac{\pi(n+1)}{4n}} k^{-\frac{n+1}{2n}} \sqrt[2n]{n!}\,\sqrt{2}}{\sqrt{\pi v}\,\sqrt[2n]{-s^{(n)}(0)}\,n} f(0)\,\Gamma\left(\frac{1}{2n}\right). \tag{3.20}$$

Integrals over the interval $[0, t_l]$ of the variable τ are therefore represented by sums of integrals

$$K_i = K_i^* + \bar{K}_i \quad (i = 1,\ 2) \tag{3.21}$$

over small intervals near the initial point $\tau = 0$ and the saddle point $\tau = \tau_0$. In accordance with (3.21) the components from (2.14) are given by

$$\left.\begin{aligned} \mathrm{u}_{l0} &= \mathrm{u}_{l01} + \mathrm{u}_{l02}, \quad \mathrm{u}_{lR} = \mathrm{u}_{lR1} + \mathrm{u}_{lR2}, \\ \mathrm{u}_{P1} &= \mathrm{u}_{P11} + \mathrm{u}_{P12}, \quad \mathrm{u}_{S\gamma} = \mathrm{u}_{S\gamma1} + \mathrm{u}_{S\gamma2}, \end{aligned}\right\} \tag{3.22}$$

where the terms whose third subscript is "1" contain the quantities (3.15) or (3.20) and the others are expressed in terms of (3.7) or (3.10). As we see from (3.7), (3.10), (3.15) and (3.20), the first components in (3.22) describe waves formed at $t = 0$, and the second components correspond to waves formed at various subsequent times. The waves $\mathbf{u}_{l01}$, $\mathbf{u}_{lR1}$, $\mathbf{u}_{P11}$ and $\mathbf{u}_{S\gamma 1}$ are therefore called primary, and the others, secondary.

§4. Investigation of primary waves

Calculation of the integrals with respect to k between the limits k_0 and ∞ reduces to an application of the asymptotic formulas

$$\left.\begin{aligned}
&\int_{k_0}^{\infty} \frac{\sin kx}{k}\, dk \sim \pi\varepsilon(x), \qquad \int_{k_0}^{\infty} \frac{\cos kx}{k}\, dk \sim -\ln|x|,\\
&\int_{k_0}^{\infty} \frac{\cos kx}{k^2}\, dk \sim \pi x\varepsilon(x), \qquad \int_{k_0}^{\infty} \frac{\sin kx}{k^2}\, dk \sim -x\ln|x|,\\
&\int_{k_0}^{\infty} \frac{\cos kx}{k^{1-\alpha}}\, dk \sim \frac{\Gamma(\alpha)}{|x|^{\alpha}} \cos\frac{\pi\alpha}{2}, \qquad \int_{k_0}^{\infty} \frac{\sin kx}{k^{1-\alpha}}\, dk \sim \frac{\Gamma(\alpha)}{|x|^{\alpha}} \operatorname{sign} x \sin\frac{\pi\alpha}{2},\\
&\int_{k_0}^{\infty} \frac{\sin kx}{k^{1-\alpha}}\, dk \sim \frac{\Gamma(\alpha)}{1-\alpha} \operatorname{sign} x\, |x|^{1-\alpha} \cos\frac{\pi\alpha}{2},\\
&\int_{k_0}^{\infty} \frac{\cos kx}{k^{2-\alpha}}\, dk \sim -\frac{\Gamma(\alpha)}{1-\alpha} |x|^{1-\alpha} \sin\frac{\pi\alpha}{2},\\
&\int_{k_0}^{\infty} \frac{e^{ikx}e^{-ky}}{k^{1-\alpha}}\, dk \sim \frac{\Gamma(\alpha)}{(x^2+y^2)^{\frac{\alpha}{2}}} e^{i\alpha \operatorname{arctg}\frac{x}{y}} \quad (y \geqslant 0, \quad 0<\alpha<1),
\end{aligned}\right\} \tag{4.1}$$

which take account of the singularities at $x = 0$. Using (2.7), (2.9), (3.7), (3.10), (3.12), (3.15), (3.20) and (4.1) for integrals with respect to η, τ and k, we can write down and study the first terms of the asymptotic form of the vectors $\mathbf{u}_{l0i}$, $\mathbf{u}_{lRi}$, $\mathbf{u}_{P1i}$ and $\mathbf{u}_{S\gamma i}$. We begin by investigating the characteristics of the primary waves.

The longitudinal primary wave $\mathbf{u}_{P01}$. The components q_{P01} and w_{P01} of the vector $\mathbf{u}_{P01}$ are given by

$$\left.\begin{aligned}
q_{P01} &= \frac{f(0)\, g(\eta_{P0})\, A_{P01}}{\sqrt{\left|\,|\eta_{P0}|^2 - b^2 s'(0)^2\right|}} \begin{cases} \pi\varepsilon(\omega_{P01}), & |\eta_{P0}| > bs'(0),\\ \ln|\omega_{P01}|, & |\eta_{P0}| < bs'(0),\end{cases}\\
w_{P01} &= |\alpha(\eta_{P0})|\, q_{P01},
\end{aligned}\right\} \tag{4.2}$$

where

$$\left.\begin{aligned} \omega_{P01} &= r - a^{-1}\sqrt{t^2 - a^2z^2}, \quad \eta_{P0} = \eta_P(0), \\ A_{P01} &= \left[\pi\mu R(\eta_{P0})\sqrt{r\,|f''_P(\eta_{P0})|}\right]^{-1}. \end{aligned}\right\} \tag{4.3}$$

The formulas (4.2) have singularities on the hemisphere

$$r^2 + z^2 = a^{-2}t^2, \quad z > 0, \tag{4.4}$$

the trace of which is depicted by the arc AP_1 in Figures 3–6. To describe the field in the neighborhood of the hemisphere (4.4) we introduce the angle of propagation $\vartheta = \operatorname{arctg}(rz^{-1}) = \arcsin(art^{-1})$ of the ray from the origin to the point (r, z) of the hemisphere (4.4). If $\sin\vartheta < [as'(0)]^{-1}$ (AC in Figures 3 and 4, and AP_1 in Figures 5 and 6), then on the front of (4.4) the field (4.2) has a discontinuity. If the reverse inequality holds: $\sin\vartheta > [as'(0)]^{-1}$ (CP_1 in Figures 3 and 4), we have logarithmic divergence.

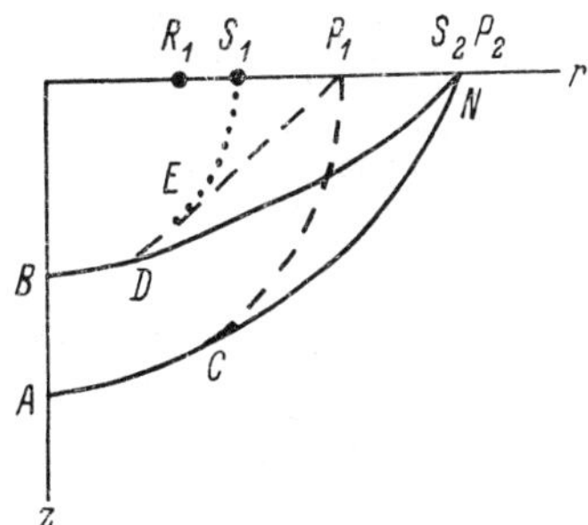

Figure 3

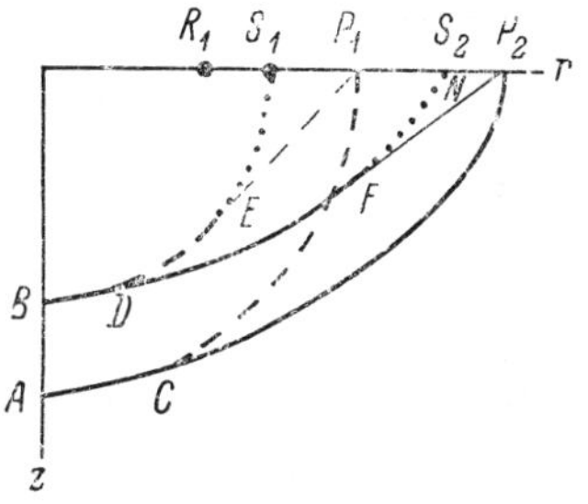

Figure 4

The transverse primary wave $u_{S01}(q_{S01}, w_{S01})$ is defined by

$$\left.\begin{aligned} w_{S01} &= -f(0)\,A_{S01}\left[\pi \operatorname{Re} B_{S01}\varepsilon(\omega_{S01}) + \operatorname{Im} B_{S01}\ln|\omega_{S01}|\right], \\ q_{S01} &= -|\beta(\eta_{S0})|\,w_{S01}, \end{aligned}\right\} \tag{4.5}$$

where

$$\begin{aligned} &\omega_{S01} = r - b^{-1}\sqrt{t^2 - a^2z^2}, \quad A_{S01} = 2\left[\pi\mu\sqrt{r\,|f''_S(\eta_{S0})|}\right]^{-1}, \\ &B_{S01} = \alpha(\eta_{S0})[\delta(\eta_{S0})R(\eta_{S0})]^{-1}, \quad \eta_{S0} = \eta_S(0), \\ &\delta(\eta_{S0}) = \begin{cases} \sqrt{b^2s'(0)^2 + \eta^2_{S0}}, & bs'(0) > |\eta_{S0}|, \\ i\sqrt{-b^2s'(0)^2 - \eta^2_{S0}}, & bs'(0) < |\eta_{S0}|. \end{cases} \end{aligned} \tag{4.6}$$

The right sides of (4.5) have singularities on the hemisphere

$$r^2 + z^2 = b^{-2}t^2, \quad z \geq 0, \tag{4.7}$$

the trace of which is depicted by BS_1 in Figures 3–6. To describe the field in the neighborhood of the hemisphere it is convenient to introduce the angle of propagation $\vartheta = \operatorname{arctg}(rz^{-1}) = \arcsin(brt^{-1})$ of the ray through the point (r, z) of the hemisphere. A singularity of the vector $\mathbf{u}_{S01}$ on the front is defined by ϑ which is related to the ordinate η_{S0} of the saddle point by $\sin\vartheta = |\eta_{S0}^{-1}|$. In the case $\sin\vartheta < \gamma$ and $\sin\vartheta < [bs'(0)]^{-1}$ (BD in Figures 3 and 4, and BE in Figures 5 and 6) the number B_{S01} is real and the functions (4.5) have discontinuities on the front (4.7). If $\gamma > \sin\vartheta > [bs'(0)]^{-1}$ (DE in Figures 3 and 4), then B_{S01} is pure imaginary and the singularity on the part of the hemisphere is logarithmic. Finally, if $\sin\vartheta > \gamma$ (ES_1 in Figures 3–6), then B_{S01} is complex and on the front the functions (4.5) have both a discontinuity and a logarithmic singularity.

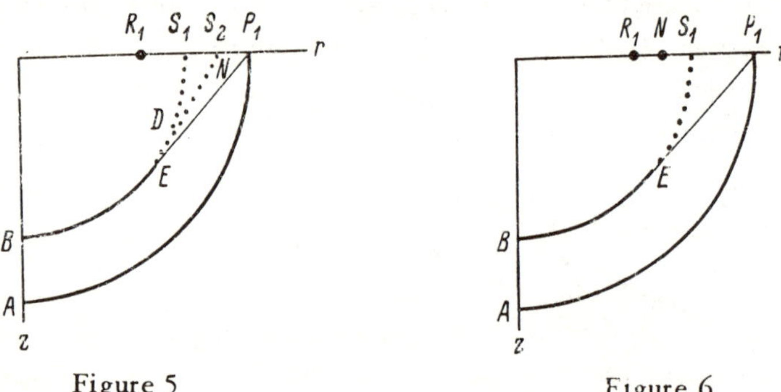

Figure 5 Figure 6

The head primary wave $\mathbf{u}_{S\gamma 1}$. When $as'(0) \neq 1$ the components $w_{S\gamma 1}$ and $q_{S\gamma 1}$ of the vector $\mathbf{u}_{S\gamma 1}$ are defined by

$$\left.\begin{aligned} w_{S\gamma 1} &= -\frac{f(0)\, A_{S\gamma 1}}{\sqrt{|a^2 s'(0)^2 - 1|}} \begin{cases} \pi \omega_{S\gamma 1} \varepsilon(\omega_{S\gamma 1}), & as'(0) < 1, \\ \omega_{S\gamma 1} \ln|\omega_{S\gamma 1}|, & as'(0) > 1, \end{cases} \\ q_{S\gamma 1} &= -\gamma^{-1}\sqrt{1-\gamma^2}\, w_{S\gamma 1}, \end{aligned}\right\} \tag{4.8}$$

where

$$\left.\begin{aligned} A_{S\gamma 1} &= 2\gamma\sqrt{\gamma}\left[\pi\mu g^2(i\gamma^{-1})\, f_S'(i\gamma^{-1})\sqrt{r f_S'(i\gamma^{-1})}\right]^{-1}, \\ \omega_{S\gamma 1} &= r - \frac{t}{a} - \frac{z\sqrt{1-\gamma^2}}{\gamma}. \end{aligned}\right\} \tag{4.9}$$

If $as'(0) = 1$, for generality we put $s''(0) = s'''(0) = \cdots = s^{(n-1)}(0) = 0$, $s^{(n)}(0) \neq 0$ $(n = 2, 3, \cdots)$. Then, by (3.16)–(3.20), in place of the first formula of (4.8) we have

$$
\begin{gathered}
w_{S\gamma 1}=-f(0)\,A_{S\gamma 1}B_{S\gamma 1}\left|\omega_{S\gamma 1}\right|^{\frac{n+1}{2n}},\\
B_{S\gamma 1}=\frac{4\Gamma\left(\frac{1}{2n}\right)\Gamma\left(\frac{n-1}{2n}\right)\sqrt[2n]{n!}}{\sqrt{2\pi a}\,(n+1)\sqrt[2n]{\left|s^{(n)}(0)\right|}}\cos\frac{\pi}{2n}\,\varepsilon(-\omega_{S\gamma 1})\left|\omega_{S\gamma 1}\right|^{\frac{n+1}{2n}}.
\end{gathered}
\tag{4.10}
$$

On the basis of (4.8) and (4.10) it is easy to see that the vector field $\mathbf{u}_{S\gamma 1}$ is continuous and the first derivatives of $\mathbf{u}_{S\gamma 1}$ (for example, the velocity $\dot{\mathbf{u}}_{S\gamma 1}$) have singularities on the conical surface

$$
ar+bz\sqrt{1-\gamma^2}=t,\quad 0<bz<t\sqrt{1-\gamma^2},
\tag{4.11}
$$

forming an angle $\tilde{\gamma}=\arcsin\gamma$ with the z-axis. The vector $\dot{\mathbf{u}}_{S\gamma 1}$ on the surface (4.11) has a discontinuity (P_1E in Figures 5 and 6) for $s'(0)<a^{-1}$ and a logarithmic singularity (P_1E in Figures 3 and 4) for $s'(0)>a^{-1}$. However, in the case $s'(0)=a^{-1}$, $s''(0)=s'''(0)=\cdots=s^{(n-1)}(0)=0$, $s^{(n)}(0)\neq 0$ the displacement velocity on the front $\omega_{S\gamma 1}=0$ tends to infinity as $|\omega_{S\gamma 1}|_{-(n-1)/2n}$.

The wave $\mathbf{u}_{P11}$ is a surface wave and has a singularity at $br=t$, $z=0$ (S_1 in Figures 3–6). If $bs'(0)=1$, the displacement velocities $\dot{\mathbf{u}}_{P11}$ have discontinuities or logarithmic singularities, and in the case $bs'(0)=1$ the singularity of $\mathbf{u}_{P11}$ is stronger. Since the front of the wave $\mathbf{u}_{P11}$ is superimposed on the hemisphere (4.7) the wave is not of great interest.

The primary Rayleigh wave $\mathbf{u}_{R1}=\mathbf{u}_{PR1}+\mathbf{u}_{SR1}$ is given by

$$
\begin{gathered}
w_{R1}=\frac{\alpha_R f(0)\sqrt{2}A_{R1}}{\sqrt{\left|s'(0)^2-v_R^2\right|}}\begin{cases}\left(g_R T_{P1}^{-1/4}\sin\psi_{P1}-2T_{S1}^{-1/4}\sin\psi_{S1}\right), & s'(0)>v_R,\\ \left(g_R T_{P1}^{-1/4}\cos\psi_{P1}-2T_{S1}^{-1/4}\cos\psi_{S1}\right), & s'(0)<v_R,\end{cases}\\
q_{R1}=\frac{f(0)\sqrt{2}A_{R1}}{\sqrt{\left|s'(0)^2-v_R^2\right|}}\begin{cases}\left(2\alpha_R\beta_R T_{S1}^{-1/4}\cos\psi_{S1}-g_R T_{P1}^{-1/4}\cos\psi_{P1}\right), & s'(0)>v_R,\\ \left(g_R T_{P1}^{-1/4}\sin\psi_{P1}-2\alpha_R\beta_R T_{S1}^{-1/4}\sin\psi_{S1}\right), & s'(0)<v_R.\end{cases}
\end{gathered}
\tag{4.12}
$$

when $s'(0)\neq v_R$. If, however, $s'(0)=v_R$, $s''(0)=s'''(0)=\cdots=s^{(n-1)}(0)=0$ and $s^{(n)}(0)\neq 0$ $(n=2,3,\cdots)$, formulas (4.12) should be replaced by

$$
\begin{aligned}
w_{R1}=\frac{\alpha_R f(0)\,A_{R1}B_R^{(n)}}{\sqrt{V_R}}\Bigg[&g_R T_{P1}^{\frac{1-2n}{4n}}\sin\left(\frac{2n-1}{2n}\varphi_{P1}+\frac{\pi}{4n}\right)\\
&-2T_{S1}^{\frac{1-2n}{4n}}\sin\left(\frac{2n-1}{2n}\varphi_{S1}+\frac{\pi}{4n}\right)\Bigg],
\end{aligned}
$$

$$g_{R1} = -\frac{f(0)\,A_{R1}B_R^{(n)}}{\sqrt{v_R}}\left[g_R T_{P1}^{\frac{1-2n}{4n}}\cos\left(\frac{2n-1}{2n}\varphi_{P1}+\frac{\pi}{4n}\right)\right.$$
$$\left. -\,2\alpha_R\beta_R T_{S1}^{\frac{1-2n}{4n}}\cos\left(\frac{2n-1}{2n}\varphi_{S1}+\frac{\pi}{4n}\right)\right]. \tag{4.13}$$

In (4.12) and (4.13) the following notation is used:

$$\begin{gathered} v_R = \tau_R b^{-1}, \quad \alpha_R = \alpha(i\tau_R), \quad \beta_R = \beta(i\tau_R), \quad g_R = g(i\tau_R), \\ T_{P1} = \alpha_R^2 z^2 + (r - v_R t)^2, \quad T_{S1} = \beta_R^2 z^2 + (r - v_R t)^2, \\ A_{R1} = (4\rho c_0 v_R \sqrt{r})^{-1}, \ c_0 = \alpha_R\beta_R^{-1} + \gamma^2\beta_R\alpha_R^{-1} - g_R, \\ \varphi_{P1} = \operatorname{arctg}\frac{r - v_R t}{\alpha_R z}, \ \varphi_{S1} = \operatorname{arctg}\frac{r - v_R t}{\beta_R z}, \quad \psi_l = \frac{\varphi_{l1}}{2} - \frac{\pi}{4}, \\ B_R^{(n)} = \sqrt[2n]{n\,!\,[-s^{(n)}(0)]^{-1}\,n^{-1}\left(\sin\frac{\pi}{2n}\right)^{-1}}. \end{gathered} \tag{4.14}$$

A study of (4.12) and (4.13) shows that w_{R1} and q_{R1} describe the surface waves and have a singularity only on the circle

$$r = v_R t, \quad z = 0. \tag{4.15}$$

If $s'(0) \neq v_R$ this singularity is of the type $|r - v_R t|^{-1/2}$, but when $s'(0) = v_R$, $s''(0) = 0, \cdots, s^{(n-1)}(0) = 0$, $s^{(n)}(0) \neq 0$ the functions (4.13) tend to infinity as $|r - v_R t|^{(1-2n)/2n}$.

Let us now compare the principal parts of the primary waves (4.2), (4.5), (4.8), (4.10), (4.12) and (4.13) with the principal parts of the primary waves caused by a uniformly expanding source

$$t_{zz}\,|_{z=0} = -\frac{\delta(r - vt)}{r}\,\varepsilon(t), \quad t_{zr}\,|_{z=0} = t_{z\theta}\,|_{z=0} = 0 \tag{4.16}$$

and investigated in [4]. When the initial source velocity $s'(0)$ is not equal to any of the propagation velocities a^{-1}, b^{-1} and v_R, the first terms of the asymptotic form of the primary waves from the sources (1.1) and (4.16) coincide, and we have $s'(0) = v$ and $f(0) = 1$. If we are interested only in the principal parts of the primary waves and assume that $s'(0) \neq a^{-1}$, $s'(0) \neq b^{-1}$ and $s'(0) \neq v_R$, then the source with variable velocity (1.1) will be equivalent to the uniformly moving source

$$t_{zz}\,|_{z=0} = -f(0)\,\varepsilon(t)\,\frac{\delta[r - s'(0)\,t]}{r}, \quad t_{zr}\,|_{z=0} = t_{z\theta}\,|_{z=0} = 0. \tag{4.17}$$

The principal parts of the primary waves from the sources (1.1) and (4.17) are identical because if $s'(0) \neq a^{-1}$, $s'(0) \neq b^{-1}$ and $s'(0) \neq v_R$, then the primary waves are formed only at the first moments of time when $s(t) = s'(0)\,t$ and $f(t) = f(0)\,\epsilon(t)$ approximately. When these relationships hold, an irregular motion of the source (1.1) can be replaced by a regular motion, and $f(t)$ is approximated by $f(0)\,\epsilon\,(t)$.

On the basis of the equivalence of the sources (1.1) and (4.17) and the results of [4] the fundamental part of the primary field of the source (1.1) can be expressed in terms of the corresponding parts of the primary field of the simplest fixed sources. If $\mathbf{u}_{l0}(0)$, $\mathbf{u}_{lR}(0)$, $\mathbf{u}_{S\gamma}(0)$ and $\mathbf{u}_{l0}(\infty)$, $\mathbf{u}_{lR}(\infty)$, $\mathbf{u}_{S\gamma}(\infty)$ denote the primary fields of the point source

$$t_{zz}\big|_{z=0} = -\frac{\delta(r)}{r}\,\varepsilon(t), \quad t_{zr}\big|_{z=0} = t_{z\theta}\big|_{z=0} = 0 \tag{4.18}$$

and of the distributed source

$$t_{zz}\big|_{z=0} = -\frac{\delta(t)}{r}, \quad t_{zr}\big|_{z=0} = t_{z\theta}\big|_{z=0} = 0, \tag{4.19}$$

we obtain

$$\begin{aligned}
\frac{\mathbf{u}_{P01}}{f(0)} &= \begin{cases} [1 - a^2 s'(0)^2 \sin^2 \tilde{\varphi}]^{-1/2}\,\mathbf{u}_{P0}(0), & as'(0)\sin\tilde{\varphi} < 1, \\ [a^2 - s'(0)^{-2}\sin^{-2}\tilde{\varphi}]^{-1/2}\,\mathbf{u}_{P0}(\infty), & as'(0)\sin\tilde{\varphi} > 1, \end{cases} \\
\frac{\mathbf{u}_{S01}}{f(0)} &= \begin{cases} [1 - b^2 s'(0)^2 \sin^2 \tilde{\psi}]^{-1/2}\,\mathbf{u}_{S0}(0), & bs'(0)\sin\tilde{\psi} < 1, \\ [b^2 - s'(0)^{-2}\sin^{-2}\tilde{\psi}]^{-1/2}\,\mathbf{u}_{S0}(\infty), & bs'(0)\sin\tilde{\psi} > 1, \end{cases} \\
\frac{\mathbf{u}_{S\gamma 1}}{f(0)} &= \begin{cases} [1 - a^2 s'(0)^2]^{-1/2}\,\mathbf{u}_{S\gamma}(0), & as'(0) < 1, \\ [s'(0)^2 - a^{-2}]^{-1/2}\,\mathbf{u}_{S\gamma}(\infty), & as'(0) > 1, \end{cases} \\
\frac{\mathbf{u}_{R1}}{f(0)} &= \begin{cases} [1 - v_R^{-2} s'(0)^2]^{-1/2}\,\mathbf{u}_{R}(0), & s'(0) < v_R, \\ [s'(0)^2 - v_R^2]^{-1/2}\,\mathbf{u}_{R}(\infty), & s'(0) > v_R. \end{cases}
\end{aligned} \tag{4.20}$$

Formulas (4.20) show that the head wave and the Rayleigh wave of an irregularly moving source are equivalent, to within a multiplier, to the correspondingly named waves in the case of fixed sources (4.18) and (4.19). As far as the longitudinal wave $\mathbf{u}_{P01}$ and the transverse wave $\mathbf{u}_{S01}$ are concerned, they are only locally equivalent to the corresponding waves $\mathbf{u}_{P0}(0)$, $\mathbf{u}_{P0}(\infty)$, $\mathbf{u}_{S0}(0)$ and $\mathbf{u}_{S0}(\infty)$.

§5. Investigation of secondary waves

The principal terms of the asymptotic form of the field of secondary waves are obtained from the exact solution on the basis of the calculations of integrals with respect to η, τ and k.

The transverse secondary wave. The basic terms of the components q_{P02} and w_{P02} of the vector $\mathbf{u}_{P02}$ are given by

$$\left.\begin{aligned}
q_{P02} &= \frac{f(\tau_0)\,A_{P02}(\tau_0)\,g\,[ibs'(\tau_0)]}{b\mu R\,[ibs'(\tau_0)]}\,\varepsilon\,[r - s(\tau_0) - \omega_P(\tau_0)], \\
w_{P02} &= q_{P02}\sqrt{a^2 s'(\tau_0)^2 - 1},
\end{aligned}\right\} \tag{5.1}$$

where $A_{P02}(\tau_0)$ is defined by

$$\left.\begin{aligned} A_{P02}(\tau_0) &= \{-rs(\tau_0)\,|\,f''_P[ibs'(\tau_0)]\,|\,[s''(\tau_0)+\omega''_P(\tau_0)]\}^{-1/2},\\ \omega_P(\tau_0) &= a^{-1}\sqrt{(t-\tau_0)^2-a^2z^2}, \end{aligned}\right\} \tag{5.2}$$

and τ_0, by (3.4), is a solution of the equation

$$as'(\tau_0)\sqrt{(t-\tau_0)^2-a^2z^2}=t-\tau_0. \tag{5.3}$$

It is easy to see that the surface on which the field (5.1) has a discontinuity is axially symmetric and described by the system of equations (5.3) and

$$a[r-s(\tau_0)]=\sqrt{(t-\tau_0)^2-a^2z^2}. \tag{5.4}$$

The points on the surface of discontinuity are conveniently described by the parameter τ_0 varying in the interval $[0, \min(t, t_1)]$, where t_1 satisfies $s'(t_1) = a^{-1}$. It is easy to see that when $\tau_0 = 0$ this surface touches the sphere (4.4) and that points of the surface corresponding to $\tau_0 = \min(t, t_1)$ lie on the plane $z = 0$. If we use the expressions for the coordinates

$$r=s(\tau_0)+a^{-2}s'(\tau_0)^{-1}(t-\tau_0),\quad z=\sqrt{1-a^{-2}s'(\tau_0)^{-2}}\,(t-\tau_0), \tag{5.5}$$

which follow from (5.3) and (5.4), we see that as τ_0 increases the distance r monotonically increases and the depth z monotonically decreases. The function $r(z)$, defining the trace of the surface of discontinuity (P_2C in Figures 3 and 4), is therefore a decreasing function. The same surface (5.3), (5.4) can be regarded as consisting of ends of rays proceeding from the points $r = s(\tau_0)$, $z = 0$ and propagated in the planes $\theta = \text{const}$ at angles $\arcsin[as'(\tau_0)]$ during the time $t - \tau_0$.

The transverse secondary wave is described by the vector $\mathbf{u}_{S02}$ and determined by the asymptotic equations

$$\left.\begin{aligned} w_{S02} &= \frac{f(\tau_0)\,A_{S02}(\tau_0)}{\pi b\mu}[\operatorname{Re} B_{S02}\ln|\omega_{S02}| - \pi\operatorname{Im} B_{S02}\varepsilon(\omega_{S02})],\\ q_{S02} &= -\sqrt{b^2s'(\tau_0)^2-1}, \end{aligned}\right\} \tag{5.6}$$

where $A_{S02}(\tau_0)$, w_{S02} and B_{S02} are given by

$$\left.\begin{aligned} A_{S02} &= 2\{-rs(\tau_0)[s''(\tau_0)+\omega''_S(\tau_0)]\,|\,f''[ibs'(\tau_0)]\,|\}^{-1/2},\\ \omega_S(\tau) &= b^{-1}\sqrt{(t-\tau)^2-b^2z^2},\quad \omega_{S02}=r-s(\tau_0)-\omega_S(\tau_0),\\ B_{S02} &= \alpha[ibs'(\tau_0)]\,R[ibs'(\tau_0)]^{-1}, \end{aligned}\right\} \tag{5.7}$$

and τ_0 is a solution of

$$bs'(\tau_0)\sqrt{(t-\tau_0)^2-b^2z^2}=t-\tau_0. \tag{5.8}$$

The surface on which u_{S02} has a singularity is described by the system of equations (5.8) and

$$b[r - s(\tau_0)] = \sqrt{(t - \tau_0)^2 - b^2z^2}. \tag{5.9}$$

The points of this surface are described by the parameter τ_0 varying in the interval $[0, \min(t, t_2)]$, where t_2 satisfies $s'(t_2) = b^{-1}$. That part of the surface where the displacement vector has only discontinuities is determined by the interval $[0, \min(t, t_1)]$ of the parameter τ_0. If $t > t_1$ the surface contains a region in which $t_1 < \tau_0 < \min(t, t_2)$ and the displacement vector has both a discontinuity and logarithmic divergence. Furthermore, the front of the wave u_{S02} touches the hemisphere (4.7) and is bounded by the plane $z = 0$. The trace of the surface in question (S_2D in Figures 3 and 4) is described by the monotonically decreasing function $r(z)$. The surface itself is the locus of ends of rays from the points $r = s(\tau_0)$, $z = 0$ and propagated in the planes $\theta = \text{const}$ at angles $\arcsin[bs'(\tau_0)]$ during the time $t - \tau_0$.

It should be noted that these investigations of longitudinal and transverse secondary waves relate to the case when the source velocity exceeds the corresponding propagation velocities. For the reverse relationships between the velocities the vectors u_{P02} and u_{S02} are represented, in accordance with (3.12), by integrals of the type

$$\int_{k_0}^{\infty} \exp\{ik[r - s(t)] - kz\sqrt{1 - s'(t)^2 v_l^{-2}}\} k^{-1}dk$$

and describe surface waves. Investigation of the singularity at $r = s(t)$, $z = 0$, directly connected with the source, is of no particular interest.

In concluding the discussion of the wave fields u_{P02} and u_{S02} we consider the special case when the source moves with constant velocity v, where $v > b^{-1}$ and $v \neq a^{-1}$. The asymptotic formulas (5.1) and (5.6) in this case turn into the corresponding equations for the vectors u_{Pv} and u_{Sv} of [4]. The systems (5.3), (5.4) and (5.8), (5.9) describing the fronts of the waves u_{P02} and u_{S02} are replaced by the equations

$$r + z\sqrt{a^2v^2 - 1} = vt, \quad r + z\sqrt{b^2v^2 - 1} = vt, \tag{5.10}$$

defining a conic surface with rectilinear generators.

The secondary head wave $u_{S\gamma 2}$ in the case $s''(t_1) \neq 0$ is described by the asymptotic equations

$$\left.\begin{aligned} w_{S\gamma 2} &= -f(t_1)A_{S\gamma 2}(t_1)[-s''(t_1)]^{-1/2}\omega_{S\gamma 2}\varepsilon(\omega_{S\gamma 2}), \\ q_{S\gamma 2} &= -\sqrt{b^2s'(t_1)^2 - 1}, \end{aligned}\right\} \tag{5.11}$$

where $A_{S\gamma 2}$ and $\omega_{S\gamma 2}$ are given by

$$\left.\begin{aligned} A_{S\gamma 2} &= 2\sqrt{\gamma}\,\{b\mu g^2\,[ibs'(t_1)]\,A_\gamma\sqrt{rA_\gamma s(t_1)}\,\}^{-1/2}, \\ A_\gamma &= |\,f_S\,[ibs'(t_1)]\,|, \quad \omega_{S\gamma 2} = r - s(t_1) - a^{-1}(t - t_1) + z\gamma^{-1}\sqrt{1-\gamma^2}, \end{aligned}\right\} \tag{5.12}$$

and t_1 satisfies $s'(t_1) = a^{-1}$. If $s''(t_1) = s'''(t_1) = \cdots = s^{(n-1)}(t_1) = 0$ and $s^{(n)}(t_1) \neq 0$, the first equation of (5.11) must be rewritten in the form

$$w_{S\gamma 2} = B_{S\gamma 2}\,|\,\omega_{S\gamma 2}\,|^{\frac{n+2}{2n}}\, f_n(\omega_{S\gamma 2}), \tag{5.13}$$

where

$$B_{S\gamma 2} = \frac{2\sqrt{2}\,\Gamma(n^{-1})\sqrt[n]{\Gamma(n+1)}}{\pi^{3/2}(n+2)\sqrt[n]{-s^{(n)}(t_1)}},$$

$$f_n(\omega_{S\gamma 2}) = \begin{cases} \cos\dfrac{\pi}{n}\,\varepsilon(-\omega_{S\gamma 2}), & n = 2m, \\ \cos\dfrac{\pi}{2n}\left[\cos\dfrac{\pi}{2n}\,\varepsilon(-\omega_{S\gamma 2}) - \sin\dfrac{\pi}{2n}\,\varepsilon(\omega_{S\gamma 2})\right], & n = 2m+1 \end{cases} \tag{5.14}$$

$$(m = 1,\ 2,\ \ldots).$$

The displacement fields (5.11)–(5.14) are continuous and the displacement velocities have singularities on the conical surface

$$r + z\gamma^{-1}\sqrt{1-\gamma^2} = s(t_1) + a^{-1}(t - t_1), \quad r - s(t_1) > \gamma b^{-1} t. \tag{5.15}$$

When $s''(t_1) \neq 0$ the vector $\dot{\mathbf{u}}_{S\gamma 2}$ has a discontinuity on passing through (5.14). If however, $s''(t_1) = s'''(t_1) = \cdots = s^{(n-1)}(t_1) = 0$ and $s^{(n)}(t_1) \neq 0$, the field $\dot{\mathbf{u}}_{S\gamma 2}$ on the front (5.14) tends to infinity as $|\omega_{S\gamma 2}|^{(2-n)/2n}$. The surface (5.14) touches the front of the wave $\mathbf{u}_{S0 2}$ and is bounded by the plane $z = 0$. Its trace is depicted in Figure 4 by P_2F. The surface itself consists of ends of rays from the points $r = s(t_1)$, $z = 0$ propagated in the planes $\theta =$ const during time $t - t_1$. These rays proceed along the plane $z = 0$ longitudinally, and then, as transverse rays, are refracted into the half-space.

The secondary screened wave $\mathbf{u}_{P12}$ is, like $\mathbf{u}_{P11}$, a surface wave. The displacements on the front $r = s(t_2) + b^{-1}(t - t_2)$, $z = 0$ are continuous, and for $s''(t_2) \neq 0$ the displacement velocities have discontinuities and logarithmic divergences. If, however, $s''(t_2) = 0$, the singularities of $\mathbf{u}_{P12}$ are stronger. As in the case of the primary screened wave, $\mathbf{u}_{P12}$ is of no great interest.

The secondary Rayleigh wave $\mathbf{u}_{R2} = \mathbf{u}_{PR2} + \mathbf{u}_{SR2}$ is formed at time $t = t_3$ when the source velocity passes through the value of the Rayleigh velocity v_R. When $s''(t_3) \neq 0$ the wave $\mathbf{u}_{R2}$ is defined by the asymptotic equations

$$\left.\begin{aligned} w_{R2} &= \alpha_R f(t_3)\,A_{R2}\,(g_R T_{P2}^{-1/4}\cos\psi_{P2} - 2T_{S2}^{-1/4}\cos\psi_{S2}), \\ q_{R2} &= f(t_3)\,A_{R2}\,(g_R T_{P2}^{-1/4}\sin\psi_{P2} - 2\alpha_R\beta_R T_{S2}^{-1/4}\sin\psi_{S2}), \end{aligned}\right\} \tag{5.16}$$

where we have used the notation

$$
\begin{aligned}
&A_{R2}=\sqrt{2}\left[4\rho v_R c_0\sqrt{-rs(t_3)\,s''(t_3)}\,\right]^{-1},\\
&T_{P2}=\omega_{R2}^2+\alpha_R^2z^2,\quad T_{S2}=\omega_{R2}^2+\beta_R^2z^2,\\
&\omega_{R2}=r-s(t_3)-v_R(t-t_3),\quad \psi_{l2}=\frac{1}{2}\varphi_{l2}-\frac{\pi}{4},\\
&\varphi_{P2}=\operatorname{arctg}\frac{\omega_{R2}}{\alpha_R z},\quad \varphi_{S2}=\operatorname{arctg}\frac{\omega_{R2}}{\beta_R z}
\end{aligned}
\tag{5.17}
$$

and (4.14). If, however, $s''(t_3)=s'''(t_3)=\dots=s^{n-1}(t_3)=0$, $s^{(n)}(t_3)\neq 0$, it is necessary to replace (5.16) by

$$
\begin{aligned}
w_{R2}&=\alpha_R f(t_3)\,B_{R2}\begin{cases}\left[g_R T_{P2}^{\frac{1-n}{2n}}\sin\chi_P^{(1)}-2T_{S2}^{\frac{1-n}{2n}}\sin\chi_S^{(1)}\right], & n=2m,\\ \cos\frac{\pi}{2n}\left[g_R T_{P2}^{\frac{1-n}{2n}}\sin\chi_P^{(2)}-2T_{S2}^{\frac{1-n}{2n}}\sin\chi_S^{(2)}\right], & n=2m+1,\end{cases}\\
q_{R2}&=f(t_3)\,B_{R2}\begin{cases}\left[g_R T_{P2}^{\frac{1-n}{2n}}\cos\chi_P^{(1)}-2\alpha_R\beta_R T_{S2}^{\frac{1-n}{2n}}\cos\chi_S^{(1)}\right], & n=2m,\\ \cos\frac{\pi}{2n}\left[g_R T_{P2}^{\frac{1-n}{2n}}\cos\chi_P^{(2)}-2\alpha_R\beta_R T_{S2}^{\frac{1-n}{2n}}\cos\chi_S^{(2)}\right], & n=2m+1\end{cases}\\
&\qquad (m=1,\ 2,\ \dots),
\end{aligned}
\tag{5.18}
$$

where

$$
\begin{aligned}
&B_{R2}=\sqrt[n]{n!}\left[2n\rho c_0 v_R\sqrt{rs(t_3)}\sqrt[n]{-s^{(n)}(t_3)}\,\sin\frac{\pi}{n}\right]^{-1},\\
&\chi_l^{(1)}=\frac{\pi}{2n}+\chi_l^{(2)},\quad \chi_l^{(2)}=\frac{n-1}{n}\varphi_{l2}.
\end{aligned}
\tag{5.19}
$$

It follows from (5.16) and (5.18) that the secondary Rayleigh waves are surface waves propagated in the direction of motion of the source. On the front

$$
r=s(t_3)+v_R(t-t_3),\quad z=0 \tag{5.20}
$$

the displacements $\mathbf{u}_{R2}$ tend to infinity as $|\omega_{R2}|^{(1-n)/n}$, where n is the least positive integer $(n\geq 2)$ satisfying $S^{(n)}(t_3)\neq 0$.

§6. Discussion of the results of investigating the wave field

Having investigated the singularities of the primary and secondary waves, we now examine the characteristics of the whole wave field and observe the formation in time of separate waves. As already indicated, the primary waves are formed only in the initial moments. The positions of the wave fronts of $\mathbf{u}_{P01}$, $\mathbf{u}_{S01}$, $\mathbf{u}_{S\gamma 1}$, $\mathbf{u}_{R1}$ and $\mathbf{u}_{P11}$ are independent of the initial source velocity $s'(0)$ and are determined by (4.4), (4.7), (4.11), (4.15) and $br=t$, $z=0$. The traces of these surfaces are depicted in Figures 3–6 by the arcs AP_1 and BS_1, the segment

P_1E and the points R_1, S_1. In studying secondary waves we consider four special cases, depending on the magnitude of the initial velocity $s'(0)$: 1) $s'(0) > a^{-1}$; 2) $b^{-1} < s'(0) \leq a^{-1}$; 3) $v_R < s'(0) \leq b^{-1}$; 4) $s'(0) \leq v_R$. We assume that the terminal source velocity $s'(\infty)$ is less than the Rayleigh velocity v_R.

In the first case only longitudinal and transverse secondary waves (P_2S and S_2D in Figure 3) are formed in the initial period $(0, t_1)$. At time $t = t_1[s'(t_1) = a^{-1}]$ a secondary head wave is formed (P_2f in Figure 4), and the formation of longitudinal waves ceases. The location of the body wave fronts for $z > 0$ and $t > t_1$ is shown in Figure 4. The traces of the fronts of the surface $z = 0$ are shown in Figure 7, in which the points P_i, S_i and $R_i (i = 1, 2)$ denote the fronts of longitudinal, transverse and Rayleigh waves, and N is the position of the source. The scheme A, B, C (Figure 7) relates to the time intervals (t_1, t_2), (t_2, t_3), $t > t_3$, respectively. At time $t = t_2[s'(t_2) = b^{-1}]$ the formation of the transverse wave $\mathbf{u}_{S02}$ ceases and the surface wave $\mathbf{u}_{P12}$ appears. Finally, when $t = t_3$ and $v_R = s'(t_3)$ a secondary Rayleigh wave is formed.

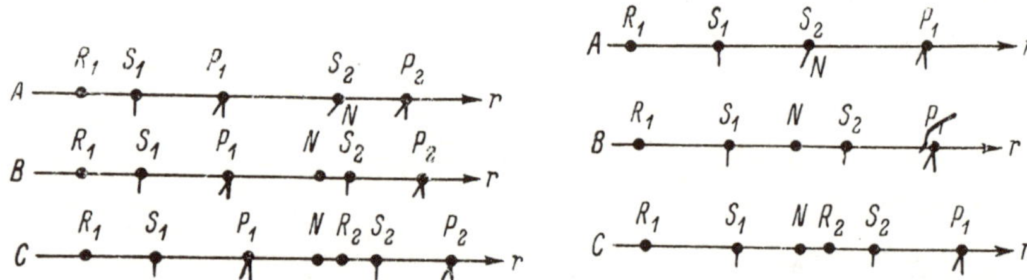

Figure 7. Figure 8.

The second case, when $b^{-1} < s'(0) \leq a^{-1}$, differs from the first by the absence of longitudinal and secondary head waves and the time interval $(0, t_1)$. The position of the body waves for $z > 0$ is shown in Figure 5; the traces of the fronts near the plane $z = 0$, in Figure 8, where A, B, and C refer respectively to the time intervals $(0, t_2)$, (t_2, t_3) and $t > t_3$. The secondary waves $\mathbf{u}_{S02}$, $\mathbf{u}_{P12}$ and $\mathbf{u}_{R2}$ are formed when $0 < t < t_2$, $t = t_2$ and $t = t_3$.

In the third case only a secondary Rayleigh wave is formed. The traces of the fronts are shown for $z > 0$ in Figure 6 and on the plane $z = 0$ in Figure 9 ($A(t < t_3)$ and $B(t > t_3)$.

Finally, in the fourth case there are, in general, no secondary waves, and the position of the fronts can be depicted as in Figure 6 ($z > 0$) and Figure 9 ($z = 0$).

We now discuss the results of the investigation into the nature of the singularities on the fronts. Figures 3–6 show the types of singularities of body waves in the case $s'(0) \neq a^{-1}$, $s''(t_1) \neq 0$. The surface of discontinuities of the displacement vector is shown by heavy continuous lines; a front with a logarithmic singularity is denoted by a heavy broken line; a surface on which the field u has both a discontinuity and a logarithmic singularity is shown by a dotted line. A front with a continuous field u is depicted by a thin continuous line in the case of a discontinuity in the displacement velocity $\dot{u}$, and by a thin broken line when $\dot{u}$ has a logarithmic singularity. On considering Figures 3–6 we deduce: 1) on fronts of the waves u_{l0i} ($l = p, s$; $i = 1, 2$) the displacements have singularities of the discontinuous type and the logarithmic type; 2) on fronts of the head waves $u_{S\gamma i}$ the displacement field is continuous and the displacement velocity undergoes a discontinuity or logarithmic divergence; 3) discontinuities of the displacement vector are connected with the first arrivals; 4) at the first arrivals of the transverse waves the vectors u_{S0i} and $u_{S\gamma i}$ undergo discontinuities; and 5) a discontinuity and a logarithmic singularity are observed simultaneously only on those parts of the fronts of u_{S0i} which are propagated behind the fronts of $u_{S\gamma i}$

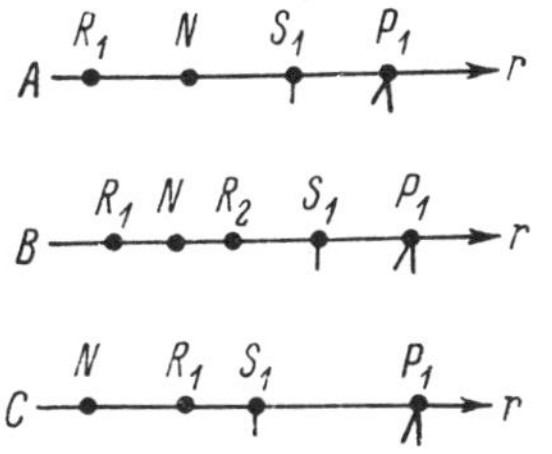

A: R_1 N S_1 P_1 r

B: R_1 N R_2 S_1 P_1 r

C: N R_1 S_1 P_1 r

Figure 9

As far as primary and secondary Rayleigh waves are concerned, when $s'(0) \neq v_R$ and $s''(t_3) \neq 0$ the displacements on the fronts $r = v_R t$ and $r = s(t_3) + v_R(t-t_3)$ ($z = 0$) tend to infinity like $\sqrt{|r - v_R t|}$ and $\sqrt{|r - s(t_3) - v_R(t-t_3)|}$. If $s'(0) = v_R$, $s''(0) = s'''(0) = \cdots = s^{(n-1)}(0) = 0$, $s^{(n)}(0) \neq 0$, then the singularity of the primary Rayleigh wave is intensified and the displacements behave as $|r - v_R t|^{(1-2n)/2n}$ close to the front. If, however, $s''(t_3) = s'''(t_3) = \cdots = s^{(m-1)}(t_3) = 0$, $s^{(m)}(t_3) \neq 0$; the singularity on the front of the secondary wave is described by $|r - s(t_3) - v_R(t-t_3)|^{(1-m)/m}$. The singularities of the primary and secondary waves are of the same nature if $m = 2n$. It should be noted that the singularity of the displacement vector on the fronts of head waves is similarly intensified if at the time of their formation $s'(0) = a^{-1}$ and $s''(t_1) = 0$.

In conclusion, let us consider the damping which is connected with the geometric divergence. The expressions in (4.2), (4.5), (4.8), (4.10), (4.12), (4.13), (5.1), (5.6), (5.11), (5.13), (5.16) and (5.18) are put in the form of products $I(r, z, t)\,\Phi\,(r, z, t)$, or the sum of such products, where Φ is a factor determining the nature of the singularity and does not vary along the ray. The dependence on time is taken into account by the factor I, which is usually called the intensity. An examination of the explicit expressions (4.2), (4.5), (5.1) and (5.6) shows that the intensity of the waves u_{l0i} ($l = P, S$; $i = 1, 2$) decreases as t^{-1}. A weakening of the intensities of the primary and secondary head waves is determined by the function t^{-2}. Finally, both Rayleigh waves decrease as $t^{-1/2}$. A study of the decrease of intensities leads to the following conclusions: 1) the geometric divergence is the same for primary and secondary like-named waves; and 2) the geometric divergence in the cases $s'(0) = a^{-1}$, $s'(0) = b^{-1}$, $s'(0) = v_R$ and $s''(t_i) = 0$ $(i = 1, 2, 3)$ does not differ from the geometric divergence in the general case.

BIBLIOGRAPHY

[1] G. I. Petrašen', *Propagation of elastic waves in stratified isotropic media separated by parallel planes*, Leningrad. Gos. Univ. Uč. Zap. 162 Ser. Mat. Nauk 26 (1952). (Russian) MR 17, 1157.

[2] M. A. Evgrafov, *Asymptotic estimates and entire functions*, 2nd rev. ed., Fizmatgiz, Moscow, 1962; English transl. of 1st ed., Gordon and Breach, New York, 1961. MR 27 #4924.

[3] A. G. Bagdoev, *Spatial non-stationary motions of a continuous medium with shock waves*, Izdat. Akad. Nauk Armjan. SSR, Erevan, 1961. (Russian)

[4] L. A. Molotkov, *On the vibrations of a homogeneous elastic half-space under the action of a source applied to a uniformly expanding circle*, Prikl. Mat. Meh. 31 (1967), no. 2, 211–221 = J. Appl. Math. Mech.

Trudy Mat. Inst. Steklov.
95 (1968)

Proc. Steklov Inst. Math.
95 (1968)

ON THE ENGINEERING EQUATIONS OF VIBRATIONS OF IMPERFECTLY ELASTIC THIN PLATES

UDC 550.341

G. I. PETRAŠEN' AND È. F. HINEN

Abstract. The paper discusses questions concerned with the derivation and proof of equations of the vibration of plates whose material can be specified by linear integral Volterra-Boltzmann operators. The investigations are based on the method of series, which is used to find exact solutions of dynamic problems in the case of an infinite plate subject to arbitrary external actions. The aim is to determine conditions which must be satisfied by the external actions so that the dynamic processes in the plate can be described approximately by the simplified equations of the engineering theory of vibrations, and to determine the form of these equations in various approximations. Special attention is given to a study of the errors of the engineering theory in describing displacement and stress fields in the plate.

Bibliography 7 items. 2 figures.

§ 0. Introduction

Our aim is to derive and prove the engineering equations of the vibration of homogeneous and isotropic plates $-h \leq z \leq h$, assuming that the processes in the plate as a three-dimensional system fit rigorously into the mathematical description of linear imperfect elasticity. In discussing such problems in perfect elasticity, most authors [1-4] start from the assumption that the mathematical description of the vibrations of thin plates will contain a small parameter arising from the static definition of the concept of a "thin plate". On this basis the assumption is made that the displacement and stress fields in the plate can be represented by segments of power series in the variable z. With the help of the Lamé equation of motion and boundary conditions for $z = \pm h$ this enables us formally to express all the coefficients of the series in terms of an auxiliary function not containing z, and also to determine a differential equation for this function. This differential equation has terms containing various powers of the "small parameter" enabling the authors mentioned above to write down the equations of motion of the engineering theory of vibrations of plates in various "approximations to reality". It should be noted that these authors have taken h as the small parameter expressed in units describing the length and width of the plate. This parameter bears no direct relationship to the dynamic processes excited in the plate as a result of the application of external forces. This method of obtaining the equations of the engineering theory of vibrations of plates, unsubstantiated by the necessary mathematical proof, cannot therefore be regarded as satisfactory from either the physical or the mathematical viewpoint. In particular, it does not answer questions concerning the conditions of applicability of the approximation theory.

Another method of deriving the equations of the engineering theory of vibrations of perfectly elastic plates, which avoids some of these shortcomings, is discussed in [5] and [6]. It is based on an analysis of exact solutions of problems for an infinite elastic sheet $-h \leq z \leq h$. Solutions of this kind are used in [7] to justify and explain the meaning of power series expansions in z of displacement and stress fields which are in fact the starting point for any of the above formal methods used to derive the equations of the engineering theory and based on the idea of a small parameter.

The results obtained in [7] seem to us to be sufficiently complete for an exhaustive solution of the mechanics problem as posed, and permit a generalization to cases of imperfectly elastic models of the medium. We shall therefore use the methods of [7] in this paper and begin the investigation with the statement and the solution of a sufficiently general dynamic problem for an infinite sheet whose material satisfies the laws of linear imperfect elasticity. Then in §2 we discuss the method of series in deriving the equations of vibrations of plates in engineering theory and clarify certain questions concerning exactness and the conditions whereby the engineering approach can be used to describe dynamic processes in plates. Finally, equations are given in §3 for longitudinal and flexural vibrations of imperfectly elastic plates, and questions concerning the application of these equations in practice are touched upon.

§1. An auxiliary problem and initial formulas

1. Suppose that in a Cartesian coordinate system (x, y, z) we have an infinite plate $-\infty < x, y < \infty, -h \leq z \leq h$ described by the density ρ and elastic lag operators

$$\left.\begin{aligned} \lambda + 2\mu &= (\lambda_0 + 2\mu_0)\left[1 - \int_0^t K_1(t-\tau)\ldots d\tau\right], \\ \mu &= \mu_0\left[1 - \int_0^t K_2(t-\tau)\ldots d\tau\right] \end{aligned}\right\} \tag{1.1}$$

in which the constants λ_0 and μ_0 and the form of the kernels $K_1(\tau)$ and $K_2(\tau)$

depend on the properties of the plate material. It is assumed that the plate is at rest for $t < 0$ and that at time $t = 0$ its boundaries are acted upon by

$$\left.\begin{array}{c}\tau_{xz}\big|_{z=\pm h}=\pm F_x^{\pm}(x,\ t),\quad \tau_{zz}\big|_{z=\pm h}=\pm F_z^{\pm}(x,\ t),\\ \tau_{yz}\big|_{z=\pm h}=0,\end{array}\right\}\tag{1.2}$$

where $F_x^{\pm}(x, t)$ and $F_z^{\pm}(x, t)$ are given functions. We require to find for $t > 0$ the plane displacement field

$$\mathbf{u}=u\,(x,\ z,\ t)\,\mathbf{i}+w\,(x,\ z,\ t)\,\mathbf{k},\tag{1.3}$$

which is excited in the plate under the action of (1.2).

Without appealing to the mathematical formulation of the problem, we note that, since the theory of imperfect elasticity is linear, the total action (1.2) can always be expressed, as in the case of the corresponding displacement field (1.3), in the form of a superposition of a symmetric and a skewsymmetric part.

The symmetric (longitudinal) part $\mathbf{u}_s$ of the displacement field in the plate is excited by actions

$$\left.\begin{array}{c}\tau_{xz}\big|_{z=\pm h}=f_x^{\pm}(x,\ t),\quad \tau_{zz}\big|_{z=\pm h}=f_z^{\pm}(x,\ t),\\ \tau_{yz}\big|_{z=\pm h}=0\end{array}\right\}\tag{1.4}$$

of the form (1.2), and the right sides contain functions

$$\begin{aligned}f_x^{+}&=-f_x^{-}=\frac{1}{2}(F_x^{+}+F_x^{-}),\\ f_z^{+}&=f_z^{-}=\frac{1}{2}(F_z^{+}+F_z^{-}).\end{aligned}\tag{1.5}$$

The skewsymmetric (flexural) part of the field $\mathbf{u}_a$ is excited by (1.4) if the functions on the right sides of (1.4) have the form

$$\begin{aligned}f_x^{+}&=f_x^{-}=\frac{1}{2}(F_x^{+}-F_x^{-}),\\ f_z^{+}&=-f_z^{-}=\frac{1}{2}(F_z^{+}-F_z^{-}).\end{aligned}\tag{1.6}$$

(The sum field $\mathbf{u} = \mathbf{u}_s + \mathbf{u}_a$ obviously satisfies the boundary conditions (1.2).)

It is natural to consider symmetric and skewsymmetric vibrations separately, since they lead to different physical effects for an exact as well as for an engineering description of the dynamic processes in the plate.

If we introduce the potentials $\phi(x, z, t)$ and $\psi(x, z, t)$, which are related to the components of the vector (1.3) by

$$u = \frac{\partial \varphi}{\partial x} - \frac{\partial \psi}{\partial z}, \quad w = \frac{\partial \varphi}{\partial z} + \frac{\partial \psi}{\partial x}, \tag{1.7}$$

the problem of determining the displacement field reduces, as we know, to the solution of the differential equations

$$\begin{gathered} (\lambda + 2\mu)\left[\frac{\partial^2 \varphi}{\partial x^2} + \frac{\partial^2 \varphi}{\partial z^2}\right] = \rho \frac{\partial^2 \varphi}{\partial t^2}, \\ \mu\left[\frac{\partial^2 \psi}{\partial x^2} + \frac{\partial^2 \psi}{\partial z^2}\right] = \rho \frac{\partial^2 \psi}{\partial t^2} \end{gathered} \tag{1.8}$$

in the region $-\infty < x < \infty$, $-h < z < h$ with zero initial data

$$\phi = \psi = \dot{\phi} = \dot{\psi} = 0 \quad (\text{for } t = 0) \tag{1.9}$$

and boundary conditions (1.4), (1.5) or (1.4), (1.6) on the planes $z = \pm h$. The left sides of (1.4) are then defined by

$$\begin{gathered} \tau_{xz} = \mu\left(2\frac{\partial^2 \varphi}{\partial x \partial z} + \frac{\partial^2 \psi}{\partial x^2} - \frac{\partial^2 \psi}{\partial z^2}\right), \\ \tau_{zz} = \lambda\left(\frac{\partial^2 \varphi}{\partial x^2} + \frac{\partial^2 \varphi}{\partial z^2}\right) + 2\mu\left(\frac{\partial^2 \varphi}{\partial z^2} + \frac{\partial^2 \psi}{\partial x \partial z}\right), \\ \tau_{yz} \equiv 0, \end{gathered} \tag{1.10}$$

which, as in the case of (1.8), contain the operators λ and μ from (1.1).

2. For definiteness we need to assign expressions to the external actions applied to the plate, that is, to the functions $f_x^+(x, t)$ and $f_z^+(x, t)$ from (1.5) and (1.6). In choosing these functions it is important to bear the following in mind:

1) The functions f_x^+ and f_z^+ must belong to a sufficiently general functional class (F) either representing arbitrary real actions (compatible with the description of the dynamics of plates within the framework of the engineering theory of vibrations) or enabling such actions to be obtained by the superposition of functions from (F).

2) Functions belonging to (F) must have the simplest possible structure, to avoid encumbering the investigations with unimportant secondary factors.

If we disregard for a moment the problems of the engineering theory of vibrations of degenerate systems and take account of mathematical considerations only, it would be natural to choose as (F) the class of functions

$$f(x, t, x_0) = \delta(x - x_0) f(t, x_0), \tag{1.11}$$

where x_0 is a parameter (the point of application of the action); $\delta(x-x_0)$ is Dirac's δ-function, and $f(t, x_0)$ is an arbitrary smooth function expressed by a Fourier or Mellin integral of the form

$$f(t, x_0)=\frac{1}{2\pi i}\int_{\sigma-i\infty}^{\sigma+i\infty}\Phi(s, x_0)e^{st}ds. \tag{1.12}$$

From the mechanics point of view, however, action concentrated at the point x_0 cannot be regarded as the "most reasonable idealization". For the actions are always applied to areas Δx_0 whose dimensions exceed a certain minimum $2\delta_0$. Even the law of distribution of forces over the area is sometimes unknown (in any event not exactly). Therefore in discussing problems of the engineering theory of vibrations of plates in sufficiently general terms for mechanics we may assume that $\delta(x-x_0)$ in (1.11) belongs to the set of functions distributed over the areas $\Delta x_0 > 2\delta_0$ and so smooth that the intervals $|x-x'| = \widetilde{\Delta x}$ of the "noticeable" change of $\delta(x-x_0)$ satisfy $\widetilde{\Delta x} \geqslant q\delta_0$.[1)] To these $\delta(x-x_0)$ belong, in particular, the functions

$$e^{-\frac{4(x-x_0)^2}{\alpha^2}}, \quad \cos(\beta x+\beta_0)e^{-\frac{4(x-x_0)^2}{\alpha^2}}, \tag{1.13}$$

and so on, if $\alpha > \delta_0$ and $|\beta| < \pi\delta_0^{-1}$.

Functions $\delta(x-x_0)$ from this set have a Fourier integral expansion with respect to $\exp[ik(x-x_0)]$ with spectral densities $\Psi(k)$ which are in practice nonzero only in an interval $|k| < k_0$. For example, in the case of (1.13) we have the expansions

$$e^{-\frac{4x^2}{\alpha^2}}=\frac{\alpha}{2\sqrt{\pi}}\int_0^\infty e^{-\left(\frac{\alpha k}{4}\right)^2}\cos kx\,dk, \tag{1.14'}$$

$$\cos\beta x e^{-\frac{4x^2}{\alpha^2}}=\frac{\alpha}{4\sqrt{\pi}}\int_0^\infty\left[e^{-\frac{\alpha^2(k-\beta)^2}{16}}+e^{-\frac{\alpha^2(k+\beta)^2}{16}}\right]\cos kx\,dk. \tag{1.14''}$$

If $k_0 = 8\delta_0^{-1}$ and $\alpha > \delta_0$, integration with respect to $k > k_0$ in (1.14′) can be ignored in any applied problems of mechanics. A similar situation holds in the case of (1.14″).

---

1) Here $0 < q < 1$ is a fixed number. The smooth function $f(x)$ given on $[a, b]$ is considered to change "noticeably" in the interval $(x, x+\Delta x_0)$ if $\Delta f \geq q_0 \max_{x\in[a,b]} f(x)$, where $\Delta f \geq \max|f(x+h) - f(x)|$ for $0 < h < \Delta x_0$ and $q_0 > 0$ is a small number chosen by the investigator.

We note further that if $\alpha > 0$ is sufficiently large (corresponding to actions distributed over a large area of the plate, which naturally must be subject to scrutiny), then the spectral function of (1.14″) has an arbitrarily sharp maximum at the point $k = \beta$, so that integration in (1.14″) is in fact taken over a small neighborhood of this point.

In consequence of this, the set of functions $\delta(x - x_0)$ from (1.11) may be assumed to be defined by [1)]

$$\delta(x) = \int_0^\infty \Psi(k) \begin{Bmatrix} \cos kx \\ \sin kx \end{Bmatrix} dk, \tag{1.15}$$

in which $\Psi(k)$ is an arbitrary smooth function in the interval $0 \le k \le k_0$, where k_0 is a (perhaps large) number; outside this interval $\Psi(k)$ has negligibly small values. For any $\delta(x - x_0)$ from this set (with a corresponding choice of k_0) the theory we develop here for the engineering equations of vibrations of plates must be valid.

Considerations which apply to the first factor in (1.11) carry over also to the second. The theory must be valid for the set of functions $f(t, x_0)$ from (1.12) whose spectral densities $\Phi(i\omega, x_0)$ are arbitrary within an interval $|\omega| < \omega_0$ and have negligibly small values outside it. One of the problems in this investigation must consist in determining k_0 and ω_0 so that the engineering theory of the vibrations of plates should be applicable as indicated for all k and ω from the intervals $0 \le k \le k_0$ and $-\omega_0 \le \omega \le \omega_0$.

In what follows we shall replace $(x - x_0)$ by x and consider the class (F) of external actions $f_x^\pm$ and $f_z^\pm$ defined by

$$\begin{aligned} f_x^\pm(x, t) &= \int_0^\infty \begin{Bmatrix} \cos kx \\ \sin kx \end{Bmatrix} \Psi_1^\pm(k)\, dk \int_{(l)} \Phi_1^\pm(s)\, e^{st} ds, \\ f_z^\pm(x, t) &= \int_0^\infty \begin{Bmatrix} \sin kx \\ -\cos kx \end{Bmatrix} \Psi_2^\pm(k)\, dk \int_{(l)} \Phi_2^\pm(s)\, e^{st} ds, \end{aligned} \tag{1.16}$$

where $\Psi_\nu^\pm(k)$ are continuous (or analytic) functions taking arbitrary values in the interval $(0, k_0)$ (or its neighborhood), decreasing as k^{-n_0} when $k \to \infty$ $(n_0 \gg 1)$ and negligibly small for $k > k_0$;[2)] (l) is the contour $\operatorname{Re} s = \sigma > 0$ on the complex

1) Here and subsequently $\begin{Bmatrix} \cos kx \\ \sin kx \end{Bmatrix}$ denotes $\cos kx$ or $\sin kx$ or some linear combination of the form $\cos kx + a(k) \sin kx$.

2) In consequence of this we can take k_0 as the upper limit of integration in (1.16).

plane (s); and finally $\Phi_\nu^\pm(s)$ are functions which are regular for $\operatorname{Re} s \geq 0$ and have a finite number of singular points (poles), take arbitrary values inside a domain $\Omega(k, s)$ containing the segment $(-i\omega_0, i\omega_0)$ of the imaginary axis, decrease as $s \to \pm i\infty$ not more slowly than $|s|^{-n_0}$ and are such that outside $\Omega(k, s)$ their values are negligibly small. We understand this last condition to mean that, for any contour (l_1) joining the points $(\sigma \pm i\infty)$ and leaving the domain $\Omega(k, s)$ to the right of itself, we have

$$\int_{(l_1)} |\Phi_\nu^\pm(s)e^{st}ds| \ll \left|\int_{(l)} \Phi_\nu^\pm(s)e^{st}ds\right|.$$

It is easy to see that, as long as there are no limitations on the choice of $\Omega(k, s)$ and the values of k_0 and ω_0, the class (F) of external actions is very general and includes practically all those actions which arise in mechanics.

We note finally that: 1) the class (F) contains functions (1.16) for which $\Psi_\nu^\pm(k)$ and $\Phi_\nu^\pm(s)$ are in practice nonzero in an arbitrarily small interval $|k - \tilde{k}| < \epsilon$ and circle $|s - s_0| < \epsilon$, where $\tilde{k}$ and s_0 are arbitrary points from $(0, k_0)$ and $\Omega(k, s)$ respectively (such functions are distributed over very large intervals of Δx and Δt); 2) if $|x| < \Delta x$ denotes the interval in which values of the functions $f_x^\pm$ and $f_z^\pm$ from (1.16) exist (that is, for $|x| > \Delta x$ the values of $f_x^\pm$ and $f_z^\pm$ can be neglected for a chosen accuracy), then for any $\Psi_\nu^\pm(k)$ we have

$$\Delta x \geqslant \frac{q}{k_0}, \tag{1.17}$$

where $q > 8$ for investigations with error up to 2% and $q > 4$ if an error of 10% is admissible.

3. It is convenient to seek solutions of the problems of §1.1 with symmetric actions (1.4), (1.5), (1.16) and skewsymmetric actions (1.4), (1.6), (1.16) in the form of integrals

$$\begin{aligned} \varphi &= \int_0^\infty \left.\begin{matrix} \sin kx \\ -\cos kx \end{matrix}\right\} dk \int_{(l)} U(z, k, s)e^{st}ds, \\ \psi &= \int_0^\infty \left.\begin{matrix} \cos kx \\ \sin kx \end{matrix}\right\} dk \int_{(l)} V(z, k, s)e^{st}ds. \end{aligned} \tag{1.18}$$

If we take account of the properties of the operators in (1.1) and the resulting relationships of the form

$$\left[1-\int_0^t K_\nu(t-\tau)\dots d\tau\right]\psi$$

$$=\int_0^\infty \left.\begin{matrix}\cos kx\\ \sin kx\end{matrix}\right\} dk \int_{(l)} [1-\varkappa_\nu(s)]\,V(z,\ k,\ s)\,e^{st}ds,$$

where

$$\varkappa_\nu(s)=\int_0^\infty K_\nu(\tau)\,e^{-s\tau}d\tau, \tag{1.19}$$

and put

$$a_0^2=\frac{\rho}{\lambda_0+2\mu_0},\qquad b_0^2=\frac{\rho}{\mu_0}, \tag{1.20}$$

then by substituting (1.18) into (1.8) we obtain

$$\left.\begin{aligned}&\frac{d^2U}{dz^2}-\left(k^2+\frac{a_0^2s^2}{1-\varkappa_1(s)}\right)U=0,\\ &\frac{d^2V}{dz^2}-\left(k^2+\frac{b_0^2s^2}{1-\varkappa_2(s)}\right)V=0.\end{aligned}\right| \tag{1.21}$$

We shall use the following convenient notation:

$$\left.\begin{aligned}&\alpha=\sqrt{k^2+\frac{a_0^2s^2}{1-\varkappa_1(s)}},\qquad \beta=\sqrt{k^2+\frac{b_0^2s^2}{1-\varkappa_2(s)}},\\ &g=k^2+\beta^2=2k^2+\frac{b_0^2s^2}{1-\varkappa_2(s)},\end{aligned}\right\} \tag{1.22}$$

where $\arg\alpha=\arg\beta=0$ for $s>0$. From (1.21) it follows that

$$\left.\begin{aligned}U(z,\ k,\ s)&=A_1(k,\ s)\operatorname{ch}\alpha z+A_2(k,\ s)\operatorname{sh}\alpha z,\\ V(z,\ k,\ s)&=B_1(k,\ s)\operatorname{sh}\beta z+B_2(k,\ s)\operatorname{ch}\beta z.\end{aligned}\right\} \tag{1.23}$$

Substituting (1.16) and (1.18) into the boundary conditions (1.4) gives

$$\left.\begin{aligned}&\{2k\alpha\,[A_1\operatorname{sh}\alpha z+A_2\operatorname{ch}\alpha z]-g\,[B_1\operatorname{sh}\beta z+B_2\operatorname{ch}\beta z]\}\,|_{z=\pm h}\\ &\qquad=\frac{\Phi_1^\pm(s)\,\Psi_1^\pm(k)}{\mu_0\,(1-\varkappa_2(s))},\\ &\{g\,[A_1\operatorname{ch}\alpha z+A_2\operatorname{sh}\alpha z]-2k\beta\,[B_1\operatorname{ch}\beta z+B_2\operatorname{sh}\beta z]\}\,|_{z=\pm h}\\ &\qquad=\frac{\Phi_2^\pm(s)\,\Psi_2^\pm(k)}{\mu_0\,(1-\varkappa_2(s))}.\end{aligned}\right| \tag{1.24}$$

When the actions are symmetric (longitudinal vibrations of the plate) we have, by (1.5),

$$\Phi_1^-\Psi_1^- = -\Phi_1^+\Psi_1^+, \quad \Phi_2^-\Psi_2^- = \Phi_2^+\Psi_2^+,$$

which forces us to put $A_2 = B_2 = 0$ in (1.23). The boundary conditions (1.24) can then be put in the form

$$\left.\begin{aligned} 2k\alpha \operatorname{sh} \alpha h A_1 - g \operatorname{sh} \beta h B_1 &= \frac{\Phi_1^+(s)\,\Psi_1^+(k)}{\mu_0(1-\varkappa_2(s))}, \\ g \operatorname{ch} \alpha h A_1 - 2k\beta \operatorname{ch} \beta h B_1 &= \frac{\Phi_2^+(s)\,\Psi_2^+(k)}{\mu_0(1-\varkappa_2(s))}. \end{aligned}\right\} \tag{1.25}$$

The components of the displacement field are given by

$$\left.\begin{aligned} u &= \int_0^\infty \left\{\begin{matrix} \cos kx \\ \sin kx \end{matrix}\right\} dk \int_{(l)} [kA_1(k,\ s) \operatorname{ch} \alpha z - \beta B_1(k,\ s) \operatorname{ch} \beta z]\, e^{st} ds, \\ w &= \int_0^\infty \left\{\begin{matrix} \sin kx \\ -\cos kx \end{matrix}\right\} dk \int_{(l)} [\alpha A_1(k,\ s) \operatorname{sh} \alpha z - kB_1(k,\ s) \operatorname{sh} \beta z]\, e^{st} ds. \end{aligned}\right\} \tag{1.26}$$

When the actions are skewsymmetric (flexural vibrations of the plate) we have, by (1.6),

$$\Phi_1^-\Psi_1^- = \Phi_1^+\Psi_1^+, \quad \Phi_2^-\Psi_2^- = -\Phi_2^+\Psi_2^+.$$

This forces us to take $A_1 = B_1 = 0$ in (1.23) and leads to the equations

$$\left.\begin{aligned} 2k\alpha \operatorname{ch} \alpha h A_2 - g \operatorname{ch} \beta h B_2 &= \frac{\Phi_1^+(s)\,\Psi_1^+(k)}{\mu_0(1-\varkappa_2(s))}, \\ g \operatorname{sh} \alpha h A_2 - 2k\beta \operatorname{sh} \beta h B_2 &= \frac{\Phi_2^+(s)\,\Psi_2^+(k)}{\mu_0(1-\varkappa_2(s))}. \end{aligned}\right\} \tag{1.27}$$

The components of the displacement field are then given by

$$\begin{aligned} u &= \int_0^\infty \left\{\begin{matrix} \cos kx \\ \sin kx \end{matrix}\right\} dk \int_{(l)} [kA_2(k,\ s) \operatorname{sh} \alpha z - \beta B_2(k,\ s) \operatorname{sh} \beta z]\, e^{st} ds, \\ w &= \int_0^\infty \left\{\begin{matrix} \sin kx \\ -\cos kx \end{matrix}\right\} dk \int_{(l)} [\alpha A_2(k,\ s) \operatorname{ch} \alpha z - kB_2(k,\ s) \operatorname{ch} \beta z]\, e^{st} ds. \end{aligned} \tag{1.28}$$

To obtain the final expressions for the components of the displacement field in the plate we need only determine the values of A_1, B_1 and A_2, B_2 from (1.25) and (1.27) respectively.

On putting [1)]

$$\left.\begin{aligned} L_1(k, s) &= s^{-2}[g^2 \operatorname{ch} \alpha h \operatorname{sh} \beta h - 4\alpha\beta k^2 \operatorname{sh} \alpha h \operatorname{ch} \beta h], \\ L_2(k, s) &= s^{-2}[g^2 \operatorname{ch} \beta h \operatorname{sh} \alpha h - 4\alpha\beta k^2 \operatorname{sh} \beta h \operatorname{ch} \alpha h], \end{aligned}\right\} \tag{1.29}$$

we have

$$\left.\begin{aligned} A_1(k, s) &= \frac{s^{-2}[g \operatorname{sh} \beta h \Phi_2^+(s) \Psi_2^+(k) - 2k\beta \operatorname{ch} \beta h \Phi_1^+(s) \Psi_1^+(k)]}{\mu_0(1 - \varkappa_2(s)) L_1(k, s)}, \\ B_1(k, s) &= \frac{s^{-2}[2k\alpha \operatorname{sh} \alpha h \Phi_2^+(s) \Psi_2^+(k) - g \operatorname{ch} \alpha h \Phi_1^+(s) \Psi_1^+(k)]}{\mu_0(1 - \varkappa_2(s)) L_1(k, s)}, \end{aligned}\right\} \tag{1.30}$$

$$\left.\begin{aligned} A_2(k, s) &= \frac{s^{-2}[g \operatorname{ch} \beta h \Phi_2^+(s) \Psi_2^+(k) - 2k\beta \operatorname{sh} \beta h \Phi_1^+(s) \Psi_1^+(k)]}{\mu_0(1 - \varkappa_2(s)) L_2(k, s)}, \\ B_2(k, s) &= \frac{s^{-2}[2k\alpha \operatorname{ch} \alpha h \Phi_2^+(s) \Psi_2^+(k) - g \operatorname{sh} \alpha h \Phi_1^+(s) \Psi_1^+(k)]}{\mu_0(1 - \varkappa_2(s)) L_2(k, s)}. \end{aligned}\right\} \tag{1.31}$$

It is easy to see that, with the given assumptions about the properties of $\Phi_\nu^+(s)$ and $\Psi_\nu^+(k)$, substituting the values of A_1, B_1 from (1.30) into (1.26), as well as the values of A_2, B_2 from (1.31) into (1.28), leads to a displacement field of symmetric and skewsymmetric vibrations of the plate strictly satisfying all the conditions of the problem.

4. The final expressions (1.26), (1.30) and (1.28), (1.31) enable us to make a thorough study of the plane displacement field in the imperfectly elastic plate $-\infty < x < \infty$, $-h \leq z \leq h$ with arbitrary actions (1.4) and (1.16). We use these formulas to clarify one particular question concerning the derivation of the equations of the engineering theory of vibrations of imperfectly elastic plates. The argument can proceed in two ways. We can start directly from the final exact formulas for the displacement field and aim to single out the "principal parts" satisfying the equations of vibrations of an engineering type which do not contain the variable z. This of course is not always possible and can only be done when the actions (1.16) satisfy certain conditions which are just those conditions for engineering theory to be applicable. The basic difficulties with this approach are connected with the singularities of the integrands in (1.26) and (1.28) on the (s)-plane which coincide with the roots $s_p = s_p(k)$ of the characteristic equations

$$L_1(k, s) = 0 \text{ and } L_2(k, s) = 0$$

of our problems. In the general case of models of imperfectly elastic media these difficulties are, generally speaking, considerable, and we therefore try to circumvent them. In constructing the engineering theory of vibrations of plates this can

1) The factor s^{-2} is introduced into (1.29)–(1.31) in order that the numerators and denominators of the explicit integrands in (1.26) and (1.28) should have no double roots at $s = 0$.

be done by employing an alternative argument similar to the method of series in the form in which it appears in [7]. In this case the assumption formulas are (1.25), (1.26) and (1.27), (1.28) for symmetric and skewsymmetric vibrations respectively, which are equivalent to the final expressions for the components of the displacement field in the plate. But the use of these formulas is based on another concept, which is discussed in the next section of the paper.

§2. The method of series in establishing the engineering theory of vibrations of plates

We know [7] that in establishing the engineering theory of the vibrations of perfectly elastic plates the components of the displacement field in the plate can be expressed as power series

$$\left.\begin{aligned} u &= \sum_{n=0}^{\infty} u_n(x,\ t)\, z^{2n}, \\ w &= \sum_{n=0}^{\infty} w_{n+1}(x,\ t)\, z^{2n+1} \end{aligned}\right\} \tag{2.1}$$

in the case of symmetric actions and

$$\left.\begin{aligned} u &= \sum_{n=0}^{\infty} \tilde{u}_{n+1}(x,\ t)\, z^{2n+1}, \\ w &= \sum_{n=0}^{\infty} \tilde{w}_n(x,\ t)\, z^{2n}, \end{aligned}\right\} \tag{2.2}$$

in the case of skewsymmetric actions (that is, for longitudinal and flexural vibrations of the plate respectively). For the coefficients of these series we obtain linear differential expressions of order not greater than $2n + 1$ and containing one or two auxiliary functions which are usually taken to be $u_0(x, t)$ (or $u_0(x, t)$ and $w_1(x, t)$) in the case of longitudinal vibrations, and $\tilde{w}_0(x, t)$ (or $\tilde{w}_0(x, t)$ and $\tilde{u}_1(x, t)$) in the case of flexural vibrations. For the auxiliary functions (when certain conditions are fulfilled) we obtain approximate differential equations which are also the equations of the engineering theory of vibrations of perfectly elastic plates.

It is shown in [7] that all the results necessary for the theory follow simply and naturally from relationships equivalent to (1.25), (1.26) and (1.27), (1.28). The traditional version of engineering theory using only one auxiliary function is shown to require more stringent conditions of applicability. In consequence of this we aim to obtain the engineering equations of the vibrations of imperfectly elastic plates containing two auxiliary functions. We discuss the case of longitudinal exhaustively. The case of flexural vibrations of plates is completely analogous, and we shall give only the necessary final results.

1) In the case of perfect elasticity considered in [7] it was found that (l) in formulas of the type (1.16) could be taken to be a finite path of integration adjacent to the segment $(-i\omega_0, i\omega_0)$ of the imaginary axis and that integration with respect to k could be taken over the interval $(0, k_0)$ only.[1] At first $k_0 > 0$ and $\omega_0 > 0$ were assumed to be arbitrary fixed numbers. In accordance with this the paths of integration in all succeeding formulas of the type (1.18), (1.26) and (1.28) were finite segments. We cannot make such assumptions because the "images" of the Boltzmann operators on the plane (s) for finite (l) do not have such a simple form as in (1.19). (In the case of the Voigt model, however, the assumptions about (l) in [7] could be retained.) It is easy to see that all the results we need are also obtained assuming infinite paths of integration in (1.16). But in formal, heuristic arguments it is convenient for us to assume that the integration in (1.16) in fact is reduced to the intervals $(0, k_0)$ and $(-i\omega_0, i\omega_0)$.

On making (temporarily) this assumption, we introduce the two auxiliary functions

$$\chi(x, t) = \int_0^\infty \left.\begin{matrix} \sin kx \\ -\cos kx \end{matrix}\right\} dk \int_{(l)} A_1(k, s)\, e^{st} ds,$$
$$\zeta(x, t) = \int_0^\infty \left.\begin{matrix} \cos kx \\ \sin kx \end{matrix}\right\} dk \int_{(l)} \beta B_1(k, s)\, e^{st} ds, \tag{2.3}$$

where $A_1(k, s)$ and $B_1(k, s)$ have the values (1.30).

On using the power series expansions for $\operatorname{ch} y$ and $\operatorname{sh} y$ (with $y = \alpha z$ and $y = \beta z$) and denoting by λ_1^n and λ_2^n the operators defined by

$$\lambda_1^n \chi(x, t) = \int_0^\infty \left.\begin{matrix} \sin kx \\ -\cos kx \end{matrix}\right\} dk \int_{(l)} \alpha^{2n} A_1(k, s)\, e^{st} ds,$$
$$\cdots\cdots\cdots\cdots\cdots\cdots \tag{2.4}$$
$$\cdots\cdots\cdots\cdots\cdots\cdots$$
$$\lambda_2^n \zeta(x, t) = \int_0^\infty \left.\begin{matrix} \cos kx \\ \sin kx \end{matrix}\right\} dk \int_{(l)} \beta^{2n} \beta B_1(k, s)\, e^{st} ds,$$

we obtain the following expressions for the components (1.26) of the displacement

1) As we noted in § 1.2, the path of integration with respect to k could here also be taken to be the interval $(0, k_0)$.

field of symmetric vibrations of a plate:

$$
\begin{aligned}
u(x, z, t) &= \sum_{n=0}^{\infty} \left\{ \lambda_1^n \frac{\partial}{\partial x} \chi(x, t) - \lambda_2^n \zeta(x, t) \right\} \frac{z^{2n}}{(2n)!}, \\
w(x, z, t) &= \sum_{n=0}^{\infty} \left\{ \lambda_1^{n+1} \chi(x, t) + \lambda_2^n \frac{\partial}{\partial x} \zeta(x, t) \right\} \frac{z^{2n+1}}{(2n+1)!},
\end{aligned} \tag{2.5}
$$

which in form coincide with (2.1), as is required in the engineering theory of vibrations. If, however, we use (1.25), then equations for the auxiliary functions $\chi(x, t)$ and $\zeta(x, t)$ can also be obtained. We need to rewrite (1.25) in the form

$$
\begin{aligned}
&\sum_{n=0}^{\infty} \frac{h^{2n+1}}{(2n+1)!} \{ 2k\alpha^{2n+2} A_1(k, s) - g\beta^{2n+1} B_1(k, s) \} = \frac{\Phi_1^+(s)\, \Psi_1^+(k)}{\mu_0 (1 - \varkappa_2(s))}, \\
&\sum_{n=0}^{\infty} \frac{h^{2n}}{(2n)!} \{ g\, \alpha^{2n} A_1(k, s) - 2k\beta^{2n+1} B_1(k, s) \} = \frac{\Phi_2^+(s)\, \Psi_2^+(k)}{\mu_0 (1 - \varkappa_2(s))}
\end{aligned} \tag{2.6}
$$

and denote by $\tilde{\mu}$ the operator transforming the "arbitrary" function

$$
f(t) = \int_{(l)} F(k, s)\, e^{st}\, ds \tag{2.7'}
$$

into the function

$$
\tilde{\mu} f(t) = \int_{(l)} \frac{F(k, s)\, e^{st}}{\mu_0 (1 - \varkappa_2(s))}\, ds. \tag{2.7''}
$$

Equations (2.6) for the integrands from (2.3) and (1.16) then turn out formally to be equivalent to the following equations for $\chi(x, t)$ and $\zeta(x, t)$:

$$
\begin{aligned}
&\sum_{n=0}^{\infty} \frac{h^{2n+1}}{(2n+1)!} \left\{ 2 \frac{\partial}{\partial x} \lambda_1^{n+1} \chi(x, t) - \left(\lambda_2 - \frac{\partial^2}{\partial x^2} \right) \lambda_2^n \zeta(x, t) \right\} = \tilde{\mu} f_x^+(x, t), \\
&\sum_{n=0}^{\infty} \frac{h^{2n}}{(2n)!} \left\{ \left(\lambda_2 - \frac{\partial^2}{\partial x^2} \right) \lambda_1^n \chi(x, t) + 2 \frac{\partial}{\partial x} \lambda_2^n \zeta(x, t) \right\} = \tilde{\mu} f_z^+(x, t).
\end{aligned} \tag{2.8}
$$

The relationships (2.8) are integro-differential equations of infinite order, and naturally cannot be the subject of study in engineering theory, where a decisive requirement is that the methematical description of the system should be the simplest possible. But if conditions could be established under which only a small number m of first terms were retained on the left sides of (2.8) and "approximate" equations for χ and ζ obtained, then these equations would be equations of the engineering theory of vibrations of imperfectly elastic plates. In letting the parameters defining deviation from a perfectly elastic medium tend to zero these equations would go over into equations of the engineering theory for longitudinal vibrations of plates, the basis of which is given in [7]. We see therefore that the problem reduces to determining conditions enabling the series in (2.5) to be terminated after $n = m_0$ terms and the series (2.8) after $n = m$ terms to give approximate (in the exact sense of this word) expressions for the components of the displacement field and approximate equations for the auxiliary functions. After this it only remains to define more exactly the meaning of the operators λ_1^n and λ_2^n from (2.4) and $\tilde{\mu}$ from (2.7), determine them explicitly and give a rational final form to the equations of vibrations of plates obtained.

2. The assumption that the paths of integration in (1.16) should be finite was necessary only in order to interchange the order of integration and summation. But in fact it is not necessary to use infinite series to obtain the equations of the engineering theory of vibrations of plates. For, assuming in accordance with §1. 2 that the integration in (1.16), (1.26), (1.28) and (2.3) has infinite limits, we choose integers $m \geq 0$ and $m_0 = m_0(m)$ which will determine subsequently the order of approximation in the equations of vibrations of plates and in the expressions for the components of the displacement field, and use the formulas

$$\left.\begin{aligned} \operatorname{ch} y &= f_m(y) + \Delta_m(y), \\ \operatorname{sh} y &= F_m(y) + \tilde{\Delta}_m(y), \end{aligned}\right\} \tag{2.9}$$

where

$$\begin{gathered} f_m(y) = \sum_{n=0}^{m} \frac{y^{2n}}{(2n)!}, \quad F_m(y) = \sum_{n=0}^{m} \frac{y^{2n+1}}{(2n+1)!}, \\ \Delta_m(y) \equiv \operatorname{ch} y - f_m(y), \\ \tilde{\Delta}_m(y) \equiv \operatorname{sh} y - F_m(y). \end{gathered} \tag{2.10}$$

Retaining the expressions (2.3) for the auxiliary functions and the definitions (2.4) of the operators λ_1^n and λ_2^n, in place of (2.5) we obtain

$$\begin{aligned} u &= \sum_{n=0}^{m_0} \left\{\lambda_1^n \frac{\partial}{\partial x} \chi(x, t) - \lambda_2^n \zeta(x, t)\right\} \frac{z^{2n}}{(2n)!} + d_{m_0}, \\ w &= \sum_{n=0}^{m_0} \left\{\lambda_1^{n+1} \chi(x, t) + \lambda_2^n \frac{\partial}{\partial x} \zeta(x, t)\right\} \frac{z^{2n+1}}{(2n+1)!} + \tilde{d}_{m_0}, \end{aligned} \tag{2.11}$$

where

$$d_{m_0}=\int_0^\infty \left.\begin{matrix}\cos kx\\ \sin kx\end{matrix}\right\} dk \int_{(l)} [kA_1(k, s)\Delta_{m_0}(\alpha z)-\beta B_1(k, s)\Delta_{m_0}(\beta z)]\, e^{st}\, ds,$$

$$\tilde{d}_{m_0}=\int_0^\infty \left.\begin{matrix}\sin kx\\ -\cos kx\end{matrix}\right\} dk \int_{(l)} [\alpha A_1(k, s)\tilde{\Delta}_{m_0}(\alpha z) - kB_1(k, s)\tilde{\Delta}_{m_0}(\beta z)]\, e^{st} ds. \qquad (2.12)$$

In exactly the same way, retaining the definition of $\tilde{\mu}$ from (2.7), in place of (2.8) we obtain

$$\sum_{n=0}^{m} \frac{h^{2n+1}}{(2n+1)!}\left\{2\frac{\partial}{\partial x}\lambda_1^{n+1}\chi(x, t)-\left(\lambda_2-\frac{\partial^2}{\partial x^2}\right)\lambda_2^{n}\zeta(x, t)\right\} + \tilde{D}_m=\tilde{\mu}f_x^+(x, t), \qquad (2.13)$$

$$\sum_{n=0}^{m} \frac{h^{2n}}{(2n)!}\left\{\left(\lambda_2-\frac{\partial^2}{\partial x^2}\right)\lambda_1^{n}\chi(x, t)+2\frac{\partial}{\partial x}\lambda_2^{n}\zeta(x, t)\right\}+D_m=\tilde{\mu}f_z^+(x, t)$$

where

$$\tilde{D}_m=\int_0^\infty \left.\begin{matrix}\cos kx\\ \sin kx\end{matrix}\right\} dk \int_{(l)} [2k\alpha A_1(k, s)\tilde{\Delta}_m(\alpha h)-gB_1(k, s)\tilde{\Delta}_m(\beta h)]\, e^{st} ds,$$

$$D_m=\int_0^\infty \left.\begin{matrix}\sin kx\\ -\cos kx\end{matrix}\right\} dk \int_{(l)} [gA_1(k, s)\Delta_m(\alpha h) - 2k\beta B_1(k, s)\Delta_m(\beta h)]\, e^{st} ds. \qquad (2.14)$$

It is clear from the above that we can arrive at the equations of motion of the engineering theory of vibrations of plates in its modern sense only if, for arbitrary actions (1.16) from class (F) and certain (small) fixed values of m and m_0, the expressions d_{m_0}, $\tilde{d}_{m_0}$ and D_m, $\tilde{D}_m$ in (2.11) and (2.13) can (in some sense) be neglected in comparison with the other terms retained on the left sides of these relations. But since the choice of actions from (F) is arbitrary, which leads to arbitrariness in the choice of values of the functions $\Psi_\nu^+(k)$ and $\Phi_\nu^+(s)$ in the whole essential part of the domain of integration of formulas (1.16), this requirement is satisfied only if the integrands in (2.12) and (2.14) are negligibly small in this domain in comparison with the integrands of the spectral expansions of the form (2.3) of all the terms retained on the left sides of (2.11) and (2.13). This, however, reduces to finding the intervals

$$0\leqslant k\leqslant k_0, \quad -\omega_0\leqslant \omega\leqslant \omega_0, \qquad (2.15)$$

and the domain $\Omega(k, s)$ containing the interval $(-i\omega_0, i\omega_0)$, for all points of which the remainder terms $\Delta_m(y)$ and $\tilde{\Delta}_m(y)$ in the expansions (2.9) can be neglected for $y = \alpha h$ and $y = \beta h$. The errors in the final formulas for the equations of vibrations of plates and the components of the displacement field will be determined in some sense with respect to the errors in (2.9). But the meaning of the concept "error" in the engineering approach to the vibrations of plates still needs to be explained.

3. Let us compare the solutions of exact equations for longitudinal vibrations of plates with the "approximate" equations of engineering theory

$$\left.\begin{aligned} \tilde{\mathscr{L}}_m(\chi, \zeta) &= \tilde{\mu} f_x^+(x, t), \\ \mathscr{L}_m(\chi, \zeta) &= \tilde{\mu} f_z^+(x, t), \end{aligned}\right\} \tag{2.16}$$

obtained from (2.13) by deleting the terms D_m and $\tilde{D}_m$ for a fixed value of $m \geq 0$. We assume that a second integer $m_0 = m_0(m) \geq m$ is chosen with respect to m such that the components of the displacement field are given by (2.11) with $d_{m_0} = \tilde{d}_{m_0} = 0$.

It follows from the above that the solutions of the exact equations for the plate can be expressed by (2.3) with the integrands $A_1(k, s)$ and $B_1(k, s)$ satisfying (1.25) and given by (1.30). In particular, from (1.25) we obtain the characteristic equation

$$L_1(k, s) = 0, \tag{2.17}$$

for the natural frequencies of the longitudinal vibrations of plates, where $L_1(k, s)$ has the value (1.29).

If the corresponding solutions of the approximate equations (2.16) are also sought in the form of integrals (2.3), it is easy to see that equations for their integrands $\tilde{A}_1(k, s)$ and $\tilde{B}_1(k, s)$ are obtained which coincide with (1.25) if ch$y$ and sh$y$ (for $y = \alpha h$ and $y = \beta h$) are replaced by $f_m(y)$ and $F_m(y)$ from (2.9). In the approximate theory, therefore, the natural frequencies are determined from the characteristic equation

$$\tilde{L}_1(k, s) = 0, \tag{2.18}$$

where

$$\tilde{L}_1(k, s) = s^{-2}[g^2 f_m(\alpha h) F_m(\beta h) - 4\alpha\beta k^2 F_m(\alpha h) f_m(\beta h)]. \tag{2.19}$$

The integrands in (2.3) now have the form

$$\left.\begin{aligned} \tilde{A}_1(k, s) &= \frac{s^{-2}\left[gF_m(\beta h)\,\Phi_2^+(s)\,\Psi_2^+(k) - 2k\beta f_m(\beta h)\,\Phi_1^+(s)\,\Psi_1^+(k)\right]}{\mu_0\,(1-\varkappa_2(s))\,\tilde{L}_1(k, s)}, \\ \tilde{B}_1(k, s) &= \frac{s^{-2}\left[2k\alpha F_m(\alpha h)\,\Phi_2^+(s)\,\Psi_2^+(k) - gf_m(\alpha h)\,\Phi_1^+(s)\,\Psi_1^+(k)\right]}{\mu_0\,(1-\varkappa_2(s))\,\tilde{L}_1(k, s)}. \end{aligned}\right\} \quad (2.20)$$

From (1.30), (2.20) and (1.26) we obtain

$$u = \int_0^\infty \left.\begin{matrix}\cos kx \\ \sin kx\end{matrix}\right\} U(z, t, k)\,dk, \qquad w = \int_0^\infty \left.\begin{matrix}\sin kx \\ -\cos kx\end{matrix}\right\} W(z, t, k)\,dk \quad (2.21)$$

for the components of the exact and approximate displacement fields, where in the case of an exact solution of the problem we have

$$\left.\begin{aligned} U &= \int_{(l)} \frac{M_1(k, s, z)\,\Phi_1^+(s)\,\Psi_1^+(k) + M_2(k, s, z)\,\Phi_2^+(s)\,\Psi_2^+(k)}{\mu_0\,(1-\varkappa_2(s))\,L_1(k, s)}\,e^{st}\,ds, \\ W &= \int_{(l)} \frac{N_1(k, s, z)\,\Phi_1^+(s)\,\Psi_1^+(k) + N_2(k, s, z)\,\Phi_2^+(s)\,\Psi_2^+(k)}{\mu_0\,(1-\varkappa_2(s))\,L_1(k, s)}\,e^{st}\,ds, \end{aligned}\right\} \quad (2.22)$$

and in the case of a solution of the approximate equations (2.16) we have

$$\begin{aligned} U = \tilde{U} &= \int_{(l)} \frac{\tilde{M}_1(k, s, z)\,\Phi_1^+(s)\,\Psi_1^+(k) + \tilde{M}_2(k, s, z)\,\Phi_2^+(s)\,\Psi_2^+(k)}{\mu_0\,(1-\varkappa_2(s))\,\tilde{L}_1(k, s)}\,e^{st}\,ds, \\ W = \tilde{W} &= \int_{(l)} \frac{\tilde{N}_1(k, s, z)\,\Phi_1^+(s)\,\Psi_1^+(k) + \tilde{N}_2(k, s, z)\,\Phi_2^+(s)\,\Psi_2^+(k)}{\mu_0\,(1-\varkappa_2(s))\,\tilde{L}_1(k, s)}\,e^{st}\,ds. \end{aligned} \quad (2.23)$$

Formulas (2.22) use the notation

$$\left.\begin{aligned} M_1 &= \beta s^{-2}\left[g\,\mathrm{ch}\,\alpha h\,\mathrm{ch}\,\beta z - 2k^2\,\mathrm{ch}\,\beta h\,\mathrm{ch}\,\alpha z\right], \\ M_2 &= ks^{-2}\left[g\,\mathrm{sh}\,\beta h\,\mathrm{ch}\,\alpha z - 2\alpha\beta\,\mathrm{sh}\,\alpha h\,\mathrm{ch}\,\beta z\right], \\ N_1 &= ks^{-2}\left[g\,\mathrm{ch}\,\alpha h\,\mathrm{sh}\,\beta z - 2\alpha\beta\,\mathrm{ch}\,\beta h\,\mathrm{sh}\,\alpha z\right], \\ N_2 &= \alpha s^{-2}\left[g\,\mathrm{sh}\,\beta h\,\mathrm{sh}\,\alpha z - 2k^2\,\mathrm{sh}\,\alpha h\,\mathrm{sh}\,\beta z\right]. \end{aligned}\right\} \quad (2.24)$$

The expressions for $\tilde{M}_\nu$ and $\tilde{N}_\nu$ are obtained from M_ν and N_ν by replacing $\mathrm{ch}\,y$ and $\mathrm{sh}\,y$ respectively by $f_m(y)$ and $F_m(y)$ from (2.9) if $y = (\alpha h$ and $\beta h)$, and by $f_{m0}(y)$ and $F_{m0}(y)$ if $y = (\alpha z$ and $\beta z)$. For $\nu = 1, 2$ we then have

$$\left.\begin{aligned} M_\nu(k, s; z) &= \tilde{M}_\nu(k, s; z) + \omega_m^{(\nu)}(k, s; z), \\ N_\nu(k, s; z) &= \tilde{N}_\nu(k, s; z) + \tilde{\omega}_m^{(\nu)}(k, s; z), \end{aligned}\right\} \quad (2.25)$$

where

$$\left.\begin{aligned} \omega_m^{(1)}(k,\ s;\ z) &= \beta s^{-2}\{g\,[f_m(\alpha h)\,\Delta_{m_0}(\beta z) + \operatorname{ch}\beta z\Delta_m(\alpha h)] \\ &\quad - 2k^2\,[f_m(\beta h)\,\dot{\Delta}_{m_0}(\alpha z) + \operatorname{ch}\alpha z\Delta_m(\beta h)]\}, \\ \omega_m^{(2)}(k,\ s;\ z) &= ks^{-2}\{g\,[F_m(\beta h)\,\Delta_{m_0}(\alpha z) + \operatorname{ch}\alpha z\tilde{\Delta}_m(\beta h)] \\ &\quad - 2\alpha\beta\,[F_m(\alpha h)\,\Delta_{m_0}(\beta z) + \operatorname{ch}\beta z\tilde{\Delta}_m(\alpha h)]\}, \end{aligned}\right\} \qquad (2.26)$$

$$\left.\begin{aligned} \tilde{\omega}_m^{(1)}(k,\ s;\ z) &= ks^{-2}\{g\,[\operatorname{sh}\beta z\Delta_m(\alpha h) + f_m(\alpha h)\,\tilde{\Delta}_{m_0}(\beta z)] \\ &\quad - 2\alpha\beta\,[f_m(\beta h)\,\tilde{\Delta}_{m_0}(\alpha z) + \operatorname{sh}\alpha z\Delta_m(\beta h)]\}, \\ \tilde{\omega}_m^{(2)}(k,\ s;\ z) &= \alpha s^{-2}\{g\,[F_m(\beta h)\,\tilde{\Delta}_{m_0}(\alpha z) + \operatorname{sh}\alpha z\tilde{\Delta}_m(\beta h)] \\ &\quad - 2k^2\,[F_m(\alpha h)\,\tilde{\Delta}_{m_0}(\beta z) + \operatorname{sh}\beta z\tilde{\Delta}_m(\alpha h)]\}. \end{aligned}\right\} \qquad (2.27)$$

On the basis of the explicit form of the expressions for the components of the displacement field we can assert that equations (2.16) of the engineering theory of longitudinal vibrations of plates will in fact be the approximate equations of longitudinal vibrations only if formulas (2.21), (2.22) and (2.21), (2.23) for arbitrary actions (1.16) from class (F) describe vibratory processes which are physically similar.[1)]

To determine when this can take place we note that the integrands in (2.22) and (2.23) are regular everywhere except for the roots of the equations (2.17) and (2.18) respectively and for common singularities of the functions $\Phi_\nu^+(s)$. The integrals in (2.22) and (2.23) can be calculated on the basis of the theorem on residues by deforming the contour of integration (l) into the left half-plane of the variable s. Therefore U and $\tilde{U}$ and also W and $\tilde{W}$ from (2.22) and (2.23) are phsically similar if and only if there exists a domain $\Omega(k,\ s)$ containing the intervals (2.15) for certain values of the numbers k_0 and ω_0 and possessing the following properties:

1) Inside $\Omega(k,\ s)$ the equations (2.17) and (2.18) have a certain number of roots s_1 and $\tilde{s}_1$ respectively, coinciding with each other with relative error $\epsilon > 0$.

2) The residues of the integrands for U and $\tilde{U}$ as well as for W and $\tilde{W}$ at these roots have approximately equal multipliers for $\exp(s_1 t)$ and $\exp(\tilde{s}_1 t)$.

3) Inside $\Omega(k,\ s)$ the residues of these integrands are approximately equal at all the poles of $\Phi_\nu^+(s)$, provided that $\Phi_\nu^+(s)$ are chosen arbitrarily from the integrands (1.16) of class (F).

---

1) We shall call vibratory processes of two systems (1 and 2) physically similar under conditions F if natural and forced vibrations with frequencies ω and amplitudes A are excited in the systems such that $|\omega_1 - \omega_2| < \epsilon|\omega_1|$ and $|A_1 - A_2| < \epsilon_1|A_1|$, where $\epsilon > 0$, $\epsilon_1 > 0$ are given small numbers.

4) Outside $\Omega(k, s)$ the functions $\Phi_\nu^+(s)$ are so small that the integrals (2.22) and (2.23) along contours (l) outside $\Omega(k, s)$ have negligibly small values.[1)]

The last (fourth) condition obviously establishes precisely the conditions of applicability of the engineering theory of vibrations of plates.

To clarify the question of the existence (and dimensions) of the domain $\Omega(k, s)$, we need to consider in the light of the above conditions the denominators (that is, the characteristic equations (2.17) and (2.18)) and also the numerators of the integrands from (2.22) and (2.23).

4. Let us consider first the characteristic equations (2.17) and (2.18) where we take account of the formula

$$L_1(k, s) = \tilde{L}_1(k, s) + \Delta_m(k, s), \tag{2.28}$$

in which

$$\begin{aligned}\Delta_m(k, s) = s^{-2}\{g^2[\operatorname{sh}\beta h\Delta_m(\alpha h) + f_m(\alpha h)\tilde{\Delta}_m(\beta h)]\\ - 4\alpha\beta k^2[\operatorname{sh}\alpha h\Delta_m(\beta h) + f_m(\beta h)\tilde{\Delta}_m(\alpha h)]\},\end{aligned} \tag{2.29}$$

and $f_m(y)$, $\Delta_m(y)$ and $\tilde{\Delta}_m(y)$ are defined in (2.10).

It follows from (1.22), (1.29), (2.19) and (2.29) that the functions in (2.28) are regular with respect to k and s in the domains

$$|\alpha h| < q, \quad |\beta h| < q, \tag{2.30}$$

where $q > 0$ is any fixed number. The inequalities (2.30) define a natural domain which contains the intervals (2.15) for suitably chosen values of k_0 and ω_0 and in which

$$\Delta_m(k, s), \quad \frac{\partial\Delta_m(k, s)}{\partial s}, \quad \frac{\partial\Delta_m(k, s)}{\partial k} \tag{2.31}$$

can be made uniformly as small as we please for a corresponding choice of the number $m > m(q)$. Thus for any $q > 0$ and sufficiently large values of m the equation (2.18) can be regarded as an approximate equation for the roots of (2.17) in the domain (2.30). To any root s_1 of (2.17) in this domain there corresponds an

---

1) It is easy to see that only these conditions will guarantee that U, $\tilde{U}$ and W, $\tilde{W}$ (together with the components of the displacement fields) are physically similar in respect of the natural frequencies of the vibrations and the amplitudes of both natural and forced vibrations of the plates.

arbitrarily close root $\tilde{s}_1$ of (2.18), and vice versa. Furthermore, if at $s = s_1$ we have[1)]

$$\frac{\partial L_1(k, s_1)}{\partial s_1} \neq 0, \tag{2.32}$$

then for the corresponding roots of (2.17) and (2.18) we have

$$L_1(k, \tilde{s}_1) - L_1(k, s_1) = \Delta_m(k, \tilde{s}_1) \approx \frac{\partial L_1(k, s_1)}{\partial s_1}(\tilde{s}_1 - s_1),$$

where for any $\epsilon > 0$, $\epsilon_1 > 0$ and arbitrary s_1 from (2.30) satisfying (2.32) the number m in (2.28) can be chosen so that

$$\left| \frac{\partial L_1(k, s_1)}{\partial s_1} - \frac{\partial \tilde{L}_1(k, \tilde{s}_1)}{\partial \tilde{s}_1} \right| < \varepsilon_1 \left| \frac{\partial \tilde{L}_1(k, \tilde{s}_1)}{\partial \tilde{s}_1} \right|, \tag{2.33}$$

$$|\tilde{s}_1 - s_1| \approx \left| \frac{\Delta_m(k, \tilde{s}_1)}{\frac{\partial \tilde{L}_1(k, \tilde{s}_1)}{\partial \tilde{s}_1}} \right| < \varepsilon |\tilde{s}_1|. \tag{2.34}$$

The numbers ϵ and ϵ_1 obviously describe the relative error in the determination of s_1 and $\partial L_1(k, s_1)/\partial s_1$ with respect to the approximate expression (2.19).

Equations (2.17) and (2.18) are of interest when $m \geq 0$ in (2.28) is fixed and small. In such cases the roots of (2.17) and (2.18) in the domains (2.30) cannot be assumed to be close for large values of q, because the quantities (2.31) are no longer small. But when the domains (2.30) are defined by sufficiently small values of $q > 0$ the quantities (2.31) are, as before, arbitrarily small. These domains will always then contain a number of roots of (2.17) and (2.18), and inequalities (2.32) will hold at the roots s_1 of (2.17).

If admissible relative errors $\epsilon > 0$ and $\epsilon_1 > 0$ are assigned in the definitions of s_1 and $\partial L_1(k, s_1)/\partial s_1$ respectively from (2.18) and (2.19), then for sufficiently small $q > 0$ inequalities (2.33) and (2.34) hold for all roots s_1 and $\tilde{s}_1$ of (2.17) and (2.18) in the domains (2.30). Therefore a least upper bound $q_m(\epsilon, \epsilon_1)$ of the numbers $q > 0$ exists such that (2.33) and (2.34) still hold in the domains (2.30) for $q < q_m(\epsilon, \epsilon_1)$.

A maximum admissible interval $(0, \tilde{k}_0(\epsilon, \epsilon_1))$ is defined with respect to the value of $q_m(\epsilon, \epsilon_1)$ and the dependence of the roots s_1 and $\tilde{s}_1$ of (2.17) and (2.18) on the parameter k. This interval is of the form (2.15) consistent with the requirement for ϵ-proximity of the exact and approximate solutions for the natural

---

1) This condition (2.32) always holds for perfect elasticity. In the case of an imperfectly elastic material it holds at all roots of (2.17) (of sufficiently small modulus) which are of interest in the engineering theory of vibrations of plates.

frequencies of vibrations of the plate and is such that inside this interval inequalities (2.33) hold.

The boundary value k_0 of the interval (2.15) which is contained in the domain $\Omega(k, s)$ mentioned at the end of §2.3 cannot exceed the value $\widetilde{k}_0(\epsilon, \epsilon_1)$.

In order to take into account the other conditions which $\Omega(k, s)$ must satisfy, we need to consider the numerators of the integrands for U, $\widetilde{U}$ and W, $\widetilde{W}$ from (2.22) and (2.23). It is necessary to remember here that the functions $\Phi_1^+(s)\Psi_1^+(k)$ and $\Phi_2^+(s)\Psi_2^+(k)$ determining the components $f_x^+(x, t)$ and $f_z^+(x, t)$ of the external action (1.16) can be assigned independently of each other.

It follows from (2.25)–(2.27) that the domains (2.30) in the case of sufficiently small $q > 0$ are, as before, natural domains for the variables k and s in which $\omega_m^{(\nu)}(k, s, z)$ and $\widetilde{\omega}_m^{(\nu)}(k, s, z)$ (describing the error in the determination of the exact values of the numerators with respect to their values from engineering theory) can be arbitrarily small. If in the domains $\Omega(k, s)$ of §2.3 the inequalities

$$\left.\begin{aligned} |\omega_m^{(\nu)}(k, s, z)| &< \varepsilon_1 M_\nu(k, s, z), \\ |\widetilde{\omega}_m^{(\nu)}(k, s, z)| &< \varepsilon_1 N_\nu(k, s, z) \end{aligned}\right\} \qquad (2.35)$$

hold simultaneously with (2.33) and (2.34), then the exact and approximate displacement fields will be physically similar not only in respect of natural frequencies but also in respect of amplitudes of both natural and forced vibrations of the plate. But it is easy to see that the inequalities (2.35) are unnecessarily stringent. For when an approximate value of the components of the field is determined the calculations in the domains Δz_0 where the components have the greatest absolute values should be carried out with the greatest accuracy. However, in the domains Δz where the components of the field are small (many times smaller in modulus than the maximum absolute value of the component in the whole domain of existence of this field) there is no necessity to carry out the calculations with great accuracy with respect to the level of the field in these domains. For the approximately calculated field to be physically similar to the true field it is sufficient here to require that the absolute error of the calculations should be small in comparison with the absolute values of the field in the domains Δz_0 in which the components have the greatest absolute values.

In accordance with this $M_\nu(k, s, z)$ and $N_\nu(k, s, z)$ on the right sides of inequalities (2.35) should be replaced by

$$\sup_z |M_\nu(k, s, z)| \text{ and } \sup_z |N_\nu(k, s, z)|$$

or

$$\frac{1}{2h}\int_{-h}^{h}|M_\nu(k, s, z)|\,dz \text{ and } \frac{1}{2h}\int_{-h}^{h}|N_\nu(k, s, z)|\,dz.$$

In practical cases of the engineering theory of longitudinal vibrations of plates, when m is always small ($m = 0$ or 1), we can take simpler expressions for the right sides of inequalities of the form (2.35), namely:

$$\sup\{|M_\nu(k, s, h)|,\quad |M_\nu(k, s, 0)|\},$$
$$\sup\{|N_\nu(k, s, h)|,\quad |N_\nu(k, s, 0)|\}$$

or even

$$\frac{1}{2}\{M_\nu(k, s, 0)| + |M_\nu(k, s, h)|\}, \tag{2.36}$$
$$|N_\nu(k, s, h)|,$$

since $N_\nu(k, s, 0) = 0$. From now on we shall use (2.36) to give the right sides for inequalities (2.35).

If we use (2.24) and write

$$\begin{aligned} Q_m^{(1)}(k, s) &= \left|\frac{\beta s^{-2}}{2}\right|\{|gf_m(\alpha h) - 2k^2 f_m(\beta h)| + \\ &+ |gf_m(\alpha h) f_{m_0}(\beta h) - 2k^2 f_m(\beta h) f_{m_0}(\alpha h)|\}, \\ Q_m^{(2)}(k, s) &= \left|\frac{ks^{-2}}{2}\right|\{|gF_m(\beta h) - 2\alpha\beta F_m(\alpha h)| + \\ &+ |gF_m(\beta h) f_{m_0}(\alpha h) - 2\alpha\beta F_m(\alpha h) f_{m_0}(\beta h)|\}, \end{aligned} \tag{2.37}$$

$$\left.\begin{aligned} \tilde{Q}_m^{(1)}(k, s) &= |ks^{-2}|\,|gf_m(\alpha h) F_{m_0}(\beta h) - 2\alpha\beta f_m(\beta h) F_{m_0}(\alpha h)|, \\ \tilde{Q}_m^{(2)}(k, s) &= |\alpha s^{-2}|\,|gF_m(\beta h) F_{m_0}(\alpha h) - 2k^2 F_m(\alpha h) F_{m_0}(\beta h)|, \end{aligned}\right\} \tag{2.38}$$

then the conditions (2.35) guaranteeing the physical similarity of the numerators of the integrands U, $\tilde{U}$ and W, $\tilde{W}$ from (2.22) and (2.23) with relative error $\epsilon_1 > 0$ can be written in the form

$$\left.\begin{aligned} |\omega_m^{(1)}(k, s, z)| &< \varepsilon_1 Q_m^{(1)}(k, s), \\ |\tilde{\omega}_m^{(1)}(k, s, z)| &< \varepsilon_1 \tilde{Q}_m^{(1)}(k, s) \end{aligned}\right\} \tag{2.39}$$

in the case $f_z^+(x, t) \equiv 0$, and in the form

$$\left.\begin{aligned} |\omega_m^{(2)}(k, s, z)| &< \varepsilon_1 Q_m^{(2)}(k, s), \\ |\tilde{\omega}_m^{(2)}(k, s, z)| &< \varepsilon_1 \tilde{Q}_m^{(2)}(k, s), \end{aligned}\right\} \tag{2.40}$$

if $f_x^+(x, t) \equiv 0$.

It is easy to see that for any fixed values of m and m_0 inequalities (2.39) and (2.40) hold in the domains (2.30) if $q > 0$ is sufficiently small. These inequalities therefore determine a nonempty domain $\Omega(k, s, \epsilon_1)$ which contains the intervals (2.15) for certain boundary values $k_0 = k_0(\epsilon_1)$ and $\omega_0 = \omega_0(\epsilon_1)$.

If for given values of $\epsilon > 0$ and $\epsilon_1 > 0$ we have $k_0(\epsilon_1) < \tilde{k}_0(\epsilon, \epsilon_1)$, where $\tilde{k}_0(\epsilon, \epsilon_1)$ is a boundary value of the interval $0 < k < \tilde{k}_0(\epsilon, \epsilon_1)$ in which (2.33) and (2.34) still hold, then it is necessary to take the domain $\Omega(k, s, \epsilon_1)$ as the unknown domain $\Omega(k, s)$. If, however, $k_0(\epsilon_1) > \tilde{k}_0(\epsilon, \epsilon_1)$, the domain $\Omega(k, s)$ must be determined from (2.39) and (2.40) together with the condition

$$0 < k < \tilde{k}_0(\varepsilon, \varepsilon_1). \tag{2.41}$$

Defining $\Omega(k, s)$ in this way and taking it to be the domain $\Omega(k, s)$ in the conditions defining the class (F) of external actions (1.16), we can assert that for arbitrary actions from the class (F) the exact and approximate solutions of our problems will be physically similar (with "exponents" ϵ and ϵ_1) with respect to the components of the displacement field. This similarity will not hold, however, for actions from the wider class of functions (1.16).

In conclusion we note that in discussing the physical similarity of the displacement fields of exact and approximate solutions it was possible earlier to take $m = m_0 \geq 0$ (that is, to retain the same number of terms in (2.11) and (2.13)). But in the analogous study of stress fields in a plate, which is of particular practical interest, it is shown that in the case $m = 0$ we cannot take $m_0 = 0$ in (2.11), since this leads to very large errors in the stresses, whatever the external actions may be. If, however, when $m = 0$ we take $m_0 = 1$, the accuracy of description of stress fields is perfectly satisfactory for actions satisfying the conditions of applicability of engineering theory.

5. In the case of flexural (skewsymmetric) vibrations of a plate the auxiliary functions are conveniently introduced by means of the formulas

$$\begin{aligned} \chi_0(x, t) &= \int_0^\infty \left.\begin{matrix} \sin kx \\ -\cos kx \end{matrix}\right\} dk \int_{(l)} \alpha A_2(k, s)\, e^{st} ds, \\ \zeta_0(x, t) &= \int_0^\infty \left.\begin{matrix} \cos kx \\ \sin kx \end{matrix}\right\} dk \int_{(l)} B_2(k, s)\, e^{st} ds, \end{aligned} \tag{2.42}$$

where $A_2(k, s)$ and $B_2(k, s)$ satisfy (1.27) and are defined by (1.31). If we choose numbers m and $m_0 = m_0(m)$ (which were referred to at the beginning of §2.2) and use (1.28), (2.9), (2.10) and the definitions (2.4) of the operators λ_ν^n, the components of the displacement field of flexural vibrations are easily obtained in the form

$$u = \sum_{n=0}^{m_0} \left\{ \lambda_1^n \frac{\partial \chi_0}{\partial x} - \lambda_2^{n+1} \zeta_0 \right\} \frac{z^{2n+1}}{(2n+1)!} + d_{m_0}^0,$$
$$w = \sum_{n=m}^{m_0} \left\{ \lambda_1^n \chi_0 + \lambda_2^n \frac{\partial \zeta_0}{\partial x} \right\} \frac{z^{2n}}{(2n)!} + \tilde{d}_{m_0}^0, \tag{2.43}$$

where the values of $d_{m\,0}^0$ and $\tilde{d}_{m\,0}^0$ are defined by (2.12) with A_1, B_1, $\Delta_{m\,0}$ and $\tilde{\Delta}_{m\,0}$ replaced by A_2, B_2, $\tilde{\Delta}_{m\,0}$ and $\Delta_{m\,0}$ respectively.

If, however, we replace $\operatorname{ch} y$ and $\operatorname{sh} y$ in the spectral relationships (1.27) (for $y = \alpha h$ and $y = \beta h$) by the right sides of (2.9) respectively, use the definitions (2.4) and (2.7) of the operators λ_ν^n and $\tilde{\mu}$ and go over to the functions χ_0, ζ_0 from (2.42), we obtain the equations

$$\sum_{n=0}^{m} \frac{h^{2n}}{(2n)!} \left\{ 2\lambda_1^n \frac{\partial \chi_0}{\partial x} + \left(\frac{\partial^2}{\partial x^2} - \lambda_2 \right) \lambda_2^n \zeta_0 \right\} + \tilde{D}_m^0 = \tilde{\mu} f_x^+(x, t),$$
$$\sum_{n=0}^{m} \frac{h^{2n+1}}{(2n+1)!} \left\{ \left(\lambda_2 - \frac{\partial^2}{\partial x^2} \right) \lambda_1^n \chi_0 + 2\lambda_2^{n+1} \frac{\partial \zeta_0}{\partial x} \right\} + D_m^0 = \tilde{\mu} f_z^+(x, t), \tag{2.44}$$

which are analogous to (2.13). Then D_m^0 and $\tilde{D}_m^0$ are defined by (2.14) if we replace A_1, B_1, Δ_m and $\tilde{\Delta}_m$ by A_2, B_2, $\tilde{\Delta}_m$ and Δ_m respectively.

Exactly as before, the equations of flexural vibrations of the engineering theory of plates can be obtained only if for arbitrary actions (1.16) from the class (F) and suitably chosen (small) numbers m and m_0 we may neglect the expressions $d_{m\,0}^0$, $\tilde{d}_{m\,0}^0$, D_m^0 and $\tilde{D}_m^0$ in (2.43) and (2.44) and thereby obtain approximate formulas for the displacement field and the equation of motion of the plate. The conditions guaranteeing this are determined exactly as in §§2.3 and 2.4. Moreover, a comparison of the formulas (1.25), (1.26) and (1.30) with (1.27), (1.28) and (1.31) shows that in such investigations all the formulas from (2.16) to (2.35) remain valid if we carry out "A-substitutions", that is, if we replace $\operatorname{ch} y$, $\operatorname{sh} y$, $f_m(y)$, $F_m(y)$, $\Delta_m(y)$ and $\tilde{\Delta}_m(y)$ (for all arguments y and subscripts m encountered) by $\operatorname{sh} y$, $\operatorname{ch} y$, $F_m(y)$, $f_m(y)$, $\tilde{\Delta}_m(y)$ and $\Delta_m(y)$ respectively. We must then of course replace L_1, A_1 and B_1 by L_2, A_2 and B_2. We can therefore write down at once all the necessary formulas relating to the conditions of applicability of engineering theory in the case of flexural vibrations of plates.

We denote by $\Delta_m^0(k, s)$; $\omega_m^{(01)}(k, s, z)$ and $\omega_m^{(02)}(k, s, z)$; and $\tilde{\omega}_m^{(01)}(k, s, z)$ and $\tilde{\omega}_m^{(02)}(k, s, z)$ the results of A-substitutions in (2.29), (2.26) and (2.27) respectively. In addition, we put

$$\begin{aligned} Q_m^{(01)}(k, s) &= |\beta s^{-2}|\,|gF_m(\alpha h)F_{m_0}(\beta h) - 2k^2F_m(\beta h)F_{m_0}(\alpha h)|, \\ Q_m^{(02)}(k, s) &= |ks^{-2}|\,|gf_m(\beta h)F_{m_0}(\alpha h) - 2\alpha\beta f_m(\alpha h)F_{m_0}(\beta h)|, \\ \tilde{Q}_m^{(01)}(k, s) &= \left|\frac{ks^{-2}}{2}\right|\{|gF_m(\alpha h) - 2\alpha\beta F_m(\beta h)| \\ &\quad + |gF_m(\alpha h)f_m(\beta h) - 2\alpha\beta F_m(\beta h)f_{m_0}(\alpha h)|\}, \\ \tilde{Q}_m^{(02)}(k, s) &= \left|\frac{\alpha s^{-2}}{2}\right|\{|gf_m(\beta h) - 2k^2f_m(\alpha h)| \\ &\quad + |gf_m(\beta h)f_{m_0}(\alpha h) - 2k^2f_m(\alpha h)f_{m_0}(\beta h)|\} \end{aligned} \tag{2.45}$$

and also write

$$\left|\frac{\Delta_m^0(k, \tilde{s}_2)}{\dfrac{\partial\tilde{L}_2(k, \tilde{s}_2)}{\partial\tilde{s}_2}}\right| < \varepsilon|\tilde{s}_2|, \tag{2.46}$$

$$\left|\frac{\partial L_2(k, s_2)}{\partial s_2} - \frac{\partial\tilde{L}_2(k, \tilde{s}_2)}{\partial\tilde{s}_2}\right| < \varepsilon_1\left|\frac{\partial\tilde{L}_2(k, \tilde{s}_2)}{\partial\tilde{s}_2}\right|,^{1)} \tag{2.47}$$

$$\left.\begin{aligned} |\omega_m^{(01)}(k, s, z)| &< \varepsilon_1 Q_m^{(01)}(k, s), \\ |\tilde{\omega}_m^{(01)}(k, s, z)| &< \varepsilon_1\tilde{Q}_m^{(01)}(k, s) \end{aligned}\right\} \quad \text{if} \quad f_z^{\pm}(x, t) \equiv 0, \tag{2.48}$$

$$\left.\begin{aligned} |\omega_m^{(02)}(k, s, z)| &< \varepsilon_1 Q_m^{(02)}(k, s), \\ |\tilde{\omega}_m^{(02)}(k, s, z)| &< \varepsilon_1\tilde{Q}_m^{(02)}(k, s) \end{aligned}\right\} \quad \text{if} \quad f_x^{\pm}(x, t) \equiv 0, \tag{2.49}$$

which are analogous to (2.34), (2.33), (2.39) and (2.40), respectively.

It can then be asserted that for chosen numbers m, m_0 and $\epsilon > 0$, $\epsilon_1 > 0$ (respectively determining the order of approximation and accuracy of description of the dynamic processes in the plate) the engineering equations of flexural vibrations of plates and the formulas for the components of the displacement field obtained from (2.44) and (2.43) will be valid only for the class (F) of actions (1.16) for which the interval $(0, k_0)$ and the domain $\Omega(k, s)$ from §1.2 are contained in domains where all the inequalities (2.46)–(2.49) still hold.

---

1) Here s_2 and $\tilde{s}_2$ denote roots of $L_2(k, s) = 0$ and $\tilde{L}_2(k, s_2) = 0$ respectively, where $L_2(k, s)$ is defined by (1.29) and $\tilde{L}_2(k, s)$ is obtained from (2.19) by means of an A-substitution.

§3. The equations of motion of the engineering theory of vibrations of imperfectly elastic plates

It follows from above that the equations of motion and formulas for the components of the displacement fields of imperfectly elastic plates contain the operators λ_ν^n from (2.4) and the operator $\tilde{\mu}$ from (2.7) which depend on the kernels $K_\nu(\tau)$ of the elastic lag from (1.1). We begin by determining the form of these operators in relation to the description of the elastic lag in the framework of the Voigt and Boltzmann models. We assume that the operators are applied to the functions

$$f(x, t) = \int_0^\infty \left.\begin{matrix}\cos kx \\ \sin kx\end{matrix}\right\} dk \int_{(l)} F(k, s) e^{st} ds. \tag{3.1}$$

Here the contour (l) coincides with the straight line $\operatorname{Re} s = \sigma > 0$; $F(k, s)$ is regular for $\operatorname{Re} s > 0$ and for any $k > 0$ and $|s| \to \infty$ in the domain $\operatorname{Re} s > 0$ decreases not more slowly than $|s|^{-n_0}$, where $n_0 >> 1$ is a sufficiently large number, and finally, for any s and $k \to \infty$, $F(k, s)$ decreases not more slowly than k^{-n_0}. Solutions of equations of interest to us are functions of just this form (3.1). These functions can be differentiated under the integral sign with respect to x and t not fewer than $n_0 - 2$ times, and they, together with these derivatives, vanish at $t = 0$.

1. **The Voigt model.** In discussing this model in place of (1.1) we usually write[1)]

$$\begin{gathered}(\lambda + 2\mu) = (\lambda_0 + 2\mu_0)\left(1 + \varepsilon \frac{\partial}{\partial t}\right), \\ \mu = \mu_0\left(1 + \Omega \frac{\partial}{\partial t}\right),\end{gathered} \tag{3.2}$$

where $\epsilon > 0$ and $\Omega > 0$ (as well as $\mu_0 > 0$ and $\lambda_0 > 0$) are constants. The transforms (1.19) of the kernels of the operators are given by

$$\varkappa_1(s) = -\varepsilon s, \quad \varkappa_2(s) = -\Omega s. \tag{3.3}$$

If we introduce the operator

$$m[p] = \int_0^t e^{-p(t-\tau)} \ldots d\tau, \tag{3.4}$$

---

1) This formally fits into the scheme (1.1) if we take Dirac's δ-function and put $K_1(\tau) = -\epsilon\,\delta'(\tau)$ and $K_2(\tau) = -\Omega\,\delta'(\tau)$.

which depends on the parameter $p > 0$, satisfies

$$\lim_{p\to\infty} pm[p] = 1$$

in the class of continuous functions, and commutes in the class of functions (3.1) with the operators $\partial^m/\partial x^m$ and $\partial^m/\partial t^m$ for $m \leq n_0 - 2$, then it is easy to see that

$$\begin{aligned} \lambda_1 &= -\frac{\partial^2}{\partial x^2} + \frac{a_0^2}{\varepsilon} m\left[\frac{1}{\varepsilon}\right]\frac{\partial^2}{\partial t^2}, \\ \lambda_2 &= -\frac{\partial^2}{\partial x^2} + \frac{b_0^2}{\Omega} m\left[\frac{1}{\Omega}\right]\frac{\partial^2}{\partial t^2}, \\ \tilde{\mu} &= \frac{1}{\mu_0 \Omega} m\left[\frac{1}{\Omega}\right]. \end{aligned} \tag{3.5}$$

For this it is sufficient merely to take account of the definitions (2.4) and (2.7) of the operators λ_ν^n and $\tilde{\mu}$ and use the notation (1.22).

The operators λ_ν^n for $2n \leq n_0 - 2$ are obtained from (3.5) by successive application of the operators λ_ν. As $\epsilon \to 0$ and $\Omega \to 0$ (corresponding to a transition to the case of perfect elasticity), we have

$$\begin{aligned} \lambda_1 &= \lambda_a = -\frac{\partial^2}{\partial x^2} + a^2\frac{\partial_2}{\partial t^2}, \\ \lambda_2 &= \lambda_b = -\frac{\partial^2}{\partial x^2} + b^2\frac{\partial^2}{\partial t^2}, \\ \tilde{\mu} &= \frac{1}{\mu}, \end{aligned} \tag{3.6}$$

where $a = a_0$, $b = b_0$ and $\mu = \mu_0$. As we know, the operators (3.6) appear in the equations of motion of the engineering theory of vibrations of perfectly elastic plates.

To conclude, we note that in the class of functions (3.1) we have the identity

$$m[p]\frac{\partial^2}{\partial t^2} = \frac{\partial}{\partial t} - p + p^2 m[p], \tag{3.7}$$

which can be used to lower the order of the operator equations of the engineering theory.

2. The Boltzmann model. The kernels of the operators from (1.1) have the form

$$K_\nu(\tau) = \sum_{i=1}^{N} \delta_i^{(\nu)} e^{-\sigma_i^{(\nu)}\tau}, \tag{3.8}$$

where the positive constants $\sigma_i^{(\nu)}$ and $\delta_i^{(\nu)}$ satisfy

$$\sum_{i=1}^{N} \frac{\delta_i^{(\nu)}}{\sigma_i^{(\nu)}} < 1 \tag{3.9}$$

and are ordered as follows:

$$0 < \sigma_1^{(\nu)} < \sigma_2^{(\nu)} < \cdots < \sigma_N^{(\nu)}. \tag{3.10}$$

The transforms (1.19) of the kernels (3.8) are given by

$$\varkappa_\nu(s) = \sum_{i=1}^{N} \frac{\delta_i^{(\nu)}}{s + \sigma_i^{(\nu)}} \qquad (\nu = 1,\ 2). \tag{3.11}$$

In order to put the operators λ_ν^n and $\tilde{\mu}$ from (2.4) and (2.7) into the most convenient form for applications, we recall that the numbers $\tilde{\delta}_i^{(\nu)}$ and $\tilde{\sigma}_i^{(\nu)}$ $(i = 1, 2, \cdots, N)$ can always be defined uniquely with respect to the given numbers $\delta_i^{(\nu)}$ and $\sigma_i^{(\nu)}$ from (3.8) satisfying (3.9) and ordered as in (3.10). We then have the identities $(\nu = 1, 2)$

$$\frac{1}{1 - \varkappa_\nu(s)} \equiv \left[1 - \sum_{i=1}^{N} \frac{\delta_i^{(\nu)}}{s + \sigma_i^{(\nu)}}\right]^{-1} \equiv 1 + \sum_{i=1}^{N} \frac{\tilde{\delta}_i^{(\nu)}}{s + \tilde{\sigma}_i^{(\nu)}}. \tag{3.12}$$

The numbers $\tilde{\sigma}_i^{(\nu)}$ are determined as roots of the equations

$$\prod_{i=1}^{N} (s + \sigma_i^{(\nu)}) - \sum_{i=1}^{N} \delta_i^{(\nu)} \prod_{j \neq i} (s + \sigma_j^{(\nu)}) = 0, \tag{3.13}$$

which are real and satisfy

$$0 < \tilde{\sigma}_1^{(\nu)} < \sigma_1^{(\nu)} < \tilde{\sigma}_2^{(\nu)} < \sigma_2^{(\nu)} < \cdots < \tilde{\sigma}_N^{(\nu)} < \sigma_N^{(\nu)}. \tag{3.14}$$

The numbers $\tilde{\delta}_i^{(\nu)}$ satisfy

$$\tilde{\delta}_i^{(\nu)} = \frac{\prod_{k=1}^{N} (\sigma_k^{(\nu)} - \tilde{\sigma}_i^{(\nu)})}{\prod_{k \neq i} (\tilde{\sigma}_k^{(\nu)} - \tilde{\sigma}_i^{(\nu)})}. \tag{3.15}$$

For, discarding the index ν in (3.12), bringing each part of the equation to a common denominator and equating separately the numerators and denominators gives

$$\prod_{i=1}^{N}(s+\sigma_i)-\sum_{i=1}^{N}\delta_i\prod_{j\neq i}(s+\sigma_j)\equiv\prod_{i=1}^{N}(s+\tilde{\sigma}_i), \tag{3.16}$$

$$\prod_{i=1}^{N}(s+\tilde{\sigma}_i)+\sum_{i=1}^{N}\tilde{\delta}_i\prod_{j\neq i}(s+\tilde{\sigma}_j)\equiv\prod_{i=1}^{N}(s+\sigma_i). \tag{3.17}$$

The identity (3.12) holds if $\tilde{\sigma}_i$ and $\tilde{\delta}_i$ can be chosen so that (3.16) and (3.17) are valid. The latter can be described as the resolution of polynomials on the left sides into their simplest factors. It follows from (3.16) that $\tilde{\sigma}_i$ are roots of the equation

$$h(s)\equiv\prod_{i=1}^{N}(s+\sigma_i)-\sum_{i=1}^{N}\delta_i\prod_{j\neq i}(s+\sigma_j)=0.$$

Taking (3.9) and (3.10) into account, we establish for this last equation that

$$h(0)>0,\ h(-\sigma_1)<0,\ h(-\sigma_2)>0,\ \ldots,\ (-)^N h(-\sigma_N)>0,$$

from which it follows that there exist N real roots which can be arranged in the order (3.14). On determining these roots we obtain (3.16).

In order that (3.17) should hold identically, it is necessary and sufficient that

$$\prod_{i=1}^{N}(\tilde{\sigma}_i-\sigma_k)+\sum_{i=1}^{N}\tilde{\delta}_i\prod_{j\neq i}(\tilde{\sigma}_j-\sigma_k)=0 \quad (k=1,\ 2,\ \ldots,\ N),$$

which are equivalent to the linear algebraic equations

$$\sum_{i=1}^{N}\frac{\tilde{\delta}_i}{\tilde{\sigma}_i-\sigma_k}=-1 \quad (k=1,\ 2,\ \ldots,\ N) \tag{3.18}$$

in $\tilde{\delta}_i$. The system (3.18) is uniquely solvable if

$$\Delta\equiv\left|\frac{1}{\tilde{\sigma}_i-\sigma_k}\right|\neq 0.$$

But this determinant is in fact nonzero, since from (3.16) with $s=-\tilde{\sigma}_k$ we have

$$\sum_{i=1}^{N}\frac{\delta_i}{\sigma_i-\tilde{\sigma}_k}=1 \quad (k=1,\ 2,\ \ldots,\ N)$$

(with the same Δ), which automatically has solutions $\delta_i>0$. Thus $\tilde{\delta}_i$ exist and are easily determined from (3.17) by putting $s=-\tilde{\sigma}_k$ which also leads to (3.15).

On determining the numbers $\tilde{\delta}_i^{(\nu)}$ and $\tilde{\sigma}_i^{(\nu)}$ from (3.14) and (3.15) and taking into account formulas (1.22), we can easily establish, on the basis of the definitions (2.4) and (2.7) of the operators λ_ν^n and $\tilde{\mu}$, that

$$\lambda_1 = -\frac{\partial^2}{\partial x^2} + a_0^2\left[1+\sum_{i=1}^{N}\tilde{\delta}_i^{(1)} m\,[\tilde{\sigma}_i^{(1)}]\right]\frac{\partial^2}{\partial t^2},$$
$$\lambda_2 = -\frac{\partial^2}{\partial x^2} + b_0^2\left[1+\sum_{i=1}^{N}\tilde{\delta}_i^{(2)} m\,[\tilde{\sigma}_i^{(2)}]\right]\frac{\partial^2}{\partial t^2}, \qquad (3.19)$$
$$\tilde{\mu} = \frac{1}{\mu_0}\left[1+\sum_{i=1}^{N}\tilde{\delta}_i^{(2)} m\,[\tilde{\sigma}_i^{(2)}]\right],$$

where the values of the operators $m\,[p]$ for $p=\tilde{\sigma}_i^{(\omega)}$ are determined from (3.4). The values of the operators λ_ν^n for $2n < n_0 - 2$ are obtained by successive application of the operators λ_ν from (3.19).

To conclude, we note that we may effect a transition from the Boltzmann model to a perfect elasticity model by assuming that $\sigma_i^{(\nu)} = \text{const}$ and $\delta_i^{(\nu)} \to 0$. As a result $\tilde{\sigma}_i^{(\nu)} \to \sigma_i^{(\nu)}$ and $\tilde{\delta}_i^{(\nu)} \to 0$, so that the operators (3.19) go into the operators (3.6) in the case of perfect elasticity. Furthermore, it is emphasized that if we introduce the operators

$$a^2 = a_0^2\left[\varkappa_0 + \int_0^t \tilde{K}_1(t-\tau)\ldots d\tau\right],$$
$$b^2 = b_0^2\left[\varkappa_0 + \int_0^t \tilde{K}_2(t-\tau)\ldots d\tau\right], \qquad (3.20)$$
$$\frac{1}{\mu} = \frac{1}{\mu_0}\left[\varkappa_0 + \int_0^t \tilde{K}_2(t-\tau)\ldots d\tau\right],$$

in which

$$\varkappa_0 = 0,\quad \tilde{K}_1(\tau) = \frac{1}{\varepsilon}e^{-\frac{t}{\varepsilon}},\quad \tilde{K}_2(\tau) = \frac{1}{\Omega}e^{-\frac{t}{\Omega}} \qquad (3.21)$$

for the Voigt model,

$$\varkappa_0 = 1,\quad \tilde{K}_1(\tau) = \sum_{i=1}^{N}\tilde{\delta}_i^{(1)}e^{-\tilde{\sigma}_i^{(1)}\tau},\quad \tilde{K}_2(\tau) = \sum_{i=1}^{N}\tilde{\delta}_i^{(2)}e^{-\tilde{\sigma}_i^{(2)}\tau} \qquad (3.22)$$

for the Boltzmann model, and

$$\varkappa_0 = 1, \quad \tilde{K}_1(\tau) = \tilde{K}_2(\tau) \equiv 0 \tag{3.23}$$

in the case of perfect elasticity, then the expressions for λ_1, λ_2 and $\tilde{\mu}$ acquire the outwardly single form (3.6) which is independent of the model of the medium. In what follows we shall take a^2, b^2 and μ^{-1} to be the operators (3.20).

3. We now discuss certain questions concerning the description of longitudinal vibrations of plates within the framework of engineering theory.

On using the notation (3.20) and putting $m = 0$ in (2.13) we obtain the equations of motion in the zero approximation:

$$\left.\begin{aligned} &2\left(a^2 \frac{\partial^3}{\partial t^2 \partial x} - \frac{\partial^3}{\partial x^3}\right)\chi + \left(2\frac{\partial^2}{\partial x^2} - b^2 \frac{\partial^2}{\partial t^2}\right)\zeta = \frac{1}{\mu}\cdot\frac{f_x^+(x,\ t)}{h}, \\ &\left(b^2 \frac{\partial^2}{\partial t^2} - 2\frac{\partial^2}{\partial x^2}\right)\chi + 2\frac{\partial}{\partial x}\zeta = \frac{1}{\mu} f_z^+(x,\ t). \end{aligned}\right\} \tag{3.24}$$

The formulas for the components of the displacement field follow from (2.11), where, putting $m_0 = 1$, we have

$$\begin{aligned} u &= \frac{\partial \chi}{\partial t} - \zeta + \frac{z^2}{2}\left[\left(a^2 \frac{\partial^2}{\partial t^2} - \frac{\partial^2}{\partial x^2}\right)\frac{\partial \chi}{\partial x} - \left(b^2 \frac{\partial^2}{\partial t^2} - \frac{\partial^2}{\partial x^2}\right)\zeta\right], \\ w &= z\left[\left(a^2 \frac{\partial^2}{\partial t^2} - \frac{\partial^2}{\partial x^2}\right)\chi + \frac{\partial \zeta}{\partial x}\right] + \frac{z^3}{6}\left[\left(a^2 \frac{\partial^2}{\partial t^2} - \frac{\partial^2}{\partial x^2}\right)^2 \chi \right. \\ &\quad \left. + \left(b^2 \frac{\partial^2}{\partial t^2} - \frac{\partial^2}{\partial x^2}\right)\frac{\partial \zeta}{\partial x}\right]. \end{aligned} \tag{3.25}$$

The conditions of applicability of equations (3.24) and formulas (3.25) follow from (2.34), (2.33), (2.39) and (2.40) with $m = 0$ and $m_0 = 1$. To avoid making special assumptions regarding the nature of the elastic lag in the material of the plate, we shall give results only in the case when $\tilde{K}_1(\tau) = \tilde{K}_2(\tau)$ in (3.21) or (3.22). We shall assume that $\lambda_0 = 2\mu_0$ (that is, the Poisson coefficient $\nu = 0.33$) and that the numbers $\epsilon > 0$ and $\epsilon_1 > 0$, which describe the relative errors in the determination of the natural frequencies and amplitudes of natural and forced vibrations of the plate, are already given and small.

When $m = 0$ the expressions for $\tilde{L}_1(k, s)$ and $\Delta_m(k, s)$ are very simple. The characteristic equation (2.18) of the system (3.24) has two complex conjugate roots for which (2.34) and (2.33) hold if

$$|kh| < 5\sqrt{\varepsilon} \text{ and } |kh| < \sqrt{2\varepsilon_1}. \tag{3.26}$$

When $m = 0$ and $m_0 = 1$ the somewhat more complex inequalities (2.39) and (2.40) give

$$\left.\begin{array}{c} h^2|\beta^2-3\alpha^2|<6\varepsilon_1,\ h^2|\beta^2-2\alpha^2|<4\varepsilon_1,\\ h^2|\alpha^2+\beta^2|<2\varepsilon_1\end{array}\right\} \tag{3.27}$$

in the case $f_z^+(x, t) \equiv 0$, and

$$\left.\begin{array}{c} h^2|\beta^2-3\alpha^2|<6\varepsilon_1,\ h^2|\beta^2+3\alpha^2|<6\varepsilon_1,\\ h^2|\alpha^2+\beta^2|<2\varepsilon_1,\end{array}\right\} \tag{3.28}$$

if $f_x^+(x, t) \equiv 0$. As a qualitative estimate of the domain of interest to us we can take

$$|\alpha h|<\sqrt{\varepsilon_1},\quad |\beta h|<\sqrt{\varepsilon_1}. \tag{3.29}$$

Conditions (3.26)–(3.28) determine the intervals (2.15) and domain $\Omega(s, k)$ on the plane (s) outside which the integrands in (1.16) of actions of class (F) must have vanishingly small values if the accuracy of the description of dynamic processes in the plate on the basis of the engineering formulas (3.24) and (3.25) is still to be characterized by ϵ and ϵ_1. It is clear from §1.2 that these conditions imply an upper bound on the frequency spectrum of the external actions and a lower bound on the dimensions of the domain over which the action is applied. The greater the accuracy required from (3.24) and (3.25), the stronger these restrictions become.

It should be noted that in the engineering theory the vibrations of plates in the zero approximation $(m = 0)$ are usually confined to the calculation in (3.25) of linear terms in z only. It would be easy to show that it is still admissible to use these formulas in describing the components of the displacement field (in a domain somewhat smaller than (3.29)), but they are unsuitable for describing the stress field in the plate. This can be seen most simply by verifying the feasibility of the boundary conditions (1.4) in the engineering description of vibrations of plates.

By using the formulas from §2 for $m = 0$ corresponding to (3.24), and displacement formulas from (2.11) for $m_0 = 0$ corresponding to linear terms in z from (3.25), we can easily obtain expressions for the left sides of the boundary conditions (1.4). It turns out that the boundary condition in τ_{zz} in approximate theory is satisfied exactly, and that for τ_{xz} when $z = \pm h$ we have

$$\tau_{xz}\Big|_{z=\pm h}=\int_0^\infty \left.\begin{array}{c}\cos kx\\ \sin kx\end{array}\right\} dk \int_{(l)} \left\{\left[1-\frac{g\beta^2-2k^2\alpha^2}{g^2-4\alpha^2k^2}\right]\Phi_1(s)\Psi_1(k) - \alpha h\frac{k\alpha\,(g-2\beta^2)}{g^2-4\alpha^2k^2}\Phi_2(s)\Psi_2(k)\right\}e^{st}ds. \tag{3.30}$$

In the domain of applicability of equations (3.24) we have $|\alpha h| \ll 1$. The contribution from the term with multiplier $\Phi_2(s)\Psi_2(k)$ to the expression τ_{xz} can therefore be assumed to be small. But the term corresponding to the integrand

$$\frac{g\beta^2 - 2k^2\alpha^2}{g^2 - 4\alpha^2k^2}\,\Phi_1(s)\,\Psi_1(k)$$

adds a contribution to τ_{xz} which is comparable with $f_x^+(x, t)$. The relative error in satisfying the boundary condition (1.4) in τ_{xz} is therefore no longer small and does not decrease as ϵ and ϵ_1 decrease. But if we use the equations of motion in the zero approximation and the displacement formulas (3.25) in the first approximation ($m_0 = 1$), the state of affairs is rectified. Each of the boundary conditions (1.4) has an error represented by integrals of the form (3.30) with integrands containing the factors $|\alpha h|$ and $|\beta h|$ to not less than the first power. In the domain of applicability of the theory these factors must be assumed small. Therefore it is in fact possible to speak of the boundary conditions being approximately satisfied.

We note in conclusion that if we introduce new auxiliary functions

$$u_0 = \frac{\partial \chi}{\partial x} - \zeta, \quad \tilde{\chi} = \chi,$$

then the equations (3.24) and displacement formulas (3.25) take the forms

$$\left.\begin{aligned} b^4\frac{\partial^2 u_0}{\partial t^2} - 4(b^2 - a^2)\frac{\partial^2 u^0}{\partial x^2} &= \frac{1}{\mu}\left[b^2\frac{f_x^+}{h} - (2a^2 - b^2)\frac{\partial f_z^+}{\partial x}\right], \\ b^2\frac{\partial^2\tilde{\chi}}{\partial t^2} - 2\frac{\partial u_0}{\partial x} &= \frac{1}{\mu}f_z^+ \end{aligned}\right\} \qquad (3.31)$$

and

$$\left.\begin{aligned} u &= u_0 + \frac{z^2}{2}\left[\left(b^2\frac{\partial^2}{\partial t^2} - \frac{\partial^2}{\partial x^2}\right)u_0 + (a^2 - b^2)\frac{\partial^3\tilde{\chi}}{\partial t^2\partial x}\right], \\ w &= z\left[a^2\frac{\partial^2\tilde{\chi}}{\partial t^2} - \frac{\partial u_0}{\partial x}\right] + \frac{z^3}{6}\left[a^4\frac{\partial^4\tilde{\chi}}{\partial t^4} - (2a^2 - b^2)\frac{\partial^4\tilde{\chi}}{\partial x^2\partial t^2} - \right. \\ &\qquad \left. - \left(b^2\frac{\partial^2}{\partial t^2} - \frac{\partial^2}{\partial x^2}\right)\frac{\partial u_0}{\partial x}\right]. \end{aligned}\right\} \qquad (3.32)$$

In the case of perfect elasticity the first equation of (3.31) has the usual form for the engineering theory of vibrations of plates and it can be assumed that (3.31) leads to a simplification of the equations of motion. However, in the case of imperfectly elastic models of the media there are no grounds for assuming equations (3.31) to be simpler than (3.24).

We shall not write out the equations of motion of longitudinal vibrations of plates in the first (and subsequent) approximations which follow from (2.13) when $m = 1$ (and $m > 1$) and $D_m = \widetilde{D}_m = 0$. We merely note that when $m = 1$ formula

(3.25) can be used for the approximate description of both the displacement field in the plate and also the stress field.

4. In the case of flexural vibrations of plates the approximate equations of motion and formulas for the components of the displacement field are obtained from (2.44) and (2.43) respectively for fixed chosen members m and m_0 and quantities $d^0_{m0} = \tilde{d}^0_{m0} = D^0_m = \tilde{D}^0_m = 0$. If we put $m = 0$ in (2.44) we obtain equations expressing the law of motion of the plate as a whole. They are not of special interest and we shall not give them here. When $m = 1$ from (2.44) we have

$$
\begin{gathered}
2\frac{\partial\chi_0}{\partial x}+\left(2\frac{\partial^2}{\partial x}-b^2\frac{\partial^2}{\partial t^2}\right)\zeta_0+\frac{h^2}{2}\left\{2\left(a^2\frac{\partial^2}{\partial t^2}-\frac{\partial^2}{\partial x^2}\right)\frac{\partial\chi}{\partial x}+\right.\\
\left.+\left(2\frac{\partial^2}{\partial x^2}-b^2\frac{\partial^2}{\partial t^2}\right)\left(b^2\frac{\partial^2}{\partial t^2}-\frac{\partial^2}{\partial x^2}\right)\zeta_0\right\}=\frac{1}{\mu}f_x^+(x,\ t),\\
\left(b^2\frac{\partial^2}{\partial t^2}-2\frac{\partial^2}{\partial x^2}\right)\chi_0+2\left(b^2\frac{\partial^2}{\partial t^2}-\frac{\partial^2}{\partial x^2}\right)\frac{\partial\zeta_0}{\partial x}\\
+\frac{h^2}{6}\left\{\left(b^2\frac{\partial^2}{\partial t^2}-2\frac{\partial^2}{\partial x^2}\right)\left(a^2\frac{\partial^2}{\partial t^2}-\frac{\partial^2}{\partial x^2}\right)\chi_0+2\left(b^2\frac{\partial^2}{\partial t^2}\right.\right.\\
\left.\left.-\frac{\partial^2}{\partial x^2}\right)^2\frac{\partial\zeta_0}{\partial x}\right\}=\frac{1}{\mu}\frac{f_z^+(x,\ t)}{h}.
\end{gathered}
\tag{3.33}
$$

From (2.43) the components of the displacement field when $m_0 = 1$ are given by

$$
\begin{gathered}
u=z\left[\frac{\partial\chi_0}{\partial x}+\left(\frac{\partial^2}{\partial x^2}-b^2\frac{\partial^2}{\partial t^2}\right)\zeta_0\right]+\frac{z^3}{6}\left\{\left(a^2\frac{\partial^2}{\partial t^2}-\frac{\partial^2}{\partial x^2}\right)\frac{\partial\chi_0}{\partial x}\right.\\
\left.-\left(b^2\frac{\partial^2}{\partial t^2}-\frac{\partial^2}{\partial x^2}\right)^2\zeta_0\right\},\\
w=\chi_0+\frac{\partial\zeta_0}{\partial x}+\frac{z^2}{2}\left\{\left(a^2\frac{\partial^2}{\partial t^2}-\frac{\partial^2}{\partial x^2}\right)\chi_0+\left(b^2\frac{\partial^2}{\partial t^2}-\frac{\partial^2}{\partial x^2}\right)\frac{\partial\zeta_0}{\partial x}\right\}.
\end{gathered}
\tag{3.34}
$$

For an approximate description of the stress field in the plate all the terms in (3.34) must, as in the case of longitudinal vibrations, be taken into account. In considering the displacement field only we may, however, confine ourselves to the linear terms in z.

Equations (3.33) are essentially different in form from those which are usually taken to be the engineering equations of flexural vibrations in the case of perfect elasticity. These are only obtained by assuming that the spectral densities of the auxiliary functions (2.42) are appreciably different from zero only in the region of "very small" values of k and $|s|$ (for $kh \ll 1$ and $|\alpha h| \ll 1$, $|\beta h| \ll 1$), corresponding to actions which are very low-frequency and very smoothly distributed with respect to x. The principal terms in (3.33) are then those on the left side of the second equation which do not contain h^2. On differentiating the first equation of (3.33) with respect to x, combining it with the second equation, and using the notation

$$w_0 = \chi_0 + \frac{\partial \zeta_0}{\partial x} \tag{3.35}$$

and the formula

$$b^2 \frac{\partial^2 \chi_0}{\partial t^2} = 2 \frac{\partial^2 w_0}{\partial x^2} - 2b^2 \frac{\partial^2}{\partial t^2} \cdot \frac{\partial \zeta_0}{\partial x} + \frac{1}{\mu} \cdot \frac{f_z^+}{h} - \frac{h^2}{6} [\ldots]$$

(which follows from the second equation of (3.33)), we obtain

$$\begin{aligned} b^4 \frac{\partial^2 w_0}{\partial t^2} + h^2 \Big[\frac{4}{3} (b^2 - a^2) \frac{\partial^4 w_0}{\partial x^4} - \frac{3b^2 - 2a^2}{2} b^2 \frac{\partial^4 w_0}{\partial x^2 \partial t^2} - \frac{2a^2 + b^2}{6} b^4 \frac{\partial^4 w_0}{\partial t^4} \Big] \\ = b^2 \frac{1}{\mu} \Big[\frac{f_z^+}{h} + \frac{\partial f_x^+}{\partial x} \Big] + \frac{h^2}{6\mu} \Big[(a^2 + b^2) b^2 \frac{\partial^2}{\partial t^2} \cdot \frac{f_z^+}{h} \\ + (4a^2 - 6b^2) \frac{\partial^2}{\partial x} \cdot \frac{f_z^+}{h} \Big] + h^4 [\ldots] \end{aligned} \tag{3.36}$$

and a second equation which we need not write out here.

Equation (3.36) together with the discarded terms containing h^2 and h^4 on the right side coincides with the generally accepted engineering equation of flexural vibrations of perfectly elastic plates. But transferring to this equation appears not to be useful even in the case of perfect elasticity because it can only result in a contraction of the domain of applicability of the theory.

5. To establish the domain of applicability of equations (3.33) and formulas (3.34) of the engineering theory of flexural vibrations of plates we need to study the inequalities (2.46), (2.47), (2.48) and (2.49) when $m = m_0 = 1$. We confine ourselves here to the simplest case when $\widetilde{K}_1(\tau) = \widetilde{K}_2(\tau)$ in (3.21) or (3.22), assuming that $\lambda_0 = 2\mu_0$ (that is, the Poisson coefficient of the plate material $\nu = 0.33$). The small numbers ϵ and ϵ_1 in (2.46)–(2.49) are assumed to be given.

We first consider the condition (2.46) containing ϵ. If we take $m = 1$ the characteristic equation $\widetilde{L}_2(k, s) = 0$ (that is, the characteristic equation of the system (3.33)) has three pairs of complex conjugate roots, for which when $kh \ll 1$ we have

$$\left. \begin{aligned} b_1^2 s_{21}^2 \approx -\frac{4}{3} (1 - \gamma^2) k^4 h^2, \\ b_1^2 s_{22}^2 = -2h^{-2}, \quad a_1^2 s_{23}^2 \approx -6h^{-2}, \end{aligned} \right\} \tag{3.37}$$

$$b_1^2 = \frac{b_0^2}{1 - \varkappa_2(s)}, \quad a_1^2 = \frac{a_0^2}{1 - \varkappa_1(s)} \text{ and } \gamma^2 = \frac{a_1^2}{b_1^2}. \tag{3.38}$$

On calculating the values of $\widetilde{\Delta}_m(k, s)$ and $\partial \widetilde{L}_2(k, s_2)/\partial s_2$ for $m = 1$, we can show that (2.46) leads to the inequality

$$k^2h^2 < 8\varepsilon \tag{3.39}$$

in the case of roots s_{21} from (3.37), to the inequality

$$\frac{1}{12} < \varepsilon$$

in the case of roots s_{22}, and to the inequality $1 < \epsilon$ in the case of roots s_{23}. Inequality (3.39) retains its qualitative meaning for $kh \leq 1$ also. The other inequalities, however, only worsen as kh increases (from the value $kh = 0$). An important corollary therefore follows from (2.46), namely that in describing dynamic processes in a plate on the basis of the engineering equation (3.33) a physical meaning can be given only to the (two) least roots (in modulus) of the characteristic equation $\widetilde{L}_2(k, s) = 0$, given by

$$b_1^2 s_{21}^2 = -\frac{4}{3}(1-\gamma^2)\, k^4h^2\left[1 - \frac{(27\lambda_0 + 34\mu_0)}{30\,(\lambda_0 + 2\mu_0)}\, k^2h^2 + \ldots\right]. \tag{3.40}$$

The values of kh which are still admissible must here satisfy (3.39). If a relative error of up to 10% is admissible in the determination of the natural frequencies we can assume that $0 \leq kh \leq 1$. For relative errors below 3%, however, we need to take $kh < 0.5$.

In order to exclude roots of the characteristic equation $\widetilde{L}_2(k, s) = 0$ with no physical meaning, it is sufficient to require that outside the domain of values (k, s) defined by (3.39) and the inequalities

$$|\beta^2h^2| < 2, \quad |\alpha^2h^2| < 1, \tag{3.41}$$

the spectral functions of the actions from (1.16) should be vanishingly small.

In the case of the roots of (3.40) condition (2.47) leads to the inequality

$$|h^2\beta^2| < 3\varepsilon_1, \tag{3.42}$$

and conditions (2.48) and (2.49) (for $m = m_0 = 1$) give

$$\left.\begin{array}{c} h^4\,|\alpha^4 + \beta^4| < 5!\,\varepsilon_1, \quad h^4\,|\alpha^4 - 6\alpha^2\beta^2 - \beta^4| < 5!\,\varepsilon_1, \\ h^4\,|2\alpha^4 + 3\alpha^2\beta^2 + 3\beta^4| < 2.5\,!\,\varepsilon_1 \end{array}\right\} \tag{3.43}$$

when $f_z^+(x, t) \equiv 0$, and

$$\left.\begin{array}{c} h^4\,|\alpha^4 + \beta^4| < 4\,!\,\varepsilon_1, \quad h^4\,|\beta^4 + 3\alpha^2\beta^2 + 4\alpha^4| < 2.4\,!\,\varepsilon_1, \\ h^4\,|2\alpha^4 - 12\alpha^2\beta^2 - 3\beta^4| < 2.5\,!\,\varepsilon_1, \end{array}\right\} \tag{3.44}$$

when $f_x^+(x, t) \equiv 0$. For a qualitative estimate of the domain (3.43) we can take

$$|\beta h| < \sqrt{5}\sqrt[4]{\varepsilon_1}, \quad |\alpha h| < \sqrt{5}\sqrt[4]{\varepsilon_1}. \tag{3.45}$$

For the domain (3.44), however, we have

$$|\beta h| < \sqrt{2.5}\sqrt[4]{\varepsilon_1}, \quad |\alpha h| < \sqrt{2.5}\sqrt[4]{\varepsilon_1}. \tag{3.46}$$

It follows from the above that in this variant of the engineering equations of flexural vibrations of plates the severest restriction on the conditions of applicability of the theory is imposed by inequality (2.46). Equations (3.33) and formulas (3.34) are applicable ("with an accuracy of ϵ, ϵ_1 with respect to the displacement field") in the class (F) of external actions (1.16) whose integrands are vanishingly small outside the intervals (2.16) in the domain $\Omega(k, s)$ defined by (3.39), (3.42), (3.43) and (3.44). A qualitative estimate of this domain can be obtained from (3.39), (3.42) and (3.45) or (3.46).

It is useful at this point to add that in describing the displacement field only by linear terms in z from (3.34) (that is, when $m = 1$ and $m_0 = 0$) in place of (3.43) and (3.44) we obtain inequalities similar to (3.27) and (3.28). Thus for a qualitative estimate of the domain $\Omega(k, s)$ in which the integrands from (1.16) are still nonzero we have in place of (3.45) and (3.46) inequalities of the form (3.29).[1] It follows that the domain of applicability of the variant of the theory when $m = 1$, $m_0 = 0$ is somewhat smaller than in the case $m = m_0 = 1$. It is easy to see that when $m = 1$, $m_0 = 2$ the accuracy of description of the displacement field, as well as the conditions of applicability of the theory, do not in practice differ from the case $m = 1$, $m_0 = 1$. In describing the stress field in the plate, however, on the basis of (3.33) the case $m_0 = 2$ gives a somewhat higher accuracy than the case $m_0 = 1$. A fuller discussion of the conditions of applicability of the equations of motion and formulas for the components of the displacement field of the engineering theory of vibrations of plates with certain specific kernels of elastic lag is given in È. F. Hinen's dissertation and will be published in the journal "Prikladnaja Matematika i Mehanika".[2]

6. For a qualitative estimate of the domains of applicability of the different variants of describing the vibrations of plates within the framework of engineering theory, it is useful to discuss briefly the form of the domains $\Omega_q(k, s)$ defined by the inequalities (2.30) for a given $q > 0$ and the boundary values k_0 and ω_0 of the intervals (2.15) contained in them.

---

1) These inequalities must of course be considered in conjunction with (3.39) and (3.42).

2) *Translator's note*: translated as "Journal of Applied Mathematics and Mechanics".

Let us consider first the case of perfect elasticity when

$$\alpha^2 = k^2 + a_0^2 s^2, \quad \beta^2 = k^2 + b_0^2 s^2, \quad 0 < a_0 < b_0.$$

Putting $s = i\omega$, it is easy to see that in the first quadrant of the (k, ω) plane the inequality

$$|k^2 + c^2 s^2| \equiv |k^2 - c^2\omega^2| < \frac{d^2}{h^2},$$

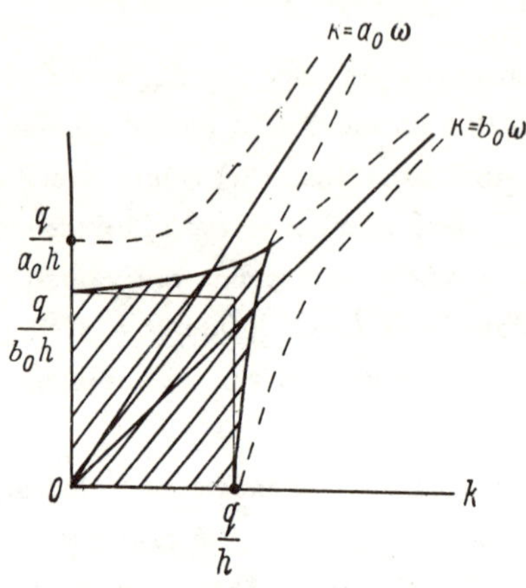

Figure 1

where $c = a_0$ or $c = b_0$, defines a region bounded by thecoordinate axes and the hyperbolas $k^2 - c^2\omega^2 = \pm q^2/h^2$ having $k = c\omega$ as asymptote. The hyperbolas intersect the coordinate ases at the points $(q/h, 0)$ and $(0, q/ch)$. We see from this that for perfect elasticity inequalities (2.30) define a region in the first quadrant of the (k, ω) plane bounded by the coordinate axes and the hyperbolas $b_0^2\omega^2 - k^2 = q^2/h^2$ and $k^2 - a_0^2\omega^2 = q^2/h^2$ (Figure 1). This region is somewhat larger than the maximum rectangle

$$0 < k < \frac{q}{h}, \quad 0 < \omega < \frac{q}{b_0 h},$$

in which k and ω can vary independently of each other. In the case of perfect elasticity, therefore, the inequalities (2.30) define admissible intervals (2.15) if we put

$$k_0 = \frac{q}{h}, \quad \omega_0 = \frac{q}{b_0 h}. \tag{3.47}$$

For $k^2 h^2 < q^2$ the domain $\Omega_q(k, s)$ is given by

$$|b_0^2 s^2| \leqslant \sqrt{\frac{q^4}{h^2} - k^4 \sin^2\varphi} - k^2 \cos\varphi, \tag{3.48}$$

where $s^2 = |s^2| \exp(i\phi)$. For a qualitative estimate of the conditions of applicability of the engineering vibrations of plates it is sufficient to consider

$$|s| < \frac{1}{b_0}\sqrt{\frac{q^2}{h^2} + k^2} < \sqrt{2}\,\frac{q}{hb_0}, \tag{3.49}$$

which contains within itself the domain $\Omega_q(k, s)$.

We now turn to the case of imperfect elasticity when

$$a^2 = k^2 + \frac{a_0^2 s^2}{1 - \varkappa_1(s)}, \quad b^2 = k^2 + \frac{b_0^2 s^2}{1 - \varkappa_2(s)},$$

where $\kappa_{1,2}(s)$ denote the images (1.19) of the kernels of the operators of elastic lag.

If we put $s = i\omega$ and take c and $\kappa(s)$ to mean a_0 and $\kappa_1(s)$ or b_0 and $\kappa_2(s)$ respectively, then in the expression

$$\frac{cs}{\sqrt{1 - \varkappa(s)}} = \frac{i\omega c}{\sqrt{1 - \varkappa(i\omega)}} = d(\omega) + i\frac{\omega}{v(\omega)}, \tag{3.50}$$

where

$$\left.\begin{aligned} d(\omega) &= \mathrm{Re}\frac{i\omega c}{\sqrt{1 - \varkappa(i\omega)}} = -\mathrm{Im}\frac{\omega c}{\sqrt{1 - \varkappa(i\omega)}}, \\ \frac{\omega}{v(\omega)} &= \mathrm{Im}\frac{i\omega c}{\sqrt{1 - \varkappa(i\omega)}} = \mathrm{Re}\frac{\omega c}{\sqrt{1 - \varkappa(i\omega)}}, \end{aligned}\right\} \tag{3.51}$$

the quantities $|d(\omega)|$ and $|v(\omega)|$ represent the damping factor and phase velocity of waves propagated in an imperfectly elastic medium with angular frequency ω. Here it is a question of the actual propagation of waves in a medium only if the excitation amplitude decreases not too rapidly over the length of the wave $\lambda = 2\pi|v(\omega)/\omega|$. If, therefore, we are interested in the processes of wave propagation along plates, we must assume that $\lambda|d(\omega)| = P$, where P is not large. On taking account of the fact that for $P = 2\pi$ the wave amplitude is attenuated over a distance λ by more than 300 times we can, without loss of generality of the physical results, assume that

$$2.5\,|d(\omega)| < \left|\frac{\omega}{v(\omega)}\right|. \tag{3.52}$$

For $s = i\omega$ it follows from (3.50) that

$$\frac{c^2 s^2}{1 - \varkappa(s)} = -\frac{c^2\omega^2}{1 - \varkappa(i\omega)} = -a_1(\omega) + ia_2(\omega), \tag{3.53}$$

where

$$a_1(\omega) = \frac{\omega^2}{v^2(\omega)} - d^2(\omega), \quad a_2(\omega) = 2d(\omega)\frac{\omega}{v(\omega)}, \tag{3.54}$$

and

$$a_1(\omega) > |a_2(\omega)|. \tag{3.55}$$

Let us consider the inequality

$$\left|k^2+\frac{c^2s^2}{1-\varkappa(s)}\right|^2=(k^2-a_1(\omega))^2+a_2^2(\omega)\equiv\chi(k,\ \omega)\leqslant\frac{q^4}{h^4}. \tag{3.56}$$

On denoting by ω_0 the upper bound of values of $\omega > 0$ for which

$$a_1^2(\omega)+a_2^2(\omega)\leqslant\frac{q^4}{h^4}, \tag{3.57}$$

we see that the curves $\chi(k, \omega) = q^4/h^4$ intersect the coordinate axes of the first quadrant of the (k, ω) plane at the points $(q/h, 0)$ and $(0, \omega_c)$. The curve passing through $(q/h, 0)$ is situated to the right of the straight line $k = q/h$ for $0 < \omega < \omega_c$ because

$$\chi\left(\frac{q}{h},\ \omega\right)=\frac{q^4}{h^4}-2\frac{q^2}{h^2}a_1(\omega)+a_1^2(\omega)+a_2^2(\omega)<\frac{q^4}{h^4}-2a_1(\omega)\left(\frac{q^2}{h^2}-a_1(\omega)\right)<\frac{q^4}{h^4}$$

by (3.55) and (3.57). The curve passing through $(0, \omega_c)$, on the other hand, lies above the line $\omega = \omega_c$ if $0 \leq k < \sqrt{2a_1(\omega_c)} = pq/h$, where $p > 1$, because

$$\begin{aligned}\chi(k,\ \omega_c)&=k^4-2k^2a_1(\omega_c)+a_1^2(\omega_c)+a_2^2(\omega_c)\\&=\frac{q^4}{h^4}-k^2(2a_1(\omega_c)-k^2)<\frac{q^4}{h^4}.\end{aligned}$$

The region of independent variation of k and ω defined by (3.56) therefore reduces to the intervals $0 < \omega < \omega_c$ and $0 < k < q/h$ (Figure 2).

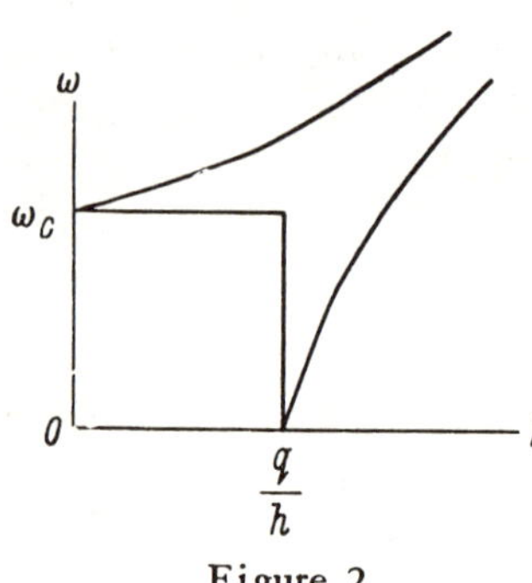

Figure 2

Taking account of the above and returning to the inequalities (2.30), we can assert that the regions of independent variation of k and ω are the intervals (2.15) if

$$k_0=\frac{q}{h},\quad \omega_0=\inf(\omega_a,\ \omega_b), \tag{3.58}$$

where ω_0 is the least positive root of

$$\left|\frac{c^2\omega^2}{1-\varkappa(i\omega)}\right|=\sqrt{a_1^2(\omega)+a_2^2(\omega)}=\frac{q^2}{h^2} \tag{3.59}$$

and c and $\kappa(s)$ are respectively equal to a_0 and $\kappa_1(s)$ or b_0 and $\kappa_2(s)$.

Finally, on putting

$$\frac{b_0^2s^2}{1-\varkappa_2(s)}=\left|\frac{b_0^2s^2}{1-\varkappa_2(s)}\right|e^{i\varphi},$$

we see that $\Omega_q(k, s)$ is defined by

$$\left|\frac{b_0^2 s^2}{\lambda - \varkappa_2(s)}\right| \leqslant \sqrt{\frac{q^4}{h^4} - k^4 \sin^2\varphi} - k^2 \cos\varphi. \tag{3.60}$$

However, for a qualitative estimate of the conditions of applicability of the engineering equations of vibrations of plates it is sufficient to consider the domain

$$\left|\frac{b_0^2 s^2}{1 - \varkappa_2(s)}\right| < \frac{q^2}{h^2} + k^2 < 2\frac{q^2}{h^2}, \tag{3.61}$$

which contains $\Omega_q(k, s)$ within itself.

BIBLIOGRAPHY

[1] N. N. Kil'čevskiĭ, *A generalization of the modern theory of shells*, Prikl. Mat. Meh. 2 (1939), 427–437; *Fundamentals of the analytical mechanics of shells*, Chap. 3, Izdat. Akad. Nauk Ukrain. SSR, Kiev, 1963. (Russian)

[2] I. T. Selezov, *Investigation of transverse vibrations of a plate*, Prikladna Meh. 6 (1960), 319–327. (Ukrainian) MR 25 #823.

[3] V. I. Uteševa, *Approximate equations of the dynamics of an elastic rod of circular cross section*, Izv. Akad. Nauk SSSR Otd. Teh. Nauk Meh. i Mašinostroenie 1963, no. 4, 154–161. (Russian)

[4] Paul S. Epstein, *On the theory of elastic vibrations in plates and shells*, J. Math. Phys. Mass. Inst. Tech. 21 (1942), 198–209. MR 4, 178.

[5] G. I. Petrašen', *On the theory of vibrations of thin plates*, Leningrad. Gos. Univ. Uč. Zap. 149 (1951), 172–249. (Russian)

[6] G. I. Petrašen' and L. A. Molotkov, *On some problems of the dynamical theory of elasticity for a medium bounded by thin layers*, Vestnik Leningrad. Univ. 13 (1958), no. 22, 137–156. (Russian) MR 21 #1043; *On vibrations of homogeneous and composite plates*, Proc. Fourth All-Union Conference Theory of Shells and Plates, Izdat. Akad. Nauk Armjan. SSR, Erevan, 1964, pp. 788–794. (Russian)

[7] G. I. Petrašen', *Problems of the engineering theory of vibrations of degenerate systems*, Investigations in Elasticity and Plasticity, no. 5, Izdat. Leningrad. Univ., Leningrad, 1966, pp. 3–33. (Russian)

Trudy Mat. Inst. Steklov. 95 (1968)

Proc. Steklov Inst. Math. 95 (1968)

ON THE PROPAGATION OF VIBRATIONS IN A MEDIUM WITH A TRANSITION LAYER

UDC 534.21

N. V. CEPELEV

Abstract. The article discusses the problem of the propagation of nonstationary vibrations in a medium consisting of two homogeneous halfspaces with different velocities of wave propagation v_0 and v_1 separated by an inhomogeneous layer. In this layer the velocity continuously and monotonically varies from v_0 to v_1. Two approaches are used to investigate the constructed solution. First, a study is made of the high-frequency part of the field and singularities on the fronts of the waves being formed are isolated. Then the roots of the characteristic equation of the problem are investigated and with their help graphs are constructed of the group and phase velocities of the interference waves.

Bibliography 11 items. 3 tables; 17 figures.

Introduction

In the investigation of wave processes in inhomogeneous media systems with transition layers are of interest. We call an inhomogeneous layer a transition layer if it separates two homogeneous media with different wave propagations velocities $v(0)$ and $v(H)$ and is such that the velocity $v(z)$ varies within it monotonically and sufficiently smoothly from $v(0)$ to $v(H)$.

Layers of this kind are very frequently encountered in real media, and it is therefore important to determine the properties of the resulting wave processes. This paper discusses a medium with a transition layer of this kind. We formulate and solve a problem on the propagation of vibrations in a system from a point source situated in a homogeneous halfspace with a smaller wave propagation velocity. An exact solution of the problem is obtained by the partial separation of variables, and the investigation is carried out by two different methods. The character of the field in the neighborhood of the wave is determined in §§1–6, and the corresponding contour integrals in the solution obtained are calculated by the stationary phase method. Similar results can be obtained by ray expansions, but in the given case a study of exact solutions is preferable if only because in the medium in question there is a head wave which originates on a weak boundary of separation of the media and has not yet been described by the ray method. The exact solutions obtained also enable us to study fields as a whole, and these questions are touched upon in §§7–9. A study is made of the roots of the characteristic equation of the problem, and with their help we obtain the graphs of the phase and group velocities of interference waves arising in the medium with a transition layer. A study of these roots is the first step in the investigation of interference phenomena in the system.

§1. Statement of the problem and its formal solution

1. Suppose that we have a medium in a cylindrical coordinate system r, θ, z consisting of two homogeneous halfspaces $z < 0$ and $z > H$ which are characterized by wave propagation velocities $v(0)$ and $v(H)$ (where $v(0) < v(H)$) respectively and are separated by an inhomogeneous transition layer $0 < z < H$ with velocity $v(z)$ (Figure 1). We assume that $v(z)$ varies from $v(0)$ to $v(H)$ and that on the boundaries $z = 0$ and $z = H$ we have $v'(z) \neq 0$ and

$$\frac{v'(0)}{v(0)} = \varepsilon_1 > 0, \quad \frac{v'(H)}{v(H)} = \varepsilon_2 > 0. \tag{1.1}$$

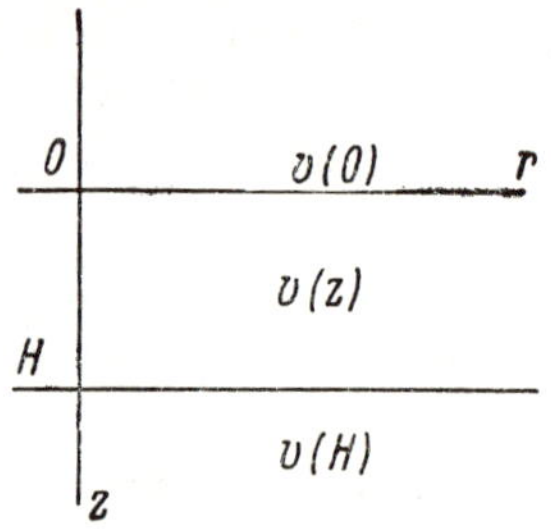

Figure 1

We assume that the propagation of vibrations in the medium is described by the wave equations:

a) for $z < 0$

$$\Delta u_0 = \frac{1}{v^2(0)} \cdot \frac{\partial^2 u_0}{\partial t^2}; \tag{1.2}$$

b) for $0 < z < H$

$$\Delta u_1 = \frac{1}{v^2(z)} \cdot \frac{\partial^2 u_1}{\partial t^2}; \tag{1.3}$$

c) for $z > H$

$$\Delta u_2 = \frac{1}{v^2(H)} \cdot \frac{\partial^2 u_2}{\partial t^2}. \tag{1.4}$$

Suppose that we are required to find functions u_0, u_1 and u_2 satisfying (1.2)–(1.4) on the assumption that the system is at rest for $t < 0$ and that a source of vibrations at the point $r = 0$, $z = -h$ is actuated at time $t = 0$ exciting a primary displacement field $\tilde{u}_0$ which for $-h < z < 0$ can be expressed in the form

$$\tilde{u}_0 = \frac{\varepsilon\,[tv(0) - R]}{R} = \int_0^\infty \left\{ \frac{1}{2\pi i} \int_{\sigma - i\infty}^{\sigma + i\infty} \frac{1}{\zeta\beta} e^{k[tv(0)\zeta - (h+z)\beta]} d\zeta \right\} J_0(kr)\, dk, \tag{1.5}$$

where

$$R=\sqrt{r^2+(h+z)^2}.$$

$J_0(kr)$ in (1.5) denotes the zero-order Bessel function, $\sigma \geq \sigma_0 > 0$ and

$$\beta=\sqrt{1+\zeta^2}. \tag{1.6}$$

The branch of the radical β on the ζ-plane cut between the points i and $-i$. is fixed by the condition $\arg \beta = 0$ for $\zeta > 0$. On the boundaries of separation $z = 0$ and $z = H$ we assume that

$$\left.\begin{aligned} u_0+\tilde{u}_0=u_1,\quad \frac{\partial u_0}{\partial z}+\frac{\partial \tilde{u}_0}{\partial z}=\frac{\partial u_1}{\partial z} \quad \text{for} \quad z=0, \\ u_1=u_2,\qquad \frac{\partial u_1}{\partial z}=\frac{\partial u_2}{\partial z} \quad \text{for} \quad z=H. \end{aligned}\right\} \tag{1.7}$$

2. The unknown functions u_ν are sought in the form

$$u_\nu=\int_0^\infty \left\{\frac{1}{2\pi i}\int_{\sigma-i\infty}^{\sigma+i\infty}\frac{1}{\zeta\beta}U_\nu(k,\ z,\ \zeta)\,e^{k[tv(0)\zeta-h\beta]}d\zeta\right\}J_0(kr)\,dk, \tag{1.8}$$

similar to (1.5). Assuming that double differentiation under the integral sign is permissible (this can be verified after finding the solution) and substituting (1.8) into (1.2)–(1.4) and (1.7), we obtain the following equations for the unknown function U_ν:

$$\begin{aligned} &\frac{d^2U_0}{dz^2}-k^2[1+\zeta^2]\,U_0=0 \quad (z<0),\\ &\frac{d^2U_1}{dz^2}-k^2\left[1+\frac{v^2(0)}{v^2(z)}\zeta^2\right]U_1=0 \quad (0<z<H),\\ &\frac{d^2U_2}{dz^2}-k^2\left[1+\frac{v^2(0)}{v^2(H)}\zeta^2\right]U_2=0 \quad (z>H) \end{aligned} \tag{1.9}$$

together with the boundary conditions

$$\left.\begin{aligned} U_0-U_1=-1,\quad \frac{dU_0}{dz}-\frac{dU_1}{dz}=k\beta \quad (z=0), \\ U_1-U_2=0,\quad \frac{dU_1}{dz}-\frac{dU_2}{dz}=0 \quad (z=H). \end{aligned}\right\} \tag{1.10}$$

Here we need take only those solutions which after substitution in (1.8) lead to expressions for u_ν which vanish for $t<0$.

If we denote two linearly independent solutions of the second equation of (1.9) by $\Phi(k,\ z,\ \zeta)$ and $\Psi(k,\ z,\ \zeta)$, it is easy to see that the unknown functions U_ν are determined by

$$U_0 = \frac{\tilde{D}_k^{(1)}(\zeta, H)\,\tilde{D}_k^{(2)}(\zeta, 0) - \tilde{D}_k^{(2)}(\zeta, H)\,\tilde{D}_k^{(1)}(\zeta, 0)}{W(k, \zeta)} e^{kz\beta}, \tag{1.11}$$

$$U_1 = 2k\beta \frac{\tilde{D}_k^{(1)}(\zeta, H)\,\Psi(k, z, \zeta) - \tilde{D}_k^{(2)}(\zeta, H)\,\Phi(k, z, \zeta)}{W(k, \zeta)}, \tag{1.12}$$

$$U_2 = -2k\beta \frac{\Phi(k, H, \zeta)\,\Psi'_z(k, H, \zeta) - \Phi'_z(k, H, \zeta)\,\Psi(k, H, \zeta)}{W(k, \zeta)} e^{k(H-z)\alpha}, \tag{1.13}$$

where

$$W(k, \zeta) = \tilde{D}_k^{(1)}(\zeta, H)\,D_k^{(2)}(\zeta, 0) - \tilde{D}_k^{(2)}(\zeta, H)\,D_k^{(1)}(\zeta, 0), \tag{1.14}$$

$$\begin{aligned} D_k^{(1)}(\zeta, z) &= k\sqrt{1 + \frac{v^2(0)}{v^2(z)}\zeta^2}\,\Phi(k, z, \zeta) + \Phi'_z(k, z, \zeta), \\ D_k^{(2)}(\zeta, z) &= k\sqrt{1 + \frac{v^2(0)}{v^2(z)}\zeta^2}\,\Psi(k, z, \zeta) + \Psi'_z(k, z, \zeta), \\ \tilde{D}_k^{(1)}(\zeta, z) &= k\sqrt{1 + \frac{v^2(0)}{v^2(z)}\zeta^2}\,\Phi(k, z, \zeta) - \Phi'_z(k, z, \zeta), \\ \tilde{D}_k^{(2)}(\zeta, z) &= k\sqrt{1 + \frac{v^2(0)}{v^2(z)}\zeta^2}\,\Psi(k, z\, \zeta) - \Psi'_z(k, z, \zeta). \end{aligned} \tag{1.15}$$

$$\alpha = \sqrt{1 + \gamma^2\zeta^2}, \qquad \gamma = \frac{v(0)}{v(H)} < 1, \tag{1.16}$$

and $\arg\,\alpha = 0$ for $\zeta > 0$.

As we know, justification of the solution which is found by the contour integral method is simple [1], and we shall not discuss it here.

3. To determine the nature of the field in the neighborhood of wave fronts, it is convenient to calculate the contour integrals in (1.8) approximately by the stationary phase method. It is necessary to expand the integrands as special series each term of which in some sense describes a definite wave arising on the boundaries of separation $z = 0$ and $z = H$ and in the inhomogeneous layer. The functions u_ν from (1.8) are conveniently expressed as the sum

$$u_\nu = u_\nu^{(1)} + u_\nu^{(2)}, \tag{1.17}$$

where

$$u_\nu^{(1)} = \int_{k_0}^{\infty} F_\nu(t, z, k)\,J_0(kr)\,dk, \tag{1.18}$$

$$u_\nu^{(2)} = \int_{0}^{k_0} F_\nu(t, z, k)\,J_0(kr)\,dk. \tag{1.19}$$

In these last two formulas $F_\nu(t, z, k)$ denotes the corresponding Mellin integrals, and $k_0 > 0$ is defined in a similar way to [2], where in addition it is shown that the term $u_\nu^{(2)}$ in (1.17) describes a smoothly varying part of the field.

If only the high-frequency part of the field is of interest (which is natural when studying the field in the neighborhood of the wave fronts), we need consider only the first term in (1.17). It is then convenient to require in addition that

$$k_0 Tv(0) \gg 1, \qquad (1.20)$$

where T is the time of propagation of vibrations along the vertical in the inhomogeneous layer. This inequality enables us to replace $D_k^{(\nu)}$ and $\tilde{D}_k^{(\nu)}$ by their asymptotic values and calculate the contour integrals in (1.8) by the stationary phase method.

The integrands in (1.8) take conjugate values at conjugate points of the ζ-plane. This allows us from now on to consider only the upper half of the ζ-plane.

§2. On the calculation of contour integrals

In studying the properties of the propagation of vibrations in the medium, the wave field in the halfspace $z < 0$, described by

$$u_0 = \int_{k_0}^{\infty} F_0(t, z, k) J_0(kr)\, dk, \qquad (2.1)$$

where

$$F_0 = \frac{1}{2\pi i} \int_{\sigma - i\infty}^{\sigma + i\infty} \frac{\tilde{D}_k^{(1)}(\zeta, H)\tilde{D}_k^{(2)}(\zeta, 0) - \tilde{D}_k^{(2)}(\zeta, H)\tilde{D}_k^{(1)}(\zeta, 0)}{\zeta\beta W(k, \zeta)} e^{k\varphi(\zeta)} d\zeta, \qquad (2.2)$$

$$\varphi(\zeta) = tv(0)\zeta - (h - z)\beta. \qquad (2.3)$$

is of the greatest interest.

To calculate this integral we need to examine the integrand in (2.2) and determine whether there is an identity transformation giving rise to terms which, for a suitable choice of contours, allow the stationary phase method to be applied. The initial contour of integration is conveniently split into three parts (l_1), (l_2) and (l_3) (Figure 2) by the points A and B, at which the integrand has an exponentially small value and whose position is determined below. We merely note here that A must lie on the boundary of a domain similar to a Watson crown for $D_k^{(\nu)}(\zeta, 0)$ and $\tilde{D}_k^{(\nu)}(\zeta, 0)$ and not too close to the point i.

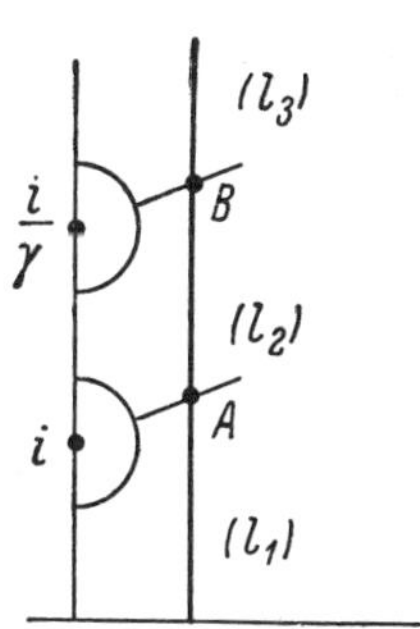

Figure 2

If we use (1.20) and replace the special functions in (2.2) by their approximate values (Appendix, Table 2), then for a suitable choice of the point [1] A we can obtain the following estimate for the integral (2.2) along (l_1):

$$|F_{0(l_1)}| < \frac{P}{k} e^{-kq}, \quad q > 0. \tag{2.4}$$

Let us consider the integral (2.2) along $(l_2) + (l_3)$. From examination of $D_k^{(\nu)}$ and $\tilde{D}_k^{(\nu)}$ we can show that for regions of the ζ-plane containing the contour $(l_2 + l_3)$ we have

$$|\tilde{D}_k^{(1)}(\zeta, H) D_k^{(2)}(\zeta, 0)| > |\tilde{D}_k^{(2)}(\zeta, H) D_k^{(1)}(\zeta, 0)|. \tag{2.5}$$

This enables us to express the integrand in (2.2) as a series [2]

$$\frac{1}{\zeta\beta} \cdot \frac{\tilde{D}_k^{(2)}(\zeta, 0)}{D_k^{(2)}(\zeta, 0)} + \frac{2k}{\zeta} \cdot \frac{V(\zeta, k)}{D_k^{(1)}(\zeta, 0) D_k^{(2)}(\zeta, 0)} \sum_{n=1}^{\infty} \left[\frac{\tilde{D}_k^{(2)}(\zeta, H) D_k^{(1)}(\zeta, 0)}{\tilde{D}_k^{(1)}(\zeta, H) D_k^{(2)}(\zeta, 0)}\right]^n, \tag{2.6}$$

where

$$V(\zeta, k) = \Phi(k, 0, \zeta) \Psi'_z(k, 0, \zeta) - \Phi'_z(k, 0, \zeta) \Psi(k, 0, \zeta). \tag{2.7}$$

In accordance with this the integral (2.2) can be expressed as the sum of integrals

$$F_0 = \tilde{F}_0 + \sum_{n=1}^{\infty} F_0^{(n)}. \tag{2.8}$$

where

$$\tilde{F}_0 = \frac{1}{2\pi i} \int_{(l_2+l_3)} \frac{1}{\zeta\beta} \cdot \frac{\tilde{D}_k^{(2)}(\zeta, 0)}{D_k^{(2)}(\zeta, 0)} e^{k\varphi(\zeta)} d\zeta, \tag{2.9}$$

$$F_0^{(n)} = \frac{k}{\pi i} \int_{(l_2+l_3)} \frac{1}{\zeta} \cdot \frac{V(\zeta, k)}{D_k^{(1)}(\zeta, 0) D_k^{(2)}(\zeta, 0)} \left[\frac{\tilde{D}_k^{(2)}(\zeta, H) D_k^{(1)}(\zeta, 0)}{\tilde{D}_k^{(1)}(\zeta, H) D_k^{(2)}(\zeta, 0)}\right]^n e^{k\varphi(\zeta)} d\zeta. \tag{2.10}$$

We note that the point i/γ is not singular for the integrand in (2.9), and we can therefore at once calculate (2.9) over the whole contour $(l_2 + l_3)$. The integral (2.10) can be conveniently calculated by splitting the contour into segments (l_2) and (l_3) by the point B (Figure 2). As in the case of point A, this must lie

1) The point A must lie within the ellipse

$$\frac{x^2}{(h - z^2)} + \frac{y^2}{t^2 v^2(0)} = \frac{1}{t^2 v^2(0) - (h - z)^2},$$

where $\zeta = x + iy$.

2) It follows from below that at any finite time t the integral of this series has only a finite number of nonzero terms.

on the boundary of a Watson crown for $D_k^{(\nu)}(\zeta, H)$ and $\tilde{D}_k^{(\nu)}(\zeta, H)$ in the region where the integrand in (2.10) and the integrands in (2.14)–(2.16) below are exponentially small. This region can always be found by investigating the phase functions of these integrals.

For the subsequent investigation of the integral (2.10) along (l_2) it is convenient to use the substitution

$$\tilde{D}_k^{(2)}(\zeta, H) = \tilde{E}_k(\zeta, H) - \tilde{D}_k^{(1)}(\zeta, H)\, e^{\frac{\pi i}{3}}, \tag{2.11}$$

where

$$\tilde{E}_k(\zeta, H) = \tilde{D}_k^{(2)}(\zeta, H) + \tilde{D}_k^{(1)}(\zeta, H)\, e^{\frac{\pi i}{3}}. \tag{2.12}$$

Simple transformations give

$$F_{0(l_2)}^{(n)} = F_{01}^{(n)} + F_{02}^{(n)} + F_{03}^{(n)}, \tag{2.13}$$

where

$$F_{01}^{(n)} = \frac{k}{\pi i}\int\limits_{(l_2)} \frac{1}{\zeta}\cdot\frac{V(\zeta, k)}{D_k^{(1)}(\zeta, 0)\, D_k^{(2)}(\zeta, 0)}\left[\frac{\tilde{E}_k(\zeta, H)\, D_k^{(1)}(\zeta, 0)}{\tilde{D}_k^{(1)}(\zeta, H)\, D_k^{(2)}(\zeta, 0)}\right]^n e^{k\varphi(\zeta)}d\zeta, \tag{2.14}$$

$$F_{02}^{(n)} = (-1)^n\frac{k}{\pi i}\int\limits_{(l_2)} \frac{1}{\zeta}\cdot\frac{V(\zeta, k)}{D_k^{(1)}(\zeta, 0)\, D_k^{(2)}(\zeta, 0)}\left[e^{\frac{\pi i}{3}}\frac{D_k^{(1)}(\zeta, 0)}{D_k^{(2)}(\zeta, 0)}\right]^n e^{k\varphi(\zeta)}d\zeta, \tag{2.15}$$

$$F_{03}^{(n)} = \frac{k}{\pi i}\int\limits_{(l_2)} \frac{1}{\zeta}\cdot\frac{V(\zeta, k)}{D_k^{(1)}(\zeta, 0)\, D_k^{(2)}(\zeta, 0)}\,\Phi_n(k, \zeta)\, e^{k\varphi(\zeta)}d\zeta, \tag{2.16}$$

and

$$\Phi_n(k, \zeta) = \sum_{m=1}^{n-1}(-1)^m e^{-\frac{\pi i}{3}(m-n)}\left[\frac{\tilde{E}_k(\zeta, H)}{\tilde{D}_k^{(1)}(\zeta, H)}\right]^m\left[\frac{D_k^{(1)}(\zeta, 0)}{D_k^{(2)}(\zeta, 0)}\right]^{n-m} C_n^m. \tag{2.17}$$

The integrals obtained enable us to use the stationary phase method of calculation. This requires that we should investigate their phase functions.

§3. Investigation of the phase functions of the integrals obtained

1. If we use the asymptotic expressions for $D_k^{(\nu)}$ and $\tilde{D}_k^{(\nu)}$ (Appendix, Table 2), the principal parts of the phase functions of the integrals (2.9), (2.10), (2.14)–(2.16) will be given respectively by the following expressions: [1)]

$$\varphi = tv(0)\zeta - (h - z)\beta, \tag{3.1}$$

1) In the case of the integrals (2.10) and (2.14), whose phase functions coincide, there is just the one formula (3.2).

$$\varphi_{n1} = tv(0)\zeta - (h-z)\beta - 2n\int_0^H \sqrt{1+\frac{v^2(0)}{v^2(z)}\zeta^2}\,dz, \tag{3.2}$$

$$\varphi_{n2} = tv(0)\zeta - (h-z)\beta - 2n\int_0^z \sqrt{1+\frac{v^2(0)}{v^2(z)}\zeta^2}\,dz, \tag{3.3}$$

$$\varphi_{n3} = tv(0)\zeta - (h-z)\beta - 2p\int_z^H \sqrt{1+\frac{v^2(0)}{v^2(z)}\zeta^2}\,dz$$

$$+ 2q\int_z^0 \sqrt{1+\frac{v^2(0)}{v^2(z)}\zeta^2}\,dz, \tag{3.4}$$

where p and q are integers and $p \neq q$. The expressions obtained have a meaning only in the domain of definition of the integrals; that is, outside the Watson crown for $D_k^{(\nu)}(\zeta, 0)$ in the case of (3.1) and (3.2) and between the crowns for $D_k^{(\nu)}(\zeta, 0)$ and $D_k^{(\nu)}(\zeta, H)$ in the case of (3.3) and (3.4).

The saddle points of the phase functions are determined by the equations

$$tv(0) - (h-z)\frac{\zeta}{\beta} = 0, \tag{3.5}$$

$$tv(0) - (h-z)\frac{\zeta}{\beta} - 2n\zeta\int_0^H \frac{\frac{v^2(0)}{v^2(z)}\,dz}{\sqrt{1+\frac{v^2(0)}{v^2(z)}\zeta^2}} = 0. \tag{3.6}$$

$$tv(0) - (h-z)\frac{\zeta}{\beta} - 2n\zeta\int_0^z \frac{\frac{v^2(0)}{v^2(z)}\,dz}{\sqrt{1+\frac{v^2(0)}{v^2(z)}\zeta^2}} = 0, \tag{3.7}$$

$$tv(0) - (h-z)\frac{\zeta}{\beta} - 2p\zeta\int_z^H \frac{\frac{v^2(0)}{v^2(z)}\,dz}{\sqrt{1+\frac{v^2(0)}{v^2(z)}\zeta^2}} + 2q\zeta\int_z^0 \frac{\frac{v^2(0)}{v^2(0)}\,dz}{\sqrt{1+\frac{v^2(0)}{v^2(z)}\zeta^2}} = 0. \tag{3.8}$$

It is easy to see that the first of these equations has two roots, which are real for $tv(0) < h - z$ and imaginary if the reverse inequality holds. In this latter case they are given by

$$\zeta_0 = \pm \frac{itv(0)}{\sqrt{t^2v^2(0) - (h-z)^2}}. \tag{3.9}$$

Equations (3.6)–(3.8) are not solvable in explicit form, and we therefore proceed as follows.

If in wave propagation problems solvable by partial separation of variables we calculate the contour integrals by the stationary phase method, the saddle points of the phase functions situated on the imaginary axis are usually related to the angles of departure of rays from the source by the formula

$$\zeta = \frac{i}{\sin \alpha_0}. \tag{3.10}$$

By means of this relation (3.6)–(3.8) can be reduced to the forms

$$t = \frac{1}{v(0)} \cdot \frac{h-z}{\cos \alpha_0} + 2n \int_0^H \frac{dz}{v(z) \cos \alpha(z)}, \tag{3.11}$$

$$t = \frac{1}{v(0)} \cdot \frac{h-z}{\cos \alpha_0} + 2n \int_0^z \frac{dz}{v(z) \cos \alpha(z)}, \tag{3.12}$$

$$t = \frac{1}{v(0)} \frac{h-z}{\cos \alpha_0} + 2p \int_0^H \frac{dz}{v(z) \cos \alpha(z)} + 2(q-p) \int_0^z \frac{dz}{v(z) \cos \alpha(z)}, \tag{3.13}$$

where

$$\sin \alpha(z) = \frac{v(z)}{v(0)} \sin \alpha_0. \tag{3.14}$$

Expressions (3.11)–(3.13) have a simple physical interpretation. The first determines the time of propagation of vibrations from the source to the point of observation in the z-plane along a ray which undergoes $(2n-1)$ reflections in the transition layer from the boundaries $z=0$ and $z=H$. These rays exist for angles α_0 between 0 and arc sin γ,[1] or, equivalently, for parameters t and $(h-z)$ satisfying

$$(h-z) + 2nv(0) \int_0^H \frac{dz}{v(z)} < tv(0) \leqslant \frac{h-z}{\sqrt{1-\gamma^2}} + 2n \int_0^H \frac{v(0)\, dz}{v(z) \sqrt{1 - \frac{v^2(z)}{v^2(H)}}}. \tag{3.15}$$

The saddle point $\zeta^{(n)}$ of the phase function (3.2) is then situated on the imaginary axis in the interval $(i/\gamma, i\infty)$.

If, however,

$$tv(0) < (h-z) + 2nv(0) \int_0^H \frac{dz}{v(z)}, \tag{3.16}$$

the point $\zeta^{(n)}$ lies on the real axis.

---

1) The angle of departure arc sin γ of the ray from the source corresponds to a ray touching the boundary $z = H$.

Further, if

$$tv(0) > \frac{h-z}{\sqrt{h-\gamma^2}} + 2n \int_0^H \frac{v(0)\,dz}{v(z)\sqrt{1-\frac{v^2(z)}{v^2(H)}}}, \tag{3.17}$$

the saddle point $\zeta^{(n)}$ departs under the cut to the second sheet of the Riemann surface ζ.

Let us consider equation (3.12), which determines the arrival time of perturbations at the point of observation along rays refracted in the inhomogeneous layer. It is necessary to remember that on the imaginary axis it is defined only in the interval $[i; i/\gamma]$ or for angles $\alpha_0 \in [\arcsin\gamma; \pi/2]$. We know that if a definite condition is satisfied in an inhomogeneous layer with an increasing refractive index, there can exist two rays along which a perturbation can reach the point of observation. In our case this condition takes the form

$$\frac{h-z}{\sqrt{1-\gamma^2}} < 2n\frac{v(0)}{v'(0)}\sqrt{1-\gamma^2} + 2n\frac{1-\gamma^2}{\gamma}\int_0^H \frac{\left[\frac{v(z)}{v'(z)}\right]' dz}{\sqrt{\frac{v^2(H)}{v^2(z)}-1}}. \tag{3.18}$$

Equation (3.12) therefore has two saddle points: $\zeta_n^{(1)}$ and $\zeta_n^{(2)}$. Let us consider values of the parameters for which, in addition to (3.7), we have

$$\frac{h-z}{\beta^3} = 2n\frac{v(0)}{v'(0)}\cdot\frac{1}{\zeta^2\beta} + \frac{2n}{\zeta^2}\int_0^z \frac{\left[\frac{v(z)}{v'(z)}\right]' dz}{\sqrt{1+\frac{v^2(0)}{v^2(z)}\zeta^2}}, \tag{3.19}$$

which is the condition for the vanishing of the second derivative of the phase function. The saddle points $\zeta_n^{(1)}$ and $\zeta_n^{(2)}$ in this case obviously coincide. We denote by $\tilde{\zeta}_n$ the point at which (3.7) and (3.19) hold simultaneously, and suppose that we always have

$$|\zeta_n^{(1)}| \geqslant |\tilde{\zeta}_n| \geqslant |\zeta_n^{(2)}|. \tag{3.20}$$

It follows that $\zeta_n^{(1)}$ can lie below i/γ only if $|\tilde{\zeta}_n| < 1/\gamma$, which is equivalent to (3.18). It is then necessary that

$$tv(0) < \frac{h-z}{\sqrt{1-\gamma^2}} + 2n\frac{v(0)}{v'(0)}\sqrt{1-\gamma^2} + 2n\gamma\int_0^H \left[\frac{v(z)}{v'(z)}\right]'\sqrt{\frac{v^2(H)}{v^2(z)}-1}\,dz. \tag{3.21}$$

If, however, (3.18) is not satisfied, only one saddle point $\zeta_n^{(2)}$ lies below i/γ, and the inequality reverse to (3.21) must hold.

It still remains for us to consider (3.8). It can be shown that there are no values of the parameters for which it has roots in the region through which (l_2) passes.

2. To calculate the integrals it is necessary to know the behavior of the stationary contours of their phase functions. They are shown for various relationships between the parameters in Figures 3–10. The stationary curves in Figures 3 and 4 correspond to equation (3.1), the curves in Figures 5–7 to (3.2) and the curves in Figures 8–10 to (3.3).

In the diagrams we have

$$P = h - z + 2nv(0) \int_0^H \frac{dz}{v(z)},$$

$$Q = \frac{h-z}{\sqrt{1-\gamma^2}} + 2nv(0) \int_0^H \frac{dz}{v(z)\sqrt{1 - \dfrac{v^2(z)}{v^2(H)}}}, \tag{3.22}$$

$$R = \frac{h-z}{\sqrt{1-\gamma^2}} - 2n\frac{\sqrt{1-\gamma^2}}{\varepsilon_1} + 2n\gamma \int_0^H \left[\frac{v(z)}{v'(z)}\right]' \sqrt{\frac{v^2(H)}{v^2(z)} - 1}\, dz.$$

Also, S is determined from the system of equations (3.19) and (3.7).

The diagrams show the stationary curves corresponding qualitatively to different moments of time. The arrows indicate the directions along which the real parts of the phase functions decrease most rapidly.

§4. Calculation of the contour integrals

1. The contour integrals (2.9) and (2.10) can be calculated independently of each other. When suitably deforming the contour of integration we need only remember that it must proceed from the same point A (Figure 2).

Let us first consider (2.9), replacing the special functions in the integrand by their asymptotic expansions (Appendix, Table 2). For different relationships between the parameters $tv(0)$ and $(h-z)$ we need to deform the contour $(l_2 + l_3)$ in accordance with Figures 3 and 4. If $tv(0) < (h-z)$ we obtain for (2.9) an exponential estimate of the form $Me^{-k\delta}$, $\delta > 0$. If the reverse inequality holds, the integral is calculated by the usual scheme of the stationary phase method to give [1)]

$$\tilde{F}_\theta = -\frac{1}{k\sqrt{2\pi k}} \cdot \frac{|\varepsilon_1 K(\zeta_0)|}{|\zeta_0 \beta_0| \sqrt{|\varphi''(\zeta_0)|}} \cos\left[k|\varphi(\zeta_0)| - \frac{\pi}{4}\right], \tag{4.1}$$

1) Here and subsequently we shall write down only the dominant terms of the asymptotic series obtained.

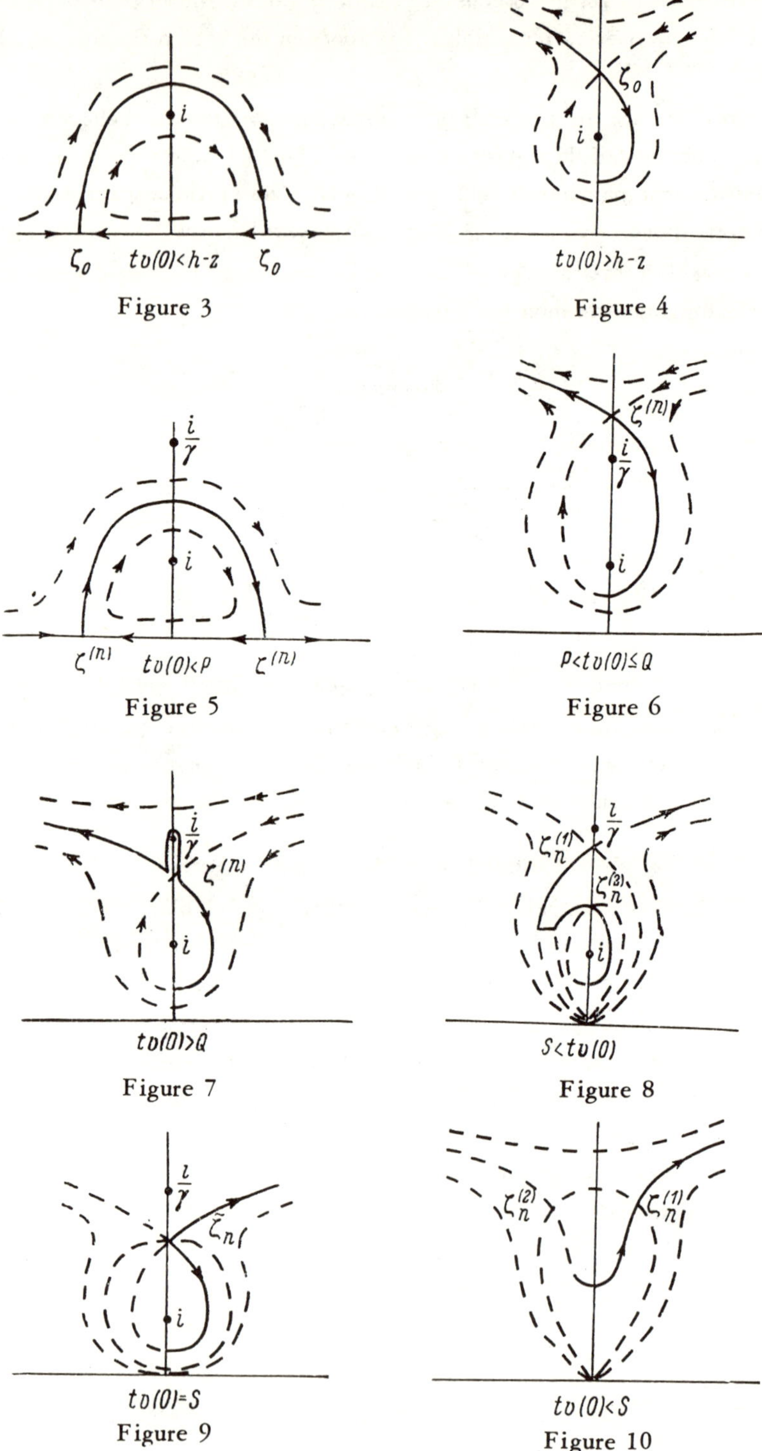

Figure 3

Figure 4

Figure 5

Figure 6

Figure 7

Figure 8

Figure 9

Figure 10

where

$$\beta_0 = \frac{i\,(h-z)}{\sqrt{t^2 v^2(0) - (h-z)^2}}, \tag{4.2}$$

$$K(\zeta) = \frac{\zeta^2}{4\zeta^3}, \tag{4.3}$$

and ϵ_1, $\phi(\zeta)$ and ζ_0 are defined by (1.1), (2.3) and (3.9) respectively.

2. As stated in §2, in calculating the integral (2.10) on (l_2) (Figure 2) we need to express it as a sum (2.13). It is then necessary to consider several cases of the relationships between the parameters.

Suppose that

$$tv(0) < P. \tag{4.4}$$

The saddle points of the phase functions of the integrals (2.10) and (2.14) then lie on the real axis. The saddle points of the phase function of the integral (2.15) lie in the right and left halfplanes. The real parts of the phase functions (3.2)–(3.4) are negative in the right half of the ζ-plane. This enables us to draw the contour $(l_2 + l_3)$ so that for the above sum of integrals we have

$$|F_0^{(n)}| < Ae^{-k\delta}. \tag{4.5}$$

Let us now consider the case in which

$$\left.\begin{aligned} P \leqslant tv(0) \leqslant Q, \\ tv(0) < S. \end{aligned}\right\} \tag{4.6}$$

These inequalities imply that a saddle point of the integrals (2.10)–(2.14) lies on the imaginary axis in the interval $(i/\gamma, i\infty)$ and a saddle point of the integral (2.15) lies in the right or left halfplanes. Integrals (2.15)–(2.16) along a contour chosen by the corresponding method have an exponential estimate, and (2.10) and (2.14) are calculated by the stationary phase method to be

$$F_0^{(n)} = \frac{(-1)^n}{k^{2n-\frac{1}{2}}} \sqrt{\frac{\pi}{2}} \frac{|\varepsilon_1 K(\zeta^{(n)})|^{n-1} |\varepsilon_2 K(\gamma\zeta^{(n)})|^n}{|\beta^{(n)}\zeta^{(n)}| \sqrt{|\varphi''_{n1}(\zeta^{(n)})|}} \cos\left[k|\varphi_{n1}(\zeta^{(n)})| - \frac{4n-3}{4}\pi\right]. \tag{4.7}$$

Here $\phi_{n1}(\zeta)$, $K(\zeta)$ and ϵ_ν are defined by (3.2), (4.3) and (1.1) respectively.

Suppose now that

$$\left.\begin{aligned} P \leqslant tv(0) \leqslant Q, \\ tv(0) \geqslant S, \\ h - z \leqslant G, \end{aligned}\right\} \tag{4.8}$$

where

$$G = 2n\frac{1-\gamma^2}{\varepsilon_1} + 2n\frac{(1-\gamma^2)^{3/2}}{\gamma}\int_0^H \frac{\left[\frac{v(z)}{v'(z)}\right]' dz}{\sqrt{\frac{v^2(H)}{v^2(z)}-1}}. \tag{4.9}$$

The first inequality guarantees that a saddle point of the integrals (2.10) and (2.14) exists in the interval $[i/\gamma, i\infty)$, and the second and third inequalities guarantee that two saddle points of the integral (2.15) exist in $[i/\gamma, i]$. The contour of integration $(l_2 + l_3)$ must be drawn in accordance with Figures 6 and 8, so that it passes through the three saddle points. The cut between i/γ and i must then be deformed into the left halfplane. The integral (2.15) needs to be calculated by different methods, depending on the mutual position of the saddle points $\zeta_n^{(1)}$ and $\zeta_n^{(2)}$. If they are far from each other the calculation follows the usual method to give

$$F_0^{(n)} = \frac{(-1)^n}{k^{2n-\frac{1}{2}}}\sqrt{\frac{\pi}{2}}\frac{|\varepsilon_1 K(\zeta^{(n)})|^{n-1}|\varepsilon_2 K(\gamma\zeta^{(n)})|^n}{|\beta^{(n)}\zeta^{(n)}|\sqrt{|\varphi_{n1}''(\zeta^{(n)})|}}\cos\left[k|\varphi_{n1}(\zeta^{(n)})| - \frac{4n-3}{4}\pi\right]$$
$$+\frac{(-1)^n}{k^{n-\frac{1}{2}}}\sqrt{\frac{\pi}{2}}\left\{\frac{|\varepsilon_1 K(\zeta_n^{(1)})|^{n-1}}{|\beta_n^{(1)}\zeta_n^{(1)}|\sqrt{|\varphi_{n2}''(\zeta_n^{(1)})|}}\cos\left[k|\varphi_{n2}(\zeta_n^{(1)})| + \frac{5-2n}{4}\pi\right]\right.$$
$$\left.+\frac{|\varepsilon_1 K(\zeta_n^{(2)})|^{n-1}}{|\beta_n^{(2)}\zeta_n^{(2)}|\sqrt{|\varphi_{n2}''(\zeta_n^{(2)})|}}\cos\left[k|\varphi_{n2}(\zeta_n^{(2)})| + \frac{3-2n}{4}\pi\right]\right\}. \tag{4.10}$$

The integral (2.16) in this case, however, has an exponential estimate. As the saddle points $\zeta_n^{(1)}$ and $\zeta_n^{(2)}$ approach each other the usual method of calculating (2.15) becomes inapplicable, and we use another method described in §5.

If

$$\left.\begin{aligned} tv(0) &> Q, \\ tv(0) &\geqslant S, \end{aligned}\right\} \tag{4.11}$$

the integrals (2.10) and (2.14) have no saddle points on the basic sheet of the ζ-plane, and (2.15) has only one saddle point $\zeta_n^{(2)}$ in the interval $[i, i/\gamma]$. The contour of integration is in this case deformed as in Figure 11. It is obvious that in addition to integrating over the neighborhood of the saddle point we must take account of the integral along the contour enclosing the cut from the point i/γ (the contour (λ)). Performing the calculations, we obtain

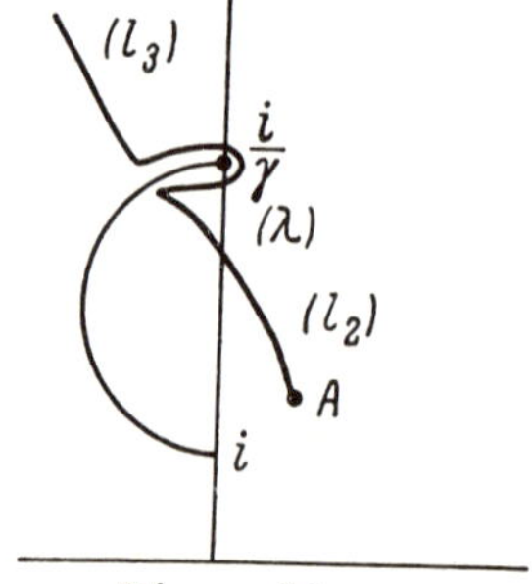

Figure 11

$$F_0^{(n)} = \frac{(-1)^n}{k^{n-\frac{1}{2}}} \sqrt{\frac{\pi}{2}} \frac{|\varepsilon_1 K(\zeta_n^{(2)})|}{|\beta_n^{(2)}\zeta_n^{(2)}| \sqrt{|\varphi''_{n2}(\zeta_n^{(2)})|}} \cos\left[k|\varphi_{n2}(\zeta_n^{(2)})| + \frac{\pi(3-2n)}{4}\right] + F_{(\lambda)}^{(n)}, \tag{4.12}$$

where $F_\lambda^{(n)}$ is the integral along the cut which is investigated below.

§5. Calculation of the contour integrals in the case of adjacent saddle points and along the cut

1. In the calculation of (2.15) we assumed that the saddle points $\zeta_n^{(1)}$ and $\zeta_n^{(2)}$ were far apart. When they approach each other the usual method of calculating the integral becomes inapplicable, and it is necessary to use another scheme [3].

The phase function (3.2) is expanded as a series in the neighborhood of the point $\tilde{\zeta}_n$ for which

$$\varphi''_{n2}(\tilde{\zeta}_n) = 0. \tag{5.1}$$

In this expansion it is necessary to retain the third derivative:

$$\varphi_{n2}(\zeta) \approx \varphi_{n2}(\tilde{\zeta}_n) + a_n(\zeta - \tilde{\zeta}_n) + b_n(\zeta - \tilde{\zeta}_n)^3, \tag{5.2}$$

where

$$a_n = \varphi'_{n2}(\tilde{\zeta}_n),$$
$$b_n = \frac{1}{6}\varphi'''_{n2}(\tilde{\zeta}_n). \tag{5.3}$$

By means of this expansion (2.15) can be expressed approximately in the form

$$F_0^{(n)} = \frac{(-1)^n}{\pi} \cdot \frac{1}{k^{n-1}} \operatorname{Re}\left\{\frac{[\varepsilon_1 K(\tilde{\zeta}_n)]^{n-1}}{\tilde{\zeta}_n \tilde{\beta}_n} e^{k\varphi_{n2}(\tilde{\zeta}_n) + \frac{3-n}{2}\pi i} J_n\right\}, \tag{5.4}$$

where

$$J_n = \int_{(\tilde{l}_2)} e^{ka_n(\zeta - \tilde{\zeta}_n) + kb_n(\zeta - \tilde{\zeta}_n)^3} d\zeta. \tag{5.5}$$

On the change of variable

$$\tau = -\sqrt[3]{3kb_n}(\zeta - \tilde{\zeta}_n), \tag{5.6}$$

this last integral becomes

$$J_n = -\frac{1}{\sqrt[3]{3kb_n}} \int_{(\tilde{l})} e^{-\frac{\tau^3}{3} - a_n \sqrt[3]{\frac{k^2}{3b_n}}\tau} d\tau. \tag{5.7}$$

The contour of integration (l) is shown in Figure 12, where

$$\tau_1 = \rho\sqrt[3]{3kb_n}\, e^{\frac{5\pi i}{4}},$$

$$\tau_2 = \rho\sqrt[3]{3kb_n}\, e^{\frac{3\pi i}{4}}.$$

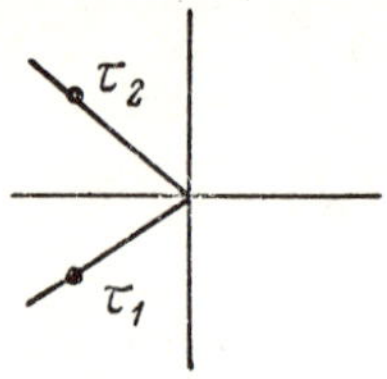

Figure 12

If we require that

$$\rho\sqrt[3]{3k\,|b_n|} \gg 1,$$

the integral (5.7) is approximately equal to the Airy function $v(\kappa_n)$ with constant multiplier $-2i\sqrt{\pi}\,(3kb_n)^{-1/3}$, where

$$v(\varkappa_n) = \frac{1}{\sqrt{\pi}}\int_0^\infty \cos\left(\frac{x^3}{3} + x\varkappa_n\right) dx, \tag{5.8}$$

$$\varkappa_n = -a_n\sqrt[3]{\frac{k^2}{3b_n}}. \tag{5.9}$$

In the case of adjacent saddle points (2.15) is therefore equal to

$$F^{(n)}_{02(l_2)} = \frac{(-1)^n}{k^{n-1}}\frac{2}{\sqrt{\pi}}\frac{v(\varkappa_n)\,|\,\varepsilon_1 K(\tilde{\zeta}_n)\,|^{n-1}}{|\,\tilde{\zeta}_n\tilde{\beta}_n\,|\sqrt[3]{3kb_n}}\sin k\,|\,\varphi_{n2}(\tilde{\zeta}_n)\,|. \tag{5.10}$$

2. As we explained in §4, if the inequalities (4.11) are satisfied the integrals (2.10), (2.14) and (2.15) are calculated over the neighborhood of the saddle point $\zeta_n^{(2)}$ and also along the contour (λ) enclosing the cut from the point i/γ (Figure 11). In calculating the integral along the cut the contour of integration goes partly into the domain requiring special asymptotic expressions for the functions $D_k^{(\nu)}(\zeta, H)$ and $\tilde{D}_k^{(\nu)}(\zeta, H)$.

The integrand then has the form

$$\Phi_n = \frac{1}{\zeta}\frac{V(\zeta, k)}{D_k^{(1)}(\zeta, 0)\,D_k^{(2)}(\zeta, 0)}\left[\frac{\tilde{D}_k^{(2)}(\zeta, H)\,D_k^{(1)}(\zeta, 0)}{\tilde{D}_k^{(1)}(\zeta, H)\,D_k^{(2)}(\zeta, 0)}\right]^n e^{k\varphi(\zeta)}. \tag{5.11}$$

Since i/γ is a branch point for $\tilde{D}_k^{(\nu)}(\zeta, H)$ only, by using the approximate expressions of $D_k^{(\nu)}(\zeta, 0)$ for large values of k we obtain

$$\Phi_n = \frac{(-1)^{n-1}}{k^{n-1}}[\varepsilon_1 K(\zeta)]^{n-1}\left\{\frac{\tilde{D}_k^{(2)}(\zeta, H)}{\tilde{D}_k^{(1)}(\zeta, H)}\right\}^n e^{k\varphi(\zeta)+2knI-\frac{5}{6}n\pi i}. \tag{5.12}$$

Here, as before,

$$K(\zeta) = \frac{\zeta^2}{4\beta^3},$$

$$I=\int\limits_{z}^{0}\sqrt{1+\frac{v^2(0)}{v^2(z)}\zeta^2}\,dz. \tag{5.13}$$

We consider the expression in (5.12) in curly brackets:

$$\left\{\frac{\tilde{D}_k^{(2)}(\zeta,H)}{\tilde{D}_k^{(1)}(\zeta,H)}\right\}^n=\left\{\frac{k\alpha\Psi(k,H,\zeta)-\Psi'_z(k,H,\zeta)}{k\alpha\Phi(k,H,\zeta)-\Phi'_z(k,H,\zeta)}\right\}^n. \tag{5.14}$$

If we rationalize the denominator and take the nth power, we obtain

$$\left\{\frac{\tilde{D}_k^{(2)}(\zeta,H)}{\tilde{D}_k^{(1)}(\zeta,H)}\right\}^n=A_n(k,\zeta)-k\alpha V(\zeta,k)B_n(k,\zeta), \tag{5.15}$$

where

$$\begin{aligned} A_n(k,\zeta)&=\sum_{m=0}^{\left[\frac{n}{2}\right]}C_n^{2m}V^{2m}(\zeta,k)\frac{[k^2\alpha^2\Phi\Psi-\Phi'_z\Psi'_z]^{n-2m}}{[k^2\alpha^2\Phi^2-\Phi'^2_z]^n}\alpha^{2m},\\ B_n(k,\zeta)&=\sum_{m=0}^{\left[\frac{n-1}{2}\right]}C_n^{2m+1}V^{2m}(\zeta,k)\frac{[k^2\alpha^2\Phi\Psi-\Phi'_z\Psi'_z]^{n-2m-1}}{[k^2\alpha^2\Phi^2-\Phi'^2_z]^n}\alpha^{2m}. \end{aligned} \tag{5.16}$$

In accordance with (5.15) the integral along the cut splits into two terms, the first of which has no branching at the point i/γ and is equal to zero. We need, therefore, consider only the integral containing the second term and, in place of the integrand Φ_n (5.12), the function

$$\alpha\frac{(-1)^{n-1}}{k^{n-2}}[\varepsilon_1K(\zeta)]^{n-1}V(\zeta,k)B_n(k,\zeta)e^{k\varphi(\zeta)+2nkI-\frac{5}{6}n\pi i}. \tag{5.17}$$

According to the usual scheme for calculating an integral along a cut, we expand the preexponential and phase functions in (5.17) in powers of $(\zeta - i/\gamma)$:

$$k\varphi(\zeta)+2nkI=k\tilde{a}_n\left(\frac{i}{\gamma}\right)+k\tilde{b}_n\left(\frac{i}{\gamma}\right)\left(\zeta-\frac{i}{\gamma}\right)+\ldots, \tag{5.18}$$

$$\begin{aligned}&\alpha\frac{(-1)^{n-1}}{k^{n-2}}[\varepsilon_1K(\zeta)]^{n-1}V(\zeta,k)B_n(k,\zeta)\\ &=\frac{3n\gamma^{5/2}\Gamma\left(\frac{1}{3}\right)}{\pi\sqrt{2}\,(1-\gamma^2)}\left[\frac{k}{6\varepsilon_2}\right]^{1/3}\left[\frac{\varepsilon_1}{k}\left|K\left(\frac{i}{\gamma}\right)\right|\right]^{n-1}\sqrt{\zeta-\frac{i}{\gamma}}\,e^{\frac{11\pi i}{12}-\frac{5n\pi i}{3}}+\ldots.\end{aligned} \tag{5.19}$$

Here

$$\begin{aligned}\tilde{a}_n\left(\frac{i}{\gamma}\right)&=\varphi\left(\frac{i}{\gamma}\right)+2nI\left(\frac{i}{\gamma}\right),\\ \tilde{b}_n\left(\frac{i}{\gamma}\right)&=\varphi'\left(\frac{i}{\gamma}\right)+2nI'\left(\frac{i}{\gamma}\right).\end{aligned} \tag{5.20}$$

We therefore need to consider the integral with the following integrand:

$$\frac{5n\gamma^{5/2}\Gamma\left(\frac{1}{3}\right)}{\pi\sqrt{2(1-\gamma^2)}}\left[\frac{k}{6\varepsilon_2}\right]^{1/3}\left[\frac{\varepsilon_1}{k}\left|K\left(\frac{i}{\gamma}\right)\right|\right]^{n-1}\sqrt{\zeta-\frac{i}{\gamma}}\,e^{k\tilde{a}_n+k\tilde{b}_n\left(\zeta-\frac{i}{\gamma}\right)+\frac{\pi i}{3}\left(\frac{11}{4}-5n\right)}. \tag{5.21}$$

By the usual method we obtain

$$F_{(\lambda)}^{(n)}=\frac{3\sqrt{2}\,n\gamma^{5/2}\Gamma\left(\frac{1}{3}\right)}{\pi^2\sqrt{2(1-\gamma^2)}}\left[\frac{k}{6\varepsilon_2}\right]^{1/3}\left[\frac{\varepsilon_1}{k}\left|K\left(\frac{i}{\gamma}\right)\right|\right]^{n-1}\frac{\cos\left[k\,|\tilde{a}_n|+\frac{\pi}{3}\left(\frac{11}{4}-5n\right)\right]}{|k\tilde{b}_n|^{3/2}}. \tag{5.22}$$

3. In the calculation of the above integral we should take account of the effect of the roots of the function $\tilde{D}_k^{(1)}(\zeta, H)$ in the denominator of (5.11). However, we can show that this may be neglected. For if we use the approximate representation of $\tilde{D}_k^{(1)}(\zeta, H)$ and consider points of the ζ-plane satisfying

$$\left|\zeta-\frac{i}{\gamma}\right|\ll 1,$$

the equation for the roots can be reduced to the form

$$e^{-z_n}=6i\varepsilon_2 z_n, \tag{5.23}$$

where

$$\zeta_n=\frac{i}{\gamma}+\frac{1}{2\gamma}\left[\frac{3\varepsilon_2 z_n}{2k}\right]^{2/3}e^{-\frac{\pi i}{2}}. \tag{5.24}$$

Since z_n depends only on the parameter ϵ_2 the distance between the roots and the point i/γ may be estimated by

$$\left|\zeta_n-\frac{i}{\gamma}\right|=O(k^{-2/3}). \tag{5.25}$$

However, in calculating the integral along the cut we require that

$$\left|\zeta_n-\frac{i}{\gamma}\right|>\frac{B}{k}. \tag{5.26}$$

On comparing (5.25) and (5.26) we see that the effect of the roots of $\tilde{D}_k^{(1)}(\zeta, H)=0$ may be neglected in the calculation of the integral along the cut.

§6. Investigation of the Fourier-Bessel integrals and physical consequences

1. After calculating the Mellin integral in the solution obtained, we can proceed to examine the Fourier-Bessel integral (2.1). Since the contour integral splits into the sum (2.8), it is convenient to split the Fourier-Bessel integral into corresponding separate terms.

Let us consider the first of these:

$$u_0^{(1)} = -\frac{1}{\sqrt{2\pi}} \cdot \frac{1}{|\zeta_0\beta_0|} \cdot \frac{|\varepsilon_1 K(\zeta_0)|}{\sqrt{|\varphi''(\zeta_0)|}} \int_{k_0}^{\infty} \frac{\cos\left[k\,|\varphi(\zeta_0)| - \frac{\pi}{4}\right]}{k^{3/2}} J_0(kr)\,dk. \qquad (6.1)$$

Let $k_0 r \gg 1$. Then the function $J_0(kr)$ can be replaced by its approximate value. If only the discontinuous part of the field is of interest, we obtain

$$u_0^{(1)} = \frac{1}{\pi} \cdot \frac{1}{|\zeta_0\beta_0|} \cdot \frac{|\varepsilon_1 K(\zeta_0)|}{\sqrt{r\,|\varphi''(\zeta_0)|}} \operatorname{Re} I_1[r - |\varphi(\zeta_0)|], \qquad (6.2)$$

where, as in [4],

$$I_n = \int_{k_0}^{\infty} e^{ik\chi} \frac{dk}{|ik|^{n+1}}. \qquad (6.3)$$

If we compare this with the analogous expression for the incident wave, the factors can be interpreted as follows: $|\zeta_0\beta_0|^{-1}$ describes the direction of the source, $(r|\phi''(\zeta_0)|)^{-1/2}$ the geometric divergence, and Re I_1 a singularity on the wave front. The factor Re I_1 has a discontinuity in the first derivative for $r = |\phi(\zeta_0)|$; that is, a singularity on the front of $u_0^{(1)}$ is weaker by an order than a singularity for the incident wave. The factor $|\epsilon_1 K(\zeta_0)|$ can be interpreted as the coefficient of reflection from the boundary $z = 0$ on which the velocity gradient is discontinuous. The entire perturbation of $u_0^{(1)}$ is a wave reflected from this boundary (Figure 13).

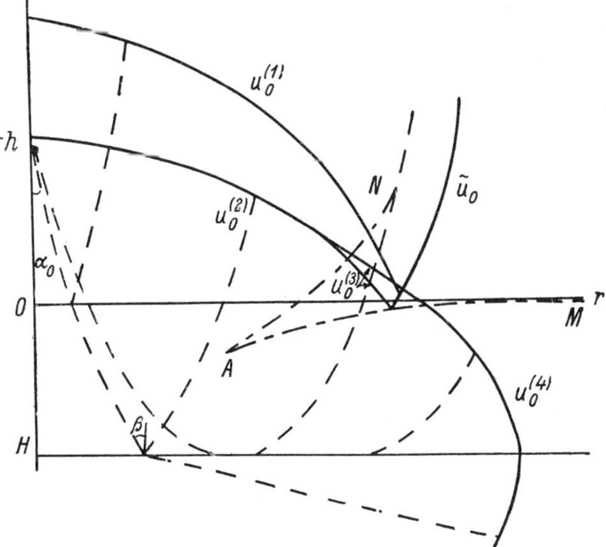

Figure 13

2. Let us now consider the term containing the integral of (4.7). After simple transformations it can be written as follows:

$$u_0^{(2)} = \frac{(-1)^n}{\pi} \cdot \frac{1}{|\zeta^{(n)}\beta^{(n)}|} \cdot \frac{|\varepsilon_1 K(\zeta^{(n)})|^{n-1} |\varepsilon_2 K(\zeta^{(n)}\gamma)|^n}{\sqrt{r}\,|\varphi''_{n1}(\zeta^{(n)})|} \operatorname{Re} I_{2n-1}[r - |\varphi_{n1}|]. \quad (6.4)$$

If first we take $n = 1$, this has the form

$$u_0^{(2)} = -\frac{1}{\pi} \cdot \frac{1}{|\zeta^{(1)}\beta^{(1)}|} \cdot \frac{|\varepsilon_2 K(\gamma\zeta^{(1)})|}{\sqrt{r}\,|\varphi''_{11}(\zeta^{(1)})|} \operatorname{Re} I_1[r - |\varphi_{11}(\zeta^{(1)})|]. \quad (6.5)$$

The equation $r = |\beta_{11}(\zeta^{(1)})|$ defines a surface of discontinuity of this expression. If we introduce the angle α_0 (Figure 13), the equation then has the explicit form

$$r = |\varphi_{11}(\zeta^{(1)})| = \frac{tv(0)}{\sin \alpha_0} - (h - z)\operatorname{ctg} \alpha_0 - 2\int_0^H \frac{dz}{\operatorname{tg} \alpha(z)} \quad (6.6)$$

and can be interpreted as a surface of discontinuity of a wave reflected from the boundary $z = H$. The multiplier $|\epsilon_2 K(\gamma\zeta^{(1)})|$ in (6.5) is then the coefficient of reflection from this boundary. For $n \neq 1$ the expression (6.4) defines a wave which undergoes $2n - 1$ reflections in the inhomogeneous layer, and the order of discontinuity on the wave front determined by $\operatorname{Re} I_{2n-1}$ decreases by unity, on each reflection.

3. Finally, we consider the Fourier-Bessel integral of expression (2.15). It was shown that the calculation of the contour integral (2.15) gives different results for different relationships between the parameters. We need to consider three fundamentally different cases: a) two saddle points far apart, b) two saddle points close together, c) one saddle point and a cut from i/γ.

a) First let $n = 1$. Then we have

$$u_0^{(3)} = -\frac{1}{\pi} \cdot \frac{1}{|\zeta_1^{(1)}\beta_1^{(1)}|} \cdot \frac{1}{\sqrt{r}\,|\varphi''_{12}(\zeta_1^{(1)})|} \operatorname{Im} I_0[r - |\varphi_{12}(\zeta_1^{(1)})|]$$
$$+ \frac{1}{\pi} \cdot \frac{1}{|\zeta_1^{(2)}\beta_1^{(2)}|} \cdot \frac{1}{\sqrt{r}\,|\varphi''_{12}(\zeta_1^{(2)})|} \operatorname{Re} I_0[r - |\varphi_{12}(\zeta_1^{(2)})|]. \quad (6.7)$$

This expression describes the singularities on the fronts of waves refracted in the inhomogeneous layer and the equations

$$\left.\begin{aligned} r &= |\varphi_{12}(\zeta_1^{(1)})|, \\ r &= |\varphi_{12}(\zeta_1^{(2)})| \end{aligned}\right\} \quad (6.8)$$

determine these fronts.

The first term in (6.7) describes the wave whose rays touch the caustic curve AM (Figure 13). In accordance with this the singularity on its front differs from the singularity on the incident wave. The second term describes the wave whose front is bounded by the limit ray and the caustic curve AN. By limit ray we mean that ray which touches the boundary of separation $z = H$.

When $n > 1$ the corresponding expression describes a wave reflected n times in the inhomogeneous layer.

b) When $n = 1$ we have

$$u_0^{(3)} = -\frac{2^{3/2}}{\pi}\cdot\frac{1}{|\tilde{\zeta}_1||\tilde{\beta}_1|\sqrt[3]{3|b_1|}}\int_{k_0}^{\infty} v(\varkappa_1)\cos\left[k(r-|\varphi_{21}(\tilde{\zeta}_1)|)-\frac{\pi}{4}\right]\frac{dk}{k^{5/6}}. \tag{6.9}$$

Here $\tilde{\zeta}_1$, b_1 and κ_1 are defined by (3.20), (5.3) and (5.9) respectively. The expression (6.9) describes the field in the neighborhood of the caustic curve AN (Figure 13). An investigation of a similar expression in [5] was sufficiently complete to establish that it has a singularity whose character is determined by the relations between the parameters.

c) It remains for us to consider the term obtained by integrating along the cut from the point i/γ. We have

$$u_0^{(4)} = \frac{6n\gamma^{5/2}\Gamma\left(\frac{1}{3}\right)}{\pi^2\sqrt{2(1-\gamma^2)}}\left[\frac{1}{6}\cdot\frac{v(H)}{v'(H)}\right]^{1/3}\frac{\left|\varepsilon_1 K\left(\frac{i}{\gamma}\right)\right|^{n-1}}{\sqrt{r}\,|\tilde{b}_n|^{3/2}}$$

$$\times\int_{k_0}^{\infty}\cos\left[k(r-|\tilde{a}_n|)-\frac{\pi(7-10n)}{6}\right]\frac{dk}{k^{n+\frac{2}{3}}}. \tag{6.10}$$

First let $n = 1$. Then

$$u_0^{(4)} = \frac{6\gamma^{5/2}\Gamma\left(\frac{1}{3}\right)}{\pi^2\sqrt{2(1-\gamma^2)}}\cdot\frac{1}{\sqrt{r}\,|\tilde{b}_1|^{3/2}}\cdot\frac{1}{(6\varepsilon_2)^{1/3}}\int_{k_0}^{\infty}\frac{\sin[k(r-|\tilde{a}_1|)]}{k^{5/3}}\,dk. \tag{6.11}$$

This expression describes the head wave formed on the boundary $z = H$. The singularity on its front is determined by the integral in (6.11). Its derivative has a singularity when $r = |\tilde{a}_1|$ of order $(r - |\tilde{a}_1|)^{-1/3}$. The expression

$$\frac{3\sqrt{2}}{\pi}\cdot\frac{\Gamma\left(\frac{1}{3}\right)}{(6\varepsilon_2)^{1/3}} = \frac{3\sqrt{2}}{\pi}\left[\frac{1}{6}\cdot\frac{v(H)}{v'(H)}\right]^{1/3}\Gamma\left(\frac{1}{3}\right) \tag{6.12}$$

can be interpreted as the coefficient of formation of a head wave on a weak boundary of separation.

We note that the singularity on the front of the head wave (6.11) is stronger than for an ordinary head wave.

For $n > 1$ the expression (6.10) describes a head wave reflected $n-1$ times in the inhomogeneous layer from the boundary $z=0$.

The complete picture of wave fronts occurring in the transition layer is shown in Figure 13.

§7. On the study of interference phenomena in the transition layer

In the study of interference phenomena the calculation of integrals (1.8) is conveniently reduced to the calculation of the sum of residues at the roots of the characteristic equation

$$W(k,\zeta)=\widetilde{D}_k^{(1)}(\zeta,H)\,D_k^{(2)}(\zeta,0)-\widetilde{D}_k^{(2)}(\zeta,H)\,D_k^{(1)}(\zeta,0)=0, \qquad (7.1)$$

intersected by contours of Mellin integrals when they are deformed into certain special contours [6]. For this we need to study the roots of (7.1) on the ζ-plane and their dependence on the parameters k and γ. We shall confine ourselves to this, since a knowledge of the behavior of the roots enables us to describe many interference properties of the transition layer.

Since a detailed study of the roots of (7.1) can be carried out only if approximate expressions for the special functions in the equation are used, it is convenient to divide the whole of the upper half of the ζ-plane into a number of regions in which these approximate expressions can be applied. We therefore divide the ζ-plane into the following regions:

I. Points which satisfy

$$|\zeta| \gg \frac{1}{\gamma} \qquad (7.2)$$

for all $k > 0$.

II, III, IV. The neighborhood of i/γ; that is,

$$\left|\zeta-\frac{i}{\gamma}\right| \ll 1, \qquad (7.3)$$

for $k > k_0$.

V. The neighborhood of i given by

$$|\zeta - i| \ll 1 \qquad (7.4)$$

for $k > k_0$.

Let (7.2) hold. In addition, we require that the following stronger inequality for small k should be satisfied:

$$|\zeta| \gg \frac{\varepsilon_2}{k\gamma}, \tag{7.5}$$

This enables the functions $D_k^{(\nu)}$ and $\widetilde{D}_k^{(\nu)}$ in (7.1) to be replaced by approximate values for a large argument (Appendix, Table 2), and the equation can then be written in the form

$$e^{2k\zeta \int_0^H \frac{v(0)}{v(z)} dz} + \frac{\varepsilon_1 \varepsilon_2}{16k^2\gamma\zeta^2} = O\left(\frac{1}{k^3\zeta^3}\right). \tag{7.6}$$

We shall use this to find the roots in region I (Figure 14).

If (7.3) holds we have to use different expressions for the functions in different regions of the neighborhood of i/γ.

Let

$$\left|\zeta - \frac{i}{\gamma}\right| \gg A\left[\frac{\varepsilon_2}{k}\right]^{2/3}, \quad A > 0, \tag{7.7}$$

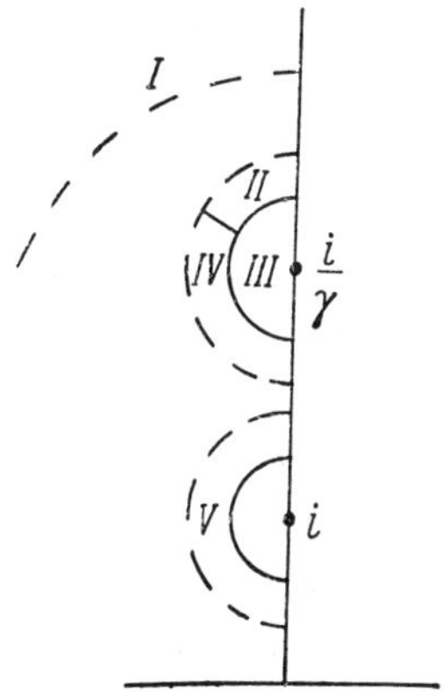

Figure 14

where

$$\frac{\pi}{2} < \arg\left(\zeta - \frac{i}{\gamma}\right) < \frac{5\pi}{6}. \tag{7.8}$$

The points ζ satisfying these relations belong to region II (Figure 14), and (7.1) takes the form

$$e^{-2ikM - 2k\frac{v^2(0)}{\gamma} N\left(\zeta - \frac{i}{\gamma}\right) + \frac{3\pi i}{4}} = 2^{11/2}\gamma^{1/2}(1-\gamma^2)^{3/2}$$

$$\times \frac{k^2}{\varepsilon_1\varepsilon_2}\left(\zeta - \frac{i}{\gamma}\right)^{3/2}\left[1 + O\left(\frac{\varepsilon_\nu}{k}\right) + O\left(\zeta - \frac{i}{\gamma}\right)\right]. \tag{7.9}$$

Here

$$M = \int_0^H \frac{\sqrt{v^2(H) - v^2(z)}}{v(z)} dz,$$
$$N = \int_0^H \frac{dz}{v(z)\sqrt{v^2(H) - v^2(z)}}. \tag{7.10}$$

If

$$\left|\zeta - \frac{i}{\gamma}\right| < A\left[\frac{\varepsilon_2}{k}\right]^{2/3} \tag{7.11}$$

(region III – Figure 14) the roots of (7.1) satisfy

$$e^{2ikM+2k\frac{v^2(0)}{\gamma}N\left(\zeta-\frac{i}{\gamma}\right)}=\frac{\varepsilon_1}{k}\cdot\frac{\gamma}{4(1-\gamma^2)^{3/2}}e^{-\frac{\pi i}{3}}\left[1+O\left(\frac{\varepsilon_2}{k}\right)\right]. \tag{7.12}$$

We now suppose that (7.7) holds and consider region IV. If

$$\frac{5\pi}{6}<\arg\left(\zeta-\frac{i}{\gamma}\right)<\frac{7\pi}{6}, \tag{7.13}$$

then (7.9) holds. If, however,

$$\frac{7\pi}{6}<\arg\left(\zeta-\frac{i}{\gamma}\right)<\frac{3\pi}{2}, \tag{7.14}$$

we see that (7.12) is satisfied.

If (7.4) holds, we can use the following approximate equation to find the roots of (7.1) in the neighborhood of i:

$$e^{-\frac{2}{3}\frac{k}{\varepsilon_1}[2i(\zeta-i)]^{3/2}}=\frac{\varepsilon_1}{4ki\,[2i(\zeta-i)]^{3/2}}\left[1+O\left(\frac{\varepsilon_1}{k}\right)\right], \tag{7.15}$$

where we assume in addition that

$$\left.\begin{aligned} &|\zeta-i|>A\left[\frac{\varepsilon_1}{k}\right]^{2/3},\\ &\frac{\pi}{2}<\arg(\zeta-i)<\frac{5\pi}{6}. \end{aligned}\right\} \tag{7.16}$$

We can show that, if the reverse inequalities hold, the equation has no roots in the neighborhood of i.

The question therefore reduces to finding roots of the approximate equations (7.6), (7.9), (7.12) and (7.15).

§8. Finding roots of the characteristic equation

Equations (7.6), (7.9) and (7.15) are of one type and are solved in the same way. We therefore consider in detail the solution of the first of them only.

We rewrite this equation in the form

$$e^{-k\zeta\int_0^H\frac{v(0)}{v(z)}dz}=\pm\frac{4ik\sqrt{\gamma}\,\zeta}{\sqrt{\varepsilon_1\varepsilon_2}} \tag{8.1}$$

and put

$$\zeta = -x + iy, \tag{8.2}$$

where x and y are positive. On separating the real and imaginary parts in (8.1), we obtain the system

$$\begin{aligned} e^{kx\int_0^H \frac{v(0)}{v(z)}dz} \cos\left[ky\int_0^H \frac{v(0)}{v(z)}dz\right] &= \mp \frac{4k\sqrt{\gamma}\,y}{\sqrt{\varepsilon_1\varepsilon_2}}, \\ e^{kx\int_0^H \frac{v(0)}{v(z)}dz} \sin\left[ky\int_0^H \frac{v(0)}{v(z)}dz\right] &= \pm \frac{4k\sqrt{\gamma}\,x}{\sqrt{\varepsilon_1\varepsilon_2}}. \end{aligned} \tag{8.3}$$

Since $\int_0^H (v(0]/v(z))dz > 0$, the trigonometric functions in (8.3) obviously have different signs. Therefore

$$y_n = \frac{\varphi_n + \frac{\pi}{2} + n\pi}{k\int_0^H \frac{v(0)}{v(z)}dz}, \tag{8.4}$$

where

$$0 < \varphi_n < \frac{\pi}{2}. \tag{8.5}$$

On dividing the equations (8.3) by each other, we obtain

$$x_n = \frac{\varphi_n + \frac{\pi}{2} + n\pi}{k\int_0^H \frac{v(0)}{v(z)}dz} \operatorname{ctg}\varphi_n. \tag{8.6}$$

The roots of (8.1) are therefore determined by

$$\zeta_n = \frac{\varphi_n + \frac{\pi}{2} + n\pi}{k\int_0^H \frac{v(0)}{v(z)}dz} \cdot \frac{e^{(\pi-\varphi_n)i}}{\sin\varphi_n}, \tag{8.7}$$

and the angle ϕ_n is found from the equation

$$e^{\left(\varphi_n+\frac{\pi}{2}+n\pi\right)\operatorname{ctg}\varphi_n} = \frac{4\sqrt{\gamma}}{\sqrt{\varepsilon_1\varepsilon_2}} \cdot \frac{\varphi_n + \frac{\pi}{2} + n\pi}{\sin\varphi_n \int_0^H \frac{v(0)}{v(z)}dz}. \tag{8.8}$$

In using (8.7) we need to remember that for small k inequality (7.5) must hold. This can be rewritten in the form

$$\frac{\varphi_n + \frac{\pi}{2} + n\pi}{\sin \varphi_n} \gg \frac{\varepsilon_2}{\gamma} \int_0^H \frac{v(0)}{v(z)} dz. \qquad (8.9)$$

To establish the dependence of ϕ_n on γ in specific terms it is helpful to consider various special cases as, for example, the exponential dependence of velocity on depth [7]. A typical example is shown in Figure 15.

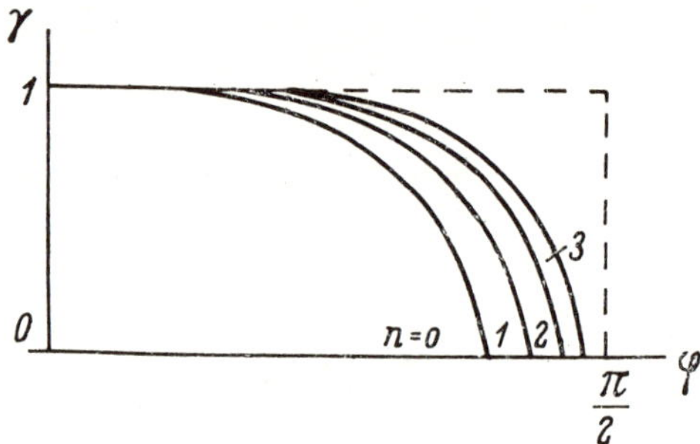

Figure 15

Let us now consider the behavior of roots in the neighborhood of i/γ. We suppose first that (7.9) is satisfied; that is, (7.7), (7.8) and (7.13) hold. Solving by a method similar to the one we have just used, we obtain

$$\zeta_n = \frac{i}{\gamma} + \frac{\frac{n\pi}{2} - kM - \frac{3}{4}\varphi_n}{kv^2(0)\,N} \gamma \frac{e^{\left(\frac{\pi}{2}+\varphi_n\right)i}}{\cos \varphi_n}, \qquad (8.10)$$

where the parameter $\phi_n \in [0; \pi/2]$ is determined by

$$e^{\frac{4}{3}\left[\frac{n\pi}{2} - kM - \frac{3}{4}\varphi_n\right]\operatorname{tg}\varphi_n} = \frac{2^{11/3}\gamma^{4/3}(1-\gamma^2)\,k^{1/3}}{(\varepsilon_1\varepsilon_2)^{2/3}} \cdot \frac{\frac{n\pi}{2} - kM - \frac{3}{4}\varphi_n}{v^2(0)\,N\cos\varphi_n}. \qquad (8.11)$$

If (7.11) is satisfied, or (7.7) and (7.14) hold simultaneously, the roots determined by (7.12) are equal to

$$\zeta_n = \frac{i}{\gamma} + \frac{\ln \frac{\varepsilon_1\gamma}{4k(1-\gamma^2)^{3/2}}}{2kv^2(0)} \gamma + \frac{\frac{n\pi}{2} - kM - \frac{\pi}{6}}{kv^2(0)\,N} \gamma i. \qquad (8.12)$$

Let us further consider the neighborhood of i in which (7.15) holds. On making the change of variable

$$\eta = \frac{2}{3} \cdot \frac{k}{\varepsilon_1} [2i(\zeta - i)]^{3/2}, \tag{8.13}$$

we obtain the following simple equation:

$$e^{\eta} = 6i\eta. \tag{8.14}$$

The unknown roots are given by

$$\zeta_n = i + \frac{1}{2}\left[\frac{3}{2}\cdot\frac{\varphi_n + \frac{3\pi}{2} + 2n\pi}{k \sin\varphi_n}\varepsilon_1\right]^{2/3} e^{\left(\frac{5\pi}{6} - \frac{2\varphi_n}{3}\right)i}, \tag{8.15}$$

and the parameter ϕ_n is found from

$$e^{\left(\varphi_n + \frac{3\pi}{2} + 2n\pi\right)\operatorname{ctg}\varphi_n} = 6\,\frac{\varphi_n + \frac{3\pi}{2} + 2n\pi}{\sin\varphi_n}. \tag{8.16}$$

In using (8.15) we need to remember that inequalities (7.16) must hold. The values of the first five roots of (8.15) are given in the following table:

n	φ_n	$\lvert \zeta_n - i \rvert \left(\frac{k}{\varepsilon_1}\right)^{2/3}$	$\arg(\zeta_n - i)$
0	0.992	2.355	1.957
1	1.231	3.616	1.796
2	1.321	4.661	1.737
3	1.372	5.665	1.704
4	1.404	6.559	1.682

§9. On the phase and group velocities of interference waves

After investigating the behavior of roots in the neighborhood of characteristic points on the complex ζ-plane, certain deductions can be made about the behavior of interference waves in the system. We know [8-10] that their phase and group velocities, depending on the value of the root, are given by

$$v^{(n)}_{\text{ph}} = v(0)\operatorname{Im}\zeta_n, \tag{9.1}$$

$$v^{(n)}_{\text{gr}} = \frac{d}{dk}(kv(0)\operatorname{Im}\zeta_n), \tag{9.2}$$

and the frequency is given by

$$\omega^{(n)} = kv(0)\operatorname{Im}\zeta_n. \tag{9.3}$$

Using these relations, we can construct the graphs of the phase and group velocities of the waves as functions of the frequency. It should, however, be noted that the use of (9.2) is admissible only if ζ_n lies on the imaginary axis. But it can also be used with sufficient accuracy in practice if the real part of the root is small. For roots with small index numbers this requirement is met only if k is such that they lie in the interval $1 < \operatorname{Im} \zeta_n \leq 1/\gamma$. For roots with large index numbers this requirement is satisfied for any k.

The real part of the root determines the damping of the interference wave the coefficient of which is found from

$$\alpha_n = -kv(0) \operatorname{Re} \zeta_n. \tag{9.4}$$

Using (8.7) and (9.3), we obtain the following value of the frequency in the case $|\zeta_n| \gg 1$:

$$\tilde{\omega}_n = \frac{\varphi_n + \frac{\pi}{2} + n\pi}{I_0}. \tag{9.5}$$

Since $\phi_n < \pi/2$, we obviously have $\tilde{\omega}_n < \tilde{\omega}_{n+1}$. It follows that the spectrum of the interference wave with index number n is contained within the limits

$$\tilde{\omega}_n < \omega_n < \infty. \tag{9.6}$$

In using a device to record a finite frequency interval in the field, we must naturally take account of only a finite number of interference waves with small index numbers.

The wave damping for $|\zeta_n| \gg 1$ is determined by

$$\alpha_n = \frac{\varphi_n + \frac{\pi}{2} + n\pi}{I_0} \operatorname{ctg} \varphi_n. \tag{9.7}$$

Since $\phi_n \longrightarrow \pi/2$ as $n \longrightarrow \infty$, it follows that

$$\alpha_n > \alpha_{n+1}. \tag{9.8}$$

It follows from (8.7) that as $\omega \longrightarrow \omega_n$ $(\omega > \omega_n)$ the phase velocities $v_{\text{ph}}^{(n)}$ tend to infinity and the group velocities $v_{\text{gr}}^{(n)}$ tend to zero.

Let us consider further the expression for the roots in the neighborhood of i/γ. Using (8.13) and (9.3), we obtain the following expression for the frequency as a function of the parameter k:

$$\omega_n = kv(H) + \frac{n\pi - \frac{\pi}{3} - 2kM}{2v(H)N}. \tag{9.9}$$

This enables us to eliminate k from the formulas for the group and phase velocities and the damping α_n. After simple transformations we have

$$v_{\text{ph}}^{(n)} = 2\omega_n \frac{v^2(H)M - N}{2\omega_n v(H)M - \left(n\pi - \frac{\pi}{3}\right)}, \tag{9.10}$$

$$v_{\text{gr}}^{(n)} = \frac{v^2(H)M - N}{v(H)M}, \tag{9.11}$$

$$\alpha_n = -\frac{\ln\left\{\dfrac{\varepsilon_1\gamma\left[v^2(H)M - N\right]}{2\left[2\omega_n v(H)M - \left(n\pi - \frac{\pi}{3}\right)\right]}\right\}}{2v(H)M}. \tag{9.12}$$

We need to remember here that (9.9)–(9.12) are valid only if (7.11) holds, which we can rewrite in the form

$$\left|\omega_n - \frac{\left(n\pi - \frac{\pi}{3}\right)v(H)}{2N}\right| < 1. \tag{9.13}$$

Suppose, further, that $|\zeta_n - i| < 1$. We then have

$$\omega_n = kv(0) + \frac{v(0)}{2}\left[\frac{3}{2}\cdot\frac{\varphi_n + \frac{3\pi}{2} + 2n\pi}{\sin\varphi_n}\varepsilon_1\right]^{2/3} k^{1/3}\sin\left(\frac{5\pi}{6} - \frac{2\varphi_n}{3}\right). \tag{9.14}$$

Using (9.1), (9.2) and (9.4), we obtain

$$v_{\text{ph}}^{(n)} - \frac{\omega_n}{k}, \tag{9.15}$$

$$v_{\text{gr}}^{(n)} = \frac{2}{3}v(0) + \frac{1}{3}\cdot\frac{\omega_n}{k}, \tag{9.16}$$

$$\alpha_n = \frac{kv(0) - \omega_n}{\operatorname{tg}\left(\frac{5\pi}{6} - \frac{2\varphi_n}{3}\right)}. \tag{9.17}$$

Since these relations are valid only as $k \to \infty$, it is easy to see that

$$\begin{gathered}\omega_n = kv(0)\left[1 + O(k^{-2/3})\right],\\ v_{\text{ph}}^{(n)} = v(0)\left[1 + O(k^{-2/3})\right],\\ v_{\text{gr}}^{(n)} = v(0)\left[1 + O(k^{-2/3})\right],\\ \alpha_n = -\frac{v^{2/3}(0)}{2}\left[\frac{3}{2}\cdot\frac{\varphi_n + \frac{3\pi}{2} + 2n\pi}{\sin\varphi_n}\varepsilon_1\right]^{2/3}\omega_n^{1/3}\cos\left(\frac{5\pi}{6} - \frac{2\varphi_n}{3}\right).\end{gathered} \tag{9.18}$$

After obtaining the expressions for the group and phase velocities and the damping of interference waves, we can depict them graphically as functions of the frequency (Figure 16). The phase velocity is shown in Figure 16 by the continuous line, the group velocity by the broken line and the damping by the dot–dash line.

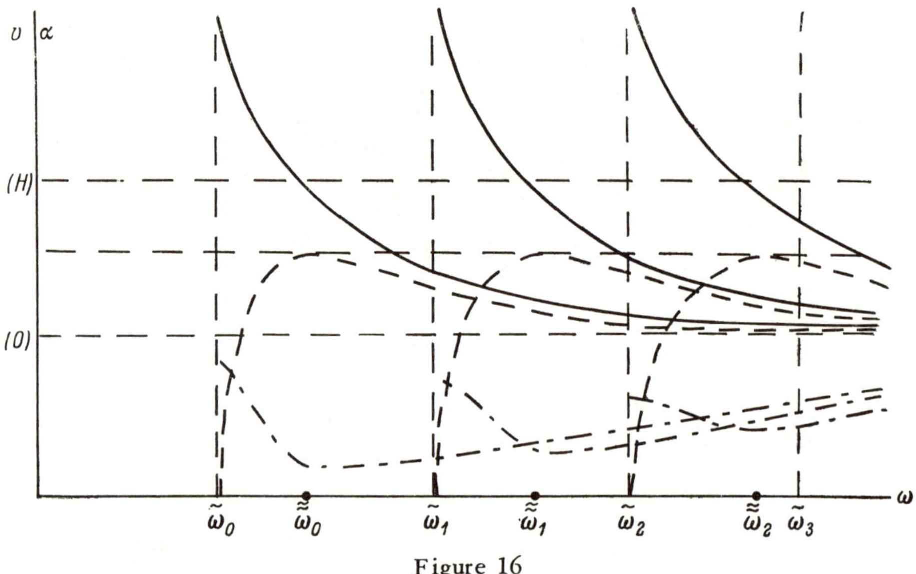

Figure 16

The following conclusion can be drawn from an examination of the behavior of these curves.

The low-frequency part of the branches of the phase velocities indicates the strong dispersion of waves at these frequencies. At higher frequencies the dispersion almost vanishes and the velocities tend to $v(0)$ as $\omega \longrightarrow \infty$.

The group velocities have a maximum at frequencies for which the phase velocities are equal to $v(H)$. The maxima are given by

$$v^{(n)}_{\mathrm{gr}} = \frac{v^2(H)\,M - N}{v(H)\,M} \tag{9.19}$$

and do not depend on the index number of the curve. This expression has a simple physical interpretation: using (7.10), we can show that it determines the velocity of the interference wave arising as a result of the superposition of waves propagated at the limit angles. The maximum lies in the interval $[v(0);\ v(H)]$.

As the frequency increases the group velocity tends to $v(0)$. We have

$$v^{(n)}_{\mathrm{ph}} - v^{(n)}_{\mathrm{gr}} = \frac{B_n}{\omega_n^{2/3}}, \tag{9.20}$$

and B_n is determined from (9.14).

The curves of the damping factor show that is is a minimum at those frequencies for which the group velocity is a maximum. As the frequency increases the damping increases as $P_n \omega_n^{1/3}$.

In conclusion, it is the author's pleasant duty to thank G. I. Petrašen′ and I. A. Molotkov for valuable assistance and discussion of the paper.

APPENDIX

Asymptotic representation of the solution of the second equation of (1.9)

The equation we are considering has the form

$$u'' = k^2 [1 + f^2(z) \zeta^2] u, \tag{1}$$

where $k > 0$ and ζ is a complex parameter. The function $f(z) = v(0)\, v(z)^{-1}$, which is given in $[0, H]$, is regular in this interval and monotonically decreases from 1 to γ $(0 < \gamma < 1)$ as z increases. We require to find the solution of (1) as $k \longrightarrow \infty$ on the ζ-plane for a fixed value of the variable z.

The problem is solved by the method of the standard equation, which in this case is written in the form

$$F''(\chi) = \chi F(\chi) \tag{2}$$

and is satisfied by the function

$$F(\chi) = \sqrt{\chi} Z_{1/3}\left(\frac{2}{3} i \chi^{3/2}\right), \tag{3}$$

where $Z_{1/3}$ is an arbitrary solution of Bessel's equation. If we take as $Z_{1/3}$ the Hankel function $H^{(\nu)}_{1/3}$, we have for large k the following asymptotic expansions for the two linearly independent solutions of (1):

$$\left.\begin{aligned} \Phi(k, z, \zeta) &= \sqrt{\frac{3}{2} \cdot \frac{I}{I'}} \{H^{(1)}_{1/3}(ikI)[1 + O(k^{-2})] + H^{(1)}_{-2/3}(ikI)\, O(k^{-1})\}, \\ \Psi(k, z, \zeta) &= \sqrt{\frac{3}{2} \cdot \frac{I}{I'}} \{H^{(2)}_{1/3}(ikI)[1 + O(k^{-2})] + H^{(2)}_{-2/3}(ikI)\, O(k^{-1})\}, \end{aligned}\right\} \tag{4}$$

which are valid in the neighborhood of the point of rotation defined by

$$f(z) = \frac{i}{\zeta_0}. \tag{5}$$

In the formulas (4) we have

$$I = \int_{\tilde{z}}^{z} \sqrt{1 + f^2(z) \zeta^2}\, dz, \tag{6}$$

where

$$f(\bar{z}) = \frac{i}{\zeta}. \tag{7}$$

It is known that for large $k|I|$ the expansions (4) become formulas which are valid at a great distance from the point of rotation. These expansions are given in Table 1:

Table 1

	$-\frac{\pi}{2} \leqslant \arg\eta \leqslant \frac{\pi}{6}$	$\frac{\pi}{6} \leqslant \arg\eta \leqslant \frac{5\pi}{6}$	$\frac{5\pi}{6} \leqslant \arg\eta \leqslant \frac{3\pi}{2}$
$\Phi(k, z, \zeta)$	$P_1 - P_2$	P_1	P_1
$\Psi(k, z, \zeta)$	$P_2 e^{\frac{\pi i}{3}}$	$P_2 e^{\frac{\pi i}{3}}$	$(P_2 - P_1) e^{\frac{\pi i}{3}}$

Here

$$\left.\begin{aligned} P_1 &= \sqrt{\frac{3}{kI'}}\, e^{-kI + \frac{\pi i}{3}} [1 + O(k^{-1})], \\ P_2 &= \sqrt{\frac{3}{kI'}}\, e^{kI + \frac{5\pi i}{6}} [1 + O(k^{-1})], \end{aligned}\right\} \tag{8}$$

$$\eta = \zeta - \zeta_0, \tag{9}$$

and ζ_0 is defined by (5).

Asymptotic expansions of the functions (1.15) can be obtained with the help of the data in Table 1, and these are given in Table 2:

Table 2

	$-\frac{\pi}{2} \leqslant \arg\eta \leqslant \frac{\pi}{6}$	$\frac{\pi}{6} \leqslant \arg\eta \leqslant \frac{5\pi}{6}$	$\frac{5\pi}{6} \leqslant \arg\eta \leqslant \frac{3\pi}{2}$
$D_k^{(1)}(\zeta, z)$	$C_k^{(1)} - C_k^{(2)}$	$C_k^{(1)}$	$C_k^{(1)}$
$\tilde{D}_k^{(1)}(\zeta, z)$	$\tilde{C}_k^{(1)} - \tilde{C}_k^{(2)}$	$\tilde{C}_k^{(1)}$	$\tilde{C}_k^{(1)}$
$D_k^{(2)}(\zeta, z)$	$C_k^{(2)} e^{\frac{\pi i}{3}}$	$C_k^{(2)} e^{\frac{\pi i}{3}}$	$(C_k^{(2)} - C_k^{(1)}) e^{\frac{\pi i}{3}}$
$\tilde{D}_k^{(2)}(\zeta, z)$	$\tilde{C}_k^{(2)} e^{\frac{\pi i}{3}}$	$\tilde{C}_k^{(2)} e^{\frac{\pi i}{3}}$	$(\tilde{C}_k^{(2)} - \tilde{C}_k^{(1)}) e^{\frac{\pi i}{3}}$

Here

$$\begin{aligned} C_k^{(1)}(\zeta, z) &= -\frac{1}{2}\cdot\frac{I''}{I'} A(k, \zeta, z)[1 + O(k^{-1})], \\ C_k^{(2)}(\zeta, z) &= 2kI'B(k, \zeta, z)[1 + O(k^{-1})], \\ \tilde{C}_k^{(1)}(\zeta, z) &= 2kI'A(k, \zeta, z)[1 + O(k^{-1})], \\ \tilde{C}_k^{(2)}(\zeta, z) &= \frac{1}{2}\cdot\frac{I''}{I'} B(k, \zeta, z)[1 + O(k^{-1})], \end{aligned} \tag{10}$$

where

$$\left.\begin{aligned} A(k, \zeta, z) &= \sqrt{\frac{3}{kI'}}\, e^{kI + \frac{\pi i}{3}}, \\ B(k, \zeta, z) &= \sqrt{\frac{3}{kI'}}\, e^{-kI + \frac{5\pi i}{6}}. \end{aligned}\right\} \qquad (11)$$

The position of the roots of $D_k^{(\nu)}$ and $\tilde{D}_k^{(\nu)}$ is of importance in the investigation of the solution of (2.1). On using the results of Table 2 we can show that in the neighborhood of the point of rotation they are situated on the rays $\arg \eta = -\pi/6$ $(\nu = 1)$ and $\arg \eta = 7\pi/6$ $(\nu = 2)$. These rays and the imaginary axis above the point of rotation divide the neighborhood of this point into regions in which the real part of the function (6) has different signs (Figure 17).

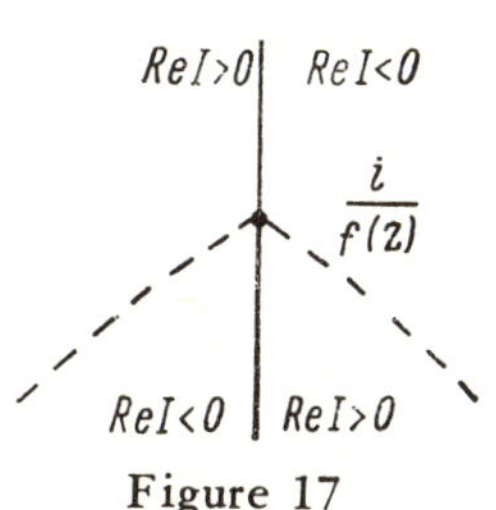

Figure 17

BIBLIOGRAPHY

[1] G. I. Petrašen′, *On a rational method of solving problems of dynamical theory of elasticity in the case of isotropic layers with plane-parallel boundaries of separation,* Leningrad. Gos. Univ. Uč. Zap. 208 Ser. Mat. Nauk 30 (1956), 5–57. (Russian) MR 19, 699.

[2] ———, *Propagation of elastic waves in stratified isotropic media separated by parallel planes,* Leningrad. Gos. Univ. Uč. Zap. 162 Ser. Mat. Nauk 26 (1952). (Russian) MR 17, 1157.

[3] V. S. Buldyrev, *Generalization of the saddle-point method to the case when the saddle points are close together,* Problems of the Dynamic Theory of the Propagation of Seismic Waves, no. 5, 136–152, Izdat. Leningrad. Univ., Leningrad, 1961. (Russian)

[4] A. S. Alekseev and B. Ja. Gel′činskiĭ, *On the ray method of calculating wave fields in the case of inhomogeneous media with curvilinear boundaries of separation,* Problems of the Dynamic Theory of the Propagation of Seismic Waves, no. 3, Izdat. Leningrad. Univ., Leningrad, 1958, pp. 12–47. (Russian)

[5] V. S. Buldyrev, *The study of a field in the neighborhood of the caustic curve,* Problems of the Dynamic Theory of the Propagation of Seismic Waves, no. 5, Izdat. Leningrad. Gos. Univ., Leningrad, 1961, pp. 153–168. (Russian)

[6] G. I. Petrašen′ and V. A. Enal′skiĭ, *On some interference phenomena in media containing thin plane-parallel layers,* Izv. Akad. Nauk SSSR Ser. Geofiz. 1956, nos. 9–11, 1009–1020, 1129–1144, 1242–1257. (Russian)

[7] N. V. Cepelev, *The propagation of waves in an acoustic medium with a transition layer,* Problems of the Dynamic Theory of the Propagation of Seismic Waves, no. 5, 169–205, Izdat. Leningrad. Univ., Leningrad, 1961. (Russian)

[8] V. S. Buldyrev and Z. A. Janson, *The propagation of interference waves SH in an elastic spherical layer.* I, II, Izv. Akad. Nauk SSSR Ser. Geofiz. (1962), no. 12, 1764–1772; (1963), no. 1, 76–89 = Bull. Acad. Sci. USSR Geophys. Ser. (USA) (1962), no. 12, 1105–1109; (1963), no. 1, 48–55.

[9] L. A. Molotkov, *On the propagation of low quality vibrations in liquid half-spaces separated by a thin elastic layer,* Problems of the Dynamic Theory of Propagation of Seismic Waves, no. 5, Izdat. Leningrad. Univ., Leningrad, 1961, pp. 281–302. (Russian)

[10] L. M. Brehovskih, *Waves in layered media,* Izdat. Akad. Nauk SSSR, Moscow, 1957. (Russian) MR 20 #1476.